Related Titles of Interest

Title	Editor/Author	ISBN
IMechE Data Book	C Matthews	1 86058 175 7
The Continuum of Design Education	N P Juster	1 86058 208 7
Industrial Sensors and Applications for Condition Monitoring	P Wild	0 85298 902 4
Managing Enterprises – Stakeholders, Engineering, Logistics, and Achievement	D T Wright, V Hanna, D Gillingwater, and N D Burns	1 86058 066 1
The Reliability of Mechanical Systems	J Davidson	0 85298 881 8
Software Quality Management V – The Quality Challenge	C Hawkins, M Ross, and G Staples	1 86058 110 2
Software Quality Management IV – Improving Quality	M Bray, M Ross, and G Staples	1 86058 031 9

For the full range of titles published by Professional Engineering Publishing contact:

Sales Department
Professional Engineering Publishing Limited
Northgate Avenue
Bury St Edmunds
Suffolk
IP32 6BW
UK

Tel: +44 (0)1284 724384
Fax: +44 (0)1284 718692

Contents

Design for Quality, Reliability, and Maintenance

Education, Training, and Medical

Foreword

Welcome to the 3rd International Conference on Quality, Reliability, and Maintenance (QRM 2000). For the past ten years or so high-tech computer visionaries have predicted many disasters in the wake of theY2K 'bug'. We were warned, indeed, by many well-wishing colleagues, of the dangers of organizing a conference in the early months of 2000. The 'end of the world brigade' foresaw doom because some computer software confused 2000 with 1900. The committee for this conference had no such foreboding and have been rewarded by the fact that the gurus who prophesied this secular apocalypse were just as unreliable as the mystical millenarians of the past. We can say that Y2K is Y2OK as far as QRM 2000 is concerned.

This conference is organized by the Universities of England Consortium for International Activities and we are privileged to have the seal of approval and the co-sponsorship of the Institution of Mechanical Engineers. The IMechE is one of the largest, most prestigious, and influential engineering institutions in the world. It combines a unique history with a dynamic approach to technological development, and the promotion of the very highest standards.

Contemporary commercial practices favour the domination of the Asset Manager, with the resultant reduced emphasis on instrument sophistication. The question now is: does it make us more competitive? The next generation of condition monitoring (CM) instruments will be dominated by simplicity and cost effectiveness. We, therefore, need to bridge the gap between the traditional refinements of detailed analytical processes and produce hardware suitable for direct and quick decisions. The relatively new technology of acoustic emission (AE) fulfils this need. Acoustic emission is based upon the detection of naturally generated high-frequency elastic waves within materials and avoids a frontal attack on the need to spend countless man-hours on diagnostic spectrum analysis, resulting in vast savings for sophisticated instrumentation. It offers a high signal-to-noise ratio, which when compared to lower frequency vibration techniques, opens the way for a more direct measure of machine faults. The instrumentation is simple, robust, and user friendly for the operators. Holroyd Instruments is a leading international company in this technology and we are pleased to welcome papers by Dr Holroyd and several others on this and related subjects. The 22 papers on *Condition Monitoring* are cogent examples of a well-balanced menu of current leading-edge practice.

It is hoped that the unhealthy distrust that has existed between management and engineers will in time be eliminated and the inclusion of six papers in this volume on *Maintenance Management* is a move towards this solution. We must also be aware of the quality and reliability theme running through these conferences and the 14 papers presented on *Quality Reliability* as well as the 14 papers on *Design for Quality, Reliability, and Maintenance* highlight this interest.

Education and training play an increasingly important role as a foundation for all practising engineers. We are pleased, therefore, to welcome 27 papers on *Education Training and Medical* relating to QRM. The range covers professional development, in-house training, education by the Internet, and many more. Several papers outlining the difficulties in the education environment are presented. Professor Chen and Dr Kruger crystallize the difficulties of education in difficult environments with unique solutions. Professor Zeitz provides a novel teaching technique with proven effectiveness for technical report writing. We shall all learn from their experiences and those of the other authors in this section, all of whom are at the sharp end of the education spectrum.

A feature of these conferences is the editorial plea that all papers must be kept to a maximum length. It is relatively easy to conceive and write a rambling, protracted document. Contemporary practice in industry is for the direct and economically sound, yet technically clear and accurate coverage of the work. This results in focussed thinking. Experience has shown that the practice induces clarity, simplicity, and directness. It also enables the reader to absorb and digest rapidly more content. We allude to the dictum 'Brevity is the soul of wit' and are grateful to the authors for the disciplined way they have observed the editorial instructions. The papers are of a high professional and technical standard, eliciting the salient points and important features of well-researched findings in the fields of Quality, Reliability, and Maintenance.

We are also pleased and privileged to welcome 83 papers in Quality, Reliability, and Maintenance by a truly international list of authors, representing the countries of Great Britain, Algeria, Hong Kong, Hungary, Italy, Malaysia, Nigeria, Poland, Romania, Russia, South Africa, Spain, Sweden, Taiwan, Thailand, Turkey, Ukraine, and USA.

G J McNulty
Organiser and Editor of QRM 2000
5 February 2000

Quality Reliability

Continual improvement in environmental quality and reliability

R AKRAM and **M J DENMAN**
School of Engineering, Sheffield Hallam University, UK
J GUBBINS
UK Health, Safety, and Environmental, Delphi Automotive Systems, UK

ABSTRACT

ISO 14001 is an international standard which imposes an Environmental management system on an organisation which enables them to assess in detail their impacts and aspects of their operations, while working towards continual improvement This paper describes a case study of the successful implementation of ISO 14001 in a manufacturing site of a large multinational automotive parts supplier.

1. INTRODUCTION

ISO 14001 is an international standard and can be described as...
"...part of the overall management system which includes the organisational structure, planning activities, responsibilities, practices, procedures, processes and resources for developing, implementing, achieving, reviewing and maintaining the environmental policy."[1]

History

ISO 14001 is an Environmental Management System (EMS) standard written by the Technical Committee 207 (TC 207). The EMS may be set-up independent to that of any other management system in the organisation. However if the organisation has a quality/Health and Safety Management system the EMS can be integrated into these with ease.

Some Common Aspects of Quality and Environmental Management Systems

Quality	EMS
Quality Policy	Environmental Policy
Adequate Resources	Adequate Resources
Responsibilities and Authorities	Responsibilities and Authorities
Training	Training
System Documentation	System Documentation
Process Controls	Operational Controls
Document Control	Document Control
System Audits	System Audits
Management Review	Management Review

2. IMPLEMENTATION

Delphi Automotive Systems Ellesmere Port is a component supplier for Vauxhall Motors' T3000 project (Astra). The site operates the most sophisticated Supply In Line Sequence (SILS) operation in Europe.[3] At the time of implantation there where 4 sites within the group applying for accreditation.

For any EMS it is necessary to have an "Environmental Champion" who must have a holistic approach and support from senior management. The EMS Project Plan was projected for completion within 10 months. This aggressive target was achieved by the setting up an Environmental Management Team (EMT) a multi-disciplined, multi-level team. This team helped with the assessing of the impacts and aspects of the company, audits and environmental awareness training amongst other items. A multi-site team was set-up, which met once a month with the aim of information sharing with other sites. A standardised approach was used for the impact risk assessments and procedures, thus saving time as each site could benefit from the work already done by the others. This information was then relayed back to the EMT of each site.

Parts of an EMS

Environmental Policy and aspects - Develop a statement of the organisation's commitment to the environment. Use this policy as a framework for planning and action. - Identify environmental attributes of the products, activities and services. Determine those that could have significant impacts on the environment.

Legal and Other Requirements - Identify and ensure access to relevant laws and regulations (and other requirements to which your organisation adheres).

Objectives and Targets - Establish environmental goals for the organisation, in line with the policy, environmental impacts, views of interested parties and other factors.

Environmental Management Program - Plan actions to achieve objectives and targets.

Structure and Responsibility - Establish roles and responsibilities and provide resources.

Training, Awareness and Competence - Ensure that employees are trained and capable of carrying out their environmental responsibilities.

Communication - Establish processes for internal and external communications on environmental management issues.

EMS Documentation - Maintain information on the EMS and related documents.

Document Control - Ensure effective management of procedures and other system documents.

Operational Control - Identify, plan and manage operations and activities in line with policy, objectives and targets.

Emergency Preparedness and Response - Identify potential emergencies and develop procedures for preventing and responding to them.
Monitoring and Measurement - Monitor key activities and track performance.
Non-conformance and corrective and preventive action - Identify and correct problems and prevent recurrences.

Records - Keep adequate records of EMS performance.

EMS Audit - Periodically verify that the EMS is operating as intended.

Management Review - Periodically review the EMS with an eye to continual improvement.

3. BENEFITS

There are major savings to be made in the implementation of an EMS, before savings can be made it is necessary to identify where the waste is, and also where the hazards are. The system helps to identify areas where saving could be made. Money is not the only reason that a company should implement a EMS, there are also environmental responsibilitiess that a company needs to address, e.g. hazardous material, oils etc.

COSTS	BENEFITS
Staff / employee time	Improved environmental performance
Possible consulting assistance	Improved compliance
Training of personnel	New customers / markets
	Increased efficiency / reduced costs
	Enhanced employee morale
	Enhanced image with public
	Reduced training effort for new employees
	Enhanced image with regulators

4. CONTINUAL IMPROVEMENT

The EMS is a continual improvement management system, thus one should address all significant issues, rather than achieve too much with EMS in one step, the EMS should be viewed as a process not a destination.

How Various Functions Can Support The EMS

Functions	How They Can Help (Possible Roles)
Purchasing	• Develop and implement controls for chemical / other material purchases
Human Resources	• Define competency requirements and job descriptions for various EMS roles • Integrate environmental management into reward, discipline and appraisal systems

Maintenance	
Finance	• Track data on environmental management costs • Prepare budgets for environmental management program • Evaluate economic feasibility of environmental projects
Engineering	• Consider environmental impacts of new or modified products and processes • Identify pollution prevention opportunities
Top Management	• Communicate importance of EMS throughout organisation • Provide necessary resources • Track and review EMS performance
Line Workers	• Provide first-hand knowledge of environmental aspects of their operations • Support training for new employees
Health & Safety	• Noise, dust, chemical monitoring • Overlapping legislation

Audits

EMS audits are a way for checking that the company is still working to control the significant environmental aspects and impacts. Ideally these audits should be carried out as to a standard of that of a third party audit.

The setting up of audit teams within the Delphi group helped with the audit process, members from different sites audited another site thus establishing a form of third party audit.

5. CONCLUSIONS

The cost of environmental management is not a consideration! Environmental management systems have been shown on numerous occasions to have paid for themselves over a short period of time, this was proven at all the Delphi sites. Many savings within the organisation can be made without significant capital investment. Examples of this were the result of awarness campaigns and training e.g. the SOS campaign (Switch Off and Save). Some projects required capital investment and management systems, such as a waste compactor to reduce the number of lifts per week and through the use of retunable packaging.
The EMT were a small dedicated team but it was necessary to instil this dedication to the rest of the workforce who where already jaded after QS9000.

Environmental Awareness training should commence before the EMS is started as this will give employees a chance to learn about what is going on and what the company intends to do, and how it WILL effect the workforce. Working practices might change with the implementation of an EMS, however it is best to have the workforce helping you with the implementation rather than them dragging their heels. Therefore it is best to keep the workforce informed of the EMS progress and get them involved in as many stages as possible. Formal awareness training may go some way to "inspiring" the workforce, but imaginative initiatives such as an environmental poster design competitions aimed at their children creates a situation where environmental issues are discussed in the home.

In many instances it would be possible to procure an "off the shelf" EMS package, to help implement ISO 14001. It was the company view to build its own system as this could be designed around its operations/processes.

As mentioned previously ISO 9000 and ISO 14001 have similarities between installation and operation, ideally it is better to run both systems in tandem thus audits (internal/External accreditation bodies) and management review etc. can take place simultaneously.

An EMS is a continuing programme, not a final destination. It concerns continuous improvement and evolution rather than a one-time measure of conformance. It cannot be over emphasised that the EMS is a process not a destination. For any EMS it is necessary to monitor and measure all activities and processes. In this way deviation from the norm can be determined using regression techniques such as CUSUM[4].

The need for management support is essential, any management system will not fully evolve unless it has the backing of senior management. The EMS should not be considered as just another certificate on the wall, the EMS and other management systems should be part of the company's business plan.

In the next couple of years there will be an integration of the two standards these being ISO 14001 and ISO 9000. Thus there will be one management standard that will cover both environmental and quality, probably in the near future we will see the integration of the health and safety standard into this one standard.

6. REFERENCES

1. ISO 14001 Environmental management systems, BSI publication, 1996
2. ISO 9000 Quality management and quality assurance standards, BSI publication 1994
3. "Driving Home - QS 9000", Quality World, 1999
4. Energy Audits for Industry, Best Practice Programme, Fuel Efficiency booklet No.1, Energy Efficiency Office of the Department of the Environment, 1993

Functional analysis: an introduction to methodologies of reliability for mechanical systems

G ARCIDIACONO and **P CITTI**
Dipartimento di Meccanicae Technologie Industriali, Universita degli Studi di Ferenze, Italy
G BANCI
Fiat Auto Dir. Tecnica Motopropulsori – Corso Settembrini, Torino, Italy

ABSTRACT

Both the market and the legislation impose strict requirements regarding quality, reliability and safety, making the management of the design phase ever more important. It must be organized with the aim of guaranteeing effectiveness, efficiency and completeness whilst also being precisely aligned to the requirements of the client.
The aim of this work is to study and describe a rational technique of systematic analysis of the product which is inserted before the prevention activities for reliability.
Starting from an analysis of reliability methodologies widely in use at the moment (e.g. FMEA), it is considered necessary to place Functional Analysis before the phase of optimizing reliability.

1. INTRODUCTION

Design has a dual role relating to concept and calculation. The concept is embraced in the drawing, the calculation is the driving force or the instrument of the design.
A market analysis is the essential step in the design planning. Market analysis is concerned with the product and optimum solutions and a flexibility is required. A knowledge of existing solutions, and a thorough knowledge of market analysis and the concomitant competition comprises the basic study plan. Customer requirement is essential and should always pervade and influence the future strategy.
The philosophy of *Functional Analysis* is an all embracing methodology of approaching the design, which enables the designer to comprehend the product and to distinguish salient aspects for improvement.
The aim therefor of this paper is to:
- outline the concept of functional analysis, highlighting its theoretical and operative procedures;
- subdivide the Functional Analysis into two discrete steps of inquiry and synthesis and introduce the concept of Correlation Matrix which is Functional Analysis in tabular form utilized by the Fiat Company.

2. FUNCTIONAL ANALYSIS: THEORETICAL PRESUMPTIONS AND AN OPERATIVE PROCEDURE

The 'Functional Analysis' "...consists in studying the product systematically, external as well as internal, in order to analyze, on one hand, all the relations that the product establishes with the environment, and, on the other hand, to analyze the solution, chosen by the examination of all the relations among the components" (1). Note that (1) highlights the necessity to point out not only the interactions among the components, but also those ones that the product or its single components have with the environment or with the user.

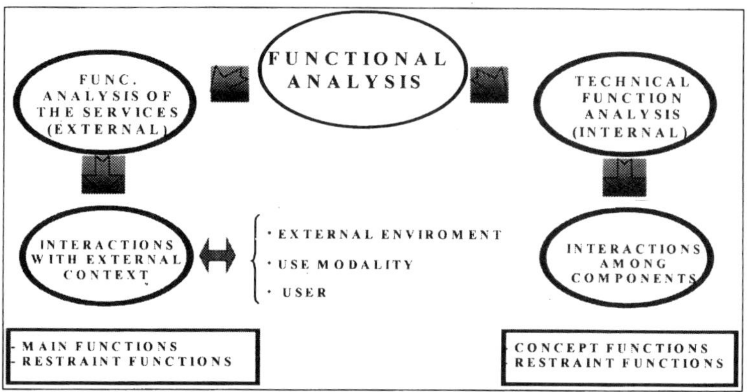

Fig 1

It is therefore important to consider (Fig. 1) a *Functional Analysis of the Systems, or External*, and a *Technical Functional Analysis*, or *Internal*. The first one wants to point out all the services requested by the customer in objective terms and independently from every technical solution, by the examine of the interactions between the product and the external context (environment, kind of use, user). This leads to point out the Principal and the Restraint Functions. The second one aims to examine the relations among the components of the product, with the double goal of observing in a detailed way how the product offers the requested services, and to review all the relations, called Concept Functions, that the components establish among themselves. By a Physical/Functional Decomposition, we determine the Concept Functions (depending on the technical solutions), the Internal Restraint Functions, and the elements useful to guarantee the concept functions in the respect of the Internal and External Restraints, and of the Main Functions.

All this aims to synthesize a functional model of the product, reporting, for each group/component/element, which and how many are the functions that it has to accomplish.

For each of these functions, it has to report which and how many are the groups/component/elements that concur to the accomplishment, or influence it by interactions that can reduce the performances. The methodological flux has been conceived on two levels (3). The first level (INQUIRY) is articulated in two steps: it starts by collecting information about the existent products, to point out the Main and Restraint Functions (*External Analysis*, or *Analysis of the Necessities*).

Then, it proceeds to select the technical solution of reference, the physical and functional decompositions are made (*Internal Analysis*), and, finally, it creates, by the *Functional Analysis System Technique* (FAST), (4), a synthetic model of the product, reproducing the hierarchical structure among the underlined functions.

This enables to correlate the product functions with the customer's requests, so that the development of the project directly follows the quality expected by the market.

All this gives the input to the second level, the creative phase (SYNTHESIS), by which, following the *Value Engineering* logic, it will be possible to study and select the technical-functional alternatives on the base of cost considerations, expected functions, and reliability. This procedure permits to obtain a complete functional idea of the new product, that will be codified in the Table of Functional Analysis.

3. CORRELATION MATRIX

The *Correlation Matrix* (C. M.) is the Table of Functional Analysis utilized by FIAT AUTO; it enables to compare in a matrix the two decompositions: the physical one (columns) and the functional one (lines). The interactions of each component/element with the other components/elements, environment, user, are expressed by interfaces (mechanical, electrical/electronic ones, with the environment, with the man). The External Restraints Functions are expressed as conditions to the rest, that can lead to a reduction of the performances linked to one or more functions, and are expressed in terms of interfaces with the environment.

To indicate which components are involved in implementing a function and, on the contrary, which and how many are the functions where a single component acts on them, we insert the small cross in the intersection cells, or we can use symbols indicating the importance of a correlation, as in the QFD technique: in Fig. 2 is shown the example of a part of the C. M. applied to the Electric Power Steering (EPS), which is a system mounted onto a steering column to help the driver to control and maneuver a car (5).

The C. M. is extremely compact and, at the same time, complete; as a matter of fact, in one only table there are:
- all the Functions, Internal and External, main and restraint ones;
- all the parts of the system;
- all the external elements that somehow interact with the system;
- all the interactions among parts and/or interfaces and functions.

Therefore, if a column has no crosses, that component/element should not have any relation with any expected function and, therefore, theoretically, could be eliminated. This fact, instead, affirms that we probably omitted to consider some function expected by the customer. Similarly, if a line is without crosses, this indicates that there is nothing that leads toward the considered function. The design should, therefore, be re-considered and modified.

Summarizing, the C. M. is:
- an alternative to the QFD for short term strategies;
- a help for the designer to consider the customer point of view during the design phase;
- a definition of the "not considered functions" on the base of which it develops the failure analysis;
- a verification regarding the completeness and consistency of the design's physical/functional structure;
- a document that establishes a Technical Standard, to add to the drafts and the technical relations of the project.

After C.M., FIAT AUTO uses the Risk Matrix (R. M.), (3), for the reliability study of the system. It comes from the FMEA logic, but its differences regard the need of a normative orientation. In the R. M., differently from the FMEA, there is not anymore the difference between actual situation and corrective actions. The best available for the firm (in terms of prescriptions for design and experimentation) is directly reported in a single draft. Moreover, the failure analysis, is directly done on the expected functions of the customer (instead on the failure modes, very often ambiguous) because it is based on the Functional Analysis results, codified in the Correlation Matrix. The R. M. is, therefore, the most natural reference document during the development of possible improvement technical solutions. The design Team will be able to refer to the information collected during the previous design activities, avoiding to study already solved problems. This strongly influences the *Time To Market*, a fundamental parameter to acquire leadership positions in a very competitive market as the one of vehicles.

Physical Decomposition / Interfaces〉〉〉 Functional Decomposition	Components / Elements / Interfaces		INTERFACES										
			COLUMN CONTAINING RPS	STEER	DEVIO GUIDA DRIVING SWITCHS	DRIVING SWITCH	ALIMENTATION	CONNECTIONS	DRIVING BOX	TRAVERSE	KNEES PROTECTION	KEYBOARD	AIRBAG
Base Functions	*Elementary Function*		A	B	C	D	E	F	G	H	I	J	K
VEHICLE HANDLING	Guarantee a good handling while parking	1	9				9	9					
VEHICLE HANDLING	Guarantee a good handling while driving	2	9				9	9					
INTERACTION WITH THE INFORMATIVE SYSTEMS	Getting signals in input	3	9					9					
INTERACTION WITH THE INFORMATIVE SYSTEMS	Transmit signals in output	4	9					9					
SAFETY	Connect the steer with the driving	5	9	9					9				
SAFETY	Guarantee the electric alimentation	6	9				9	9					
SAFETY	Guarantee the mechanical handling in case of EPS system disconnected	7	9	9					9				
VEHICLE PERFORMANCE	Contain the driving effort	8	9									1	
COMFORT	Guarantee the silence	9	9		3								
COMFORT	Not vibrate	10	9	9	1	1			9	9			
FUNCTIONALITY/ ERGONOMIC	Contribute to the ergonomy of driver's seat	11	9								1		

Fig 2

4. CONCLUSIONS

Functional Analysis is very important to achieve the objectives stated above. It was initially thought necessary to distinguish two types of action: the first aims, in objective terms and independent from other technical solutions, to identify the whole range of services expected by the client and the interactions which are established between the product and the external environment (Functional Analysis of Needs or External); the second aims to identify all the relations which are established among the components of the product by virtue of the particular technical solutions adopted (Functional Analysis Technical or Internal).

The results of the analysis were then codified into a Correlation Matrix which constitutes a concise and compact functional chart for the specific product and has been adopted, together with the Risk Matrix, by the Technical Management of FIAT AUTO as the standard form of technical rules.

REFERENCES

(1) Vigier M.G., *Pratique de la maitrise statistique des procédés: M.S.P. ou S.P.C. (Statistical Process Control)*, Les édition d'organisation.

(2) Cloarec J.M., *Analyse fonctionnelle ou la préparation de l'AMDE(C)*, Maintenance & entreprise, novenbre 1993.

(3) Banci G., *L'Analisi Funzionale come premessa delle metodologie per l'affidabilità dei sistemi meccanici*, Tesi di Laurea, Università degli Studi di Firenze, 1999.

(4) Fox J., *Quality Through Design*, McGrow-Hill, 1993.

(5) Arcidiacono G., Avenatti R., Capitani R., *Definizione delle linee guida per la stesura della normativa tecnica di un sistema di guida elettrica*, XXVIII Convegno Nazionale AIAS, Vicenza, 8-11 Settembre 1999.

Methodologies for the development of high quality, reliable steels

J R PATEL, P H BATESON, and **K RANDERSON**
British Steel Limited, Rotherham, UK

ABSTRACT

British Steel recently supplied a high strength-high toughness weldable plate steel for the construction of the Oresund Bridge, which connects Malmo in Sweden to Copenhagen in Denmark across the Oresund Strait. The Bridge will carry both rail and road traffic, and safety is of paramount importance with one important feature being the use of high strength, high toughness, weldable steel for the construction of the box sections which support the approach bridges. One of the steel grades supplied required a minimum yield strength of 460 N/mm^2 and a guaranteed Charpy impact toughness of 27 J at -50°C. British Steel had not previously supplied a steel of this strength and toughness using a thermomechanical controlled rolled (TMCR) route, i.e. a steel directly usable off the mill, without the need for further heat treatment to develop the properties. This paper describes the various methodologies used by British Steel in the development of steels for safety critical situations such as the Oresund Bridge.

BACKGROUND

The demands on modern steels have increased over the last 20 years, and continue to become more severe in nearly every sector in which they are used. In certain market sectors such as offshore platforms and oil and gas pipelines this demand has taken the form of a requirement for increased strength whilst maintaining toughness and weldability. The driving force for the increase in strength is to allow designers to utilise thinner gauge steels and thus save on structure weight and the volume of welding required during fabrication. The need to maintain the toughness is to ensure fitness for purpose and adequate resistance to fracture. The need to maintain weldability is linked to producing a sound structure, free of defects and at minimum cost to the fabricator. There is pressure therefore on the steelmaker to produce steels to comply with these requirements at a cost to the user which does not negate the benefits gained from the other changes. Structural steels for bridges is another area which is following this trend with a move to higher strength, notch tough steels. An example of such a bridge is the Oresund bridge which links Copenhagen in Denmark to Malmo in Sweden. The Oresund link consists of:- a 3.7 km sunken-tube tunnel, a 3.9 km reclaimed island and a 7.47 km bridge which connects the reclaimed island to Sweden. The Bridge will carry both rail and road traffic, and the Oresund Strait is a busy shipping route. Safety is of paramount importance and one important feature was the use of high strength, high toughness, weldable steel for the construction of the box sections which support the approach bridges. A large part of the order was for conventional steels however a significant tonnage was for a high strength steel of 40 mm thickness with a guaranteed Charpy impact toughness of 27 J at a test temperature of -50°C (460ML grade).

"

Achievement of increased strength and toughness properties is often difficult especially in micro alloyed plate steels because most of the factors which increase strength such as solid solution and precipitation strengthening have a detrimental effect on toughness. The single factor which can be influenced to improve both toughness and strength is the ferrite grain size (Fig. 1). Water can be used to improve grain size by increasing cooling rate and reducing transformation temperatures however the challenge for British Steel was to produce the required steel without the use of water cooling after rolling.

It is not possible to simply generate a best guess and make full casts of steel, the cost of failure is large and therefore prior to the production of full scale casts, British Steel utilise a variety of methods for determining an optimum composition for a steel needed for applications such as the Oresund Bridge. These methods include:- Empirical relationships based on historical practice, Neural networks, Microstructural modelling and Pilot mill trials. The remainder of this paper describes how these methods are used to develop steels such as those used for the Oresund Bridge.

METHODOLOGIES USED FOR STEEL DEVELOPMENTS

It is important to understand some metallurgical principles prior to describing the various options available for steel development. The strength and toughness of a structural steel are basically controlled by the relative contributions of solid solution strengthening, precipitation strengthening, and ferrite grain size. The contribution from each depends on the composition and the rolling schedule, however in general the first two parameters have a detrimental effect on toughness whereas refinement of the third has a beneficial effect on both strength and toughness. Empirical equations have been developed over the last twenty years which allows the strength of a structural type steel to be predicted from the composition and a knowledge of the steel microstructure. A typical equation is given below:-

Yield Strength = 70 + (32% Mn+84% Si+33% Ni-30% Cr+38% Cu+11% Mo) + k_y $d^{-\frac{1}{2}}$ where k_y = 18.1 and d is the ferrite grain size in $mm^{-\frac{1}{2}}$ (Reference (1))

The terms for Mn, Si, Ni, Cr, Cu, Mo, relate to solid solution strengthening. The relative effect of precipitation cannot be simply predicted from the composition because it is strongly influenced by the rolling schedule, however it will be related to the amount of Niobium (Nb) and Vanadium (V) added to the steel and to the rolling schedule. Figure 2 indicates the basic contribution to strength from the solid solution strengthening mechanism for the steel used for the 460ML grade. It is evident from the figure that to achieve the minimum strength specification requirement using a conventional chemistry it was necessary to ensure a significant contribution from grain size/precipitation strengthening. Contributions from theses mechanisms can be simulated by the use of metallurgical modelling, neural networks and trials on the pilot mill.

Several forms of data assessment are used on a routine basis in the process of developing new steels either singly or in combination. In many cases the starting point is a comparison of available steel grades and the new specification from which will follow a definition of the problem in terms of factors such as:- increase in strength required, increase in toughness required, increase in gauge range required, composition restrictions, often linked to weldability needs etc.

Most new steels are developed as an extension and improvement of an existing grade and the Oresund Bridge steel was no exception. In the past a review of the relevant literature would be extended to include a more detailed statistical assessment based on available commercial or laboratory data. This step is important because much of the data in the literature, while providing a good indication of trends, is not always a good indicator of absolute value. Statistical assessments can be based on the important variables identified in the literature and in this case can give clear guidance about the required composition and/or process envelope. If the assessment is done with a proper understanding of the limits within which processing must be restricted then one outcome is a fuller appreciation of the areas of difficulty. This in the past would have been the focus for some pilot plant laboratory work.

A modern development of the statistical assessment is the use of neural networks. Neural networks is a relatively new technique for British Steel to use to assist in the development of steels. A neural network technique has been employed which can recognise complex relationships and develop a quantitative method for estimating the yield and tensile strength as a function of steel composition and rolling parameters.

The traditional assessment of structure properties referred to above has involved the detection of individual effects, each of which has then been studied in detail. The most recent technique in steel development has been to reverse this trend and use metallurgical understanding to construct a model of overall behaviour. British Steel are developing a through process model to be able to predict strength and toughness from an understanding of metallurgical principles for a variety of products including plate. Although not complete the model is sufficiently developed to be able to predict trends in strength and toughness from a knowledge of composition and processing details. It is intended that the model will account for reheating (including solution of precipitation and grain growth), rolling (to include the effects of the applied temperatures and strains on recrystallisation, grain growth and precipitation), and cooling to finally predict microstructure and properties. The model can give a prediction of how the austenite grain size varies through the rolling schedule (Fig. 3) to the end of rolling and through transformation to the final ferrite grain size, this allows a prediction of final properties. The real advantage of the modelling method is that it can cope more easily with a move outside the previously available data range, e.g. by considering new rolling schedules or composition ranges. The model needs to be based on an adequate metallurgical understanding and to be backed by a proper validation process which could be a combination of commercial and laboratory data.

The final stage of the development prior to commercial production is to produce laboratory casts and roll on the pilot mill. British Steel utilise a pilot plate mill situated at its product research centre at Swinden Technology Centre to dynamically develop new compositions and existing rolling schedules in a cost effective manner. The various facilities available include a reheating furnace capable of temperatures upto 1300°C, a two high reversing mill stand, and water cooling facilities which can be applied both during and after rolling if required.

Development of steels can only be accomplished if the results obtained on the pilot mill are readily transferable to the commercial mill. Several factors are accounted for in the pilot mill trials. These include:- (a) Accurate simulation of reheating practice for either commercial slab sections or laboratory produced melts, (b) The commercial rolling schedule can be copied or varied to determine the effect of changing parameters such as finish rolling temperatures, (FRT) hold strategies, (delays in the schedule), and drafting pattern.

For the Oresund Bridge development the results from laboratory mill trials (in terms of mechanical properties) are in good agreement to those obtained from a commercial rolling, and are demonstrated in Fig. 4. The figure shows the average strength and toughness properties from the commercial order together with the results from a laboratory melt processed on the pilot mill. It can be seen that the laboratory processed plate has given similar results to that obtained from commercial processing.

The demands on steel and steel makers continue to become more severe. To increase the speed of developments British Steel has various methodologies which can be used individually or together and it is hoped this note has given an insight into these methodologies.

REFERENCES

1. W.B. Morrison, B. Mintz and R.C. Cochrane: 'Structure Property Relationships in Controlled Processed Steels', Controlled Processing of HSLA Steels, University of York, 1976.

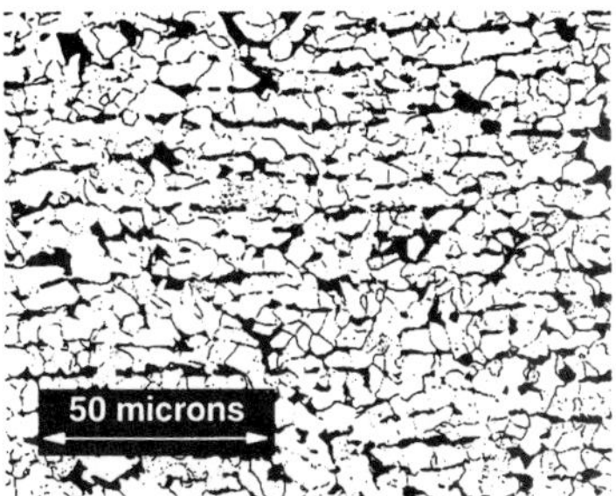

Fig. 1 Typical Ferrite/Pearlite Microstructure

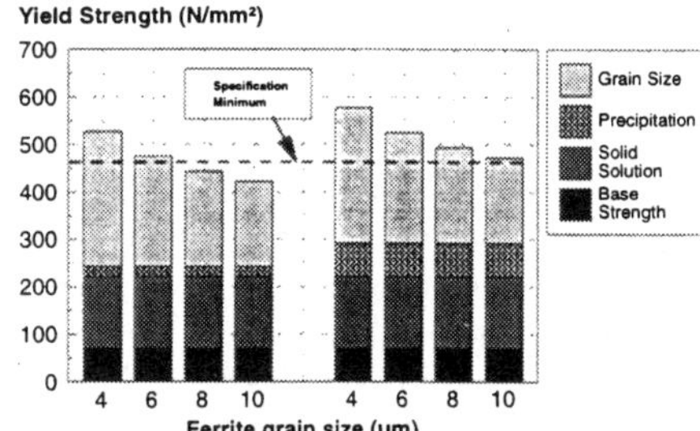

Fig. 2 Factors Contributing to Overall Plate Strength for 460ML Grade

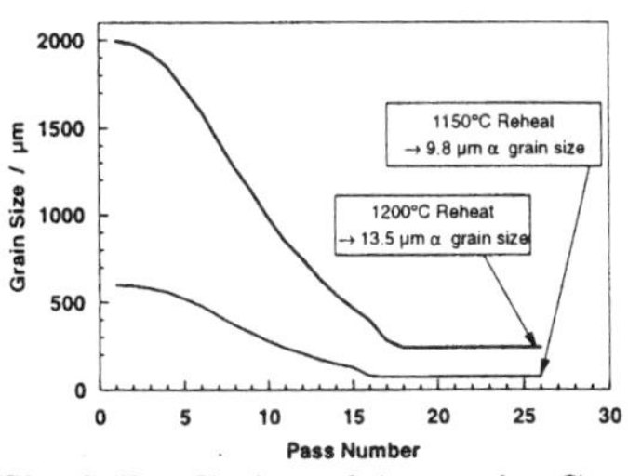

Fig. 3 Prediction of Austenite Grain Size Through the Rolling Schedule

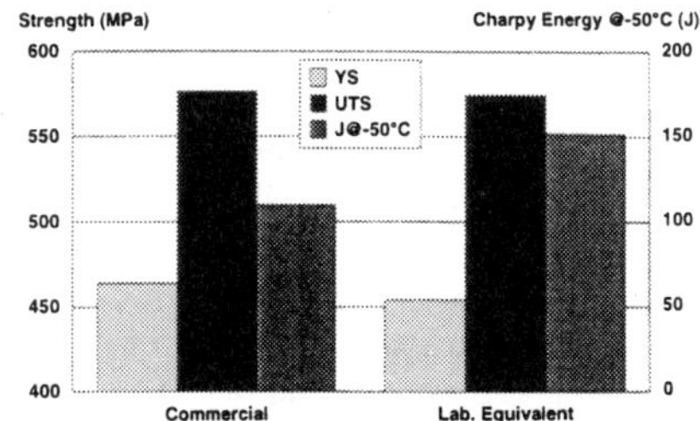

Fig. 4 Comparison of Commercial and Laboratory Rolled Plate

Quality achievement in toy manufacturing through a multimedia expert system

S F CHAN
School of Science and Technology, The Open University of Hong Kong, China
T T WONG
Department of Mechanical Engineering, The Hong Kong Polytechnic University, China

Abstract
Transfer technology is demonstrated in this work in which Universities and Industry combine to advance the development of computer aided toy manufacture. Toy manufacture is inherently small scale business initiative which inhibits the investment of research .The system uses Visual Basic due to its ability to integrate pictures, animation, text and sound. The system is built upon four major elements: knowledge database, expert mode, training mode and test mode. The proposed system provides a low cost, , robust environment for the plastic industry in diagnosing design defects. It can also be used as a learning and training tool to enhance productivity of the operating personnel. Experts from local companies were invited to evaluate the proposed system and comments received were positive.

1. Introduction

Emerging technologies in compounding, blending and alloying have led to a large number of commercially available plastic materials. This provides designers with a tremendous opportunity to develop new products from a wide range of choices of material. However the rapid growth in the number of available plastic materials can also lead to considerable confusion in the determination of the most appropriate material for a given application. Thus there are increasing difficulties in identifying product defects and characteristic behaviour of the plastic material. Consequently, many companies have to carry out expensive redesign work, product defect analysis or in-house testing in order to find out possible solutions to rectify the problem.

Companies can easily retrieve a great deal of information from material suppliers that describe the behaviour of specified plastic material both during processing and in its final shape. Customised databases fall short in the provision of a complted materials management system. There is a major need to generalise and standardise the techniques applicable to the industry. This initiative will provide a strong and coherent approach to producing a synergetic effort to maintain a competitive edge for the local toy industry.

Hong Kong has started its light manufacturing industry in textile, plastic flowers, wigs, toys and light electronics and electrical goods since 1960s. With the benefit of relatively inexpensive labour, free import of raw materials and the increasing popularity of plastic products in the worldwide market, Hong Kong expanded rapidly from simple plastic products to complex mechanical toys in recent years. Whilst the export value of Hong Kong toys was only HK$120 million in 1960, it has grown to HK$2 billion in 1981. In 1995, Hong Kong became the world's largest toys exporter (including re-exports of goods manufactured in China) with a record value of HK$77 billion [1]. Its products are now distributed all over the world and the products became more competitive due to its flexibility and adaptability in a dynamic and changing environment. Since toy manufacturing is the most popular category in local industry, its manufacturing processes are therefore an ideal candidate for investigation. Traditionally technical personnel working in the toy industry address problems in the manufacturing process by such primitive methods as past experience, direct sensation, and trial and error. Although references summarizing the major sources and remedial measures of different kinds of defects in manufacturing processes have been published[2-10], they are still rarely referred to in real life situations. The subsequent misinterpreted causes of defects would likely lead to inappropriate remedial actions, which in turn could result in a waste of resources.

This is one of the major reasons for upbringing the quality and smoothness of operation in the plastic product manufacturing industry, and in particular, the toy manufacturing industry. Recent advances in multi-media technology shed light on the development of computerised trouble shooting systems [2] to provide an effective tool in the toy industry. This re-engineering process can be achieved through effective application of low cost multimedia technology.

A development project in computerised plastic product diagnosis (CPPD) system has been carried out at the Open University of Hong Kong. It was part of the quality assurance process to capture a more representative view of what a diagnostic system should feature. In order to ensure the validity of the conceptual design of the system, a questionnaire survey had been conducted to analyse the users' expectations at the initial stage. During the process of developing the pilot model, a system evaluation was conducted at the completion stage to ensure the robustness of the system. The design of the system was aimed to achieve three main objectives:

(a) to generalise trouble-shooting techniques for the toys manufacturing processes;
(b) to design and develop a robust and expandable computerised diagnosis system incorporating generalised features to function in a multi-media environment;
(c) to ensure that the knowledge of experienced technical staff are retained in a computerised system for future retrieval.

Demonstration of the proposed system would reveal that it provides a low cost, robust environment for the industry in diagnosing design defects, incorporating industry-wide trouble-shooting techniques and procedures. It can also be used as a learning/training tool to enhance productivity of the operating personnel. Experts from local companies were invited to evaluate the proposed system and comments received were quite positive.

2. Features incorporated in the CPPD system

Organisations such as toys manufacturers that deal with technical information for plastics are facing a common operation problem: the specifications and characteristics of various plastic product defects which occur during and after processing are often spread within among

Improving measurement precision of laser displacement sensors for surface inspection

J CHEN, R WARD, and G WATERWORTH
Faculty of Information and Engineering Systems, Leeds Metropolitan University, UK

ABSTRACT:

A new application of laser displacement sensors in measuring surface texture has been reported recently from industry (1). For low-resolution sensors, internal noise is the main factor affecting measurement precision. This paper proposes methods dedicated to in-process measurement to reduce the internal noise in the sensor output signal. An artificial neural network (ANN) has been trained to evaluate suitable values for the average sample number (N). Experimentation o surface profile measurement shows that the proposed methods can dramatically improve the measurement precision whilst maintaining a suitable dynamic response.

1. PRINCIPLES OF INTERNAL NOISE REMOVAL.

A LA40HR laser sensor (resolution $10mV/20\mu m$) (2) is used in this research. When measuring a fixed distance, the output voltage of the sensor shows variation due to its random internal noise Fig. 1(a) shows a sample profile measurement of a surface with roughness average $Ra = 3.2\mu m$ which is less than the sensor resolution ($20\mu m$). The desired surface profile shown in Fig. 1(b) estimated with a filter, is hidden by the internal noise.

1.1 Statistical principles of improving measurement precision.

In general, the internal noise could be a combination of a random component n_{ri} and a periodica compon0ent n_{pi} (3), and the output signal d_i is in the form of equation 1.

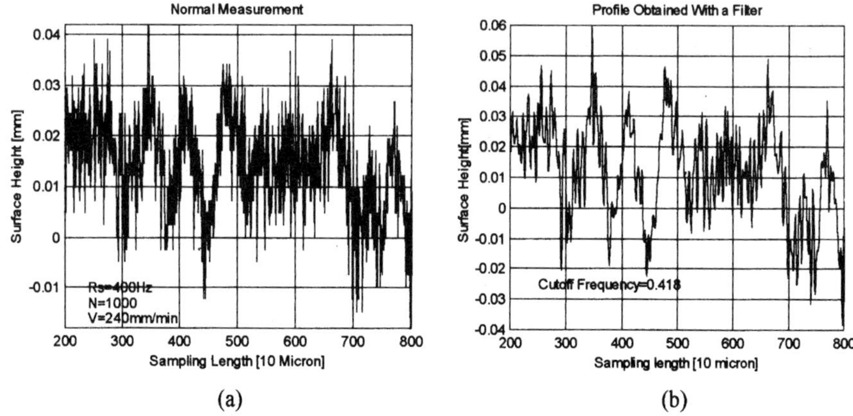

Fig. 1 Surface Profile Measurement with Normal Method

$$d_i = d_0 + n_{ri} + n_{pi}, \qquad i = 1, \dots , N \tag{1}$$

where d_0 represents the true distance D_0 between the sensor and the surface, and $(n_{ri} + n_{pi})$ produces the signal variation. The mean M of this signal can be approximated in equation 2 (4)

$$M \approx M^* = d_0 + M_p \tag{2}$$

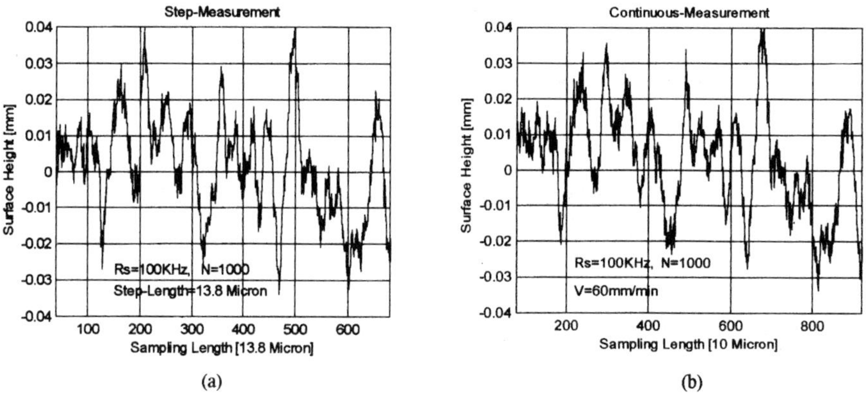

Fig. 2 Measurement of Surface Profile

where M_p is the mean of n_{pi}. Since M_p Can be considered as a shift, M^* can then represent the true distance variation accurately if the average sample number N is large enough.

This conclusion can be directly used for off-line surface profile measurement. The laser sensor is moved step by step longitudinally along a round metal bar (the step length is 13.8μm for the devices used). At step j, the mean of N measurements is calculated after using M_j , M_j^*, d_{oj} instead of M, M^*, and d_o in equation 2. Fig. 2(a) is a sample surface profile measured with this ehthod.

1.2 Methods to Achieve High Precision for In-Process Measurement.

The key idea of the proposed methods is to keep the sensor moving while the measurement is taken continuously, so in-cycle or -process measurement becomes possible. At the same time, the internal noise could be reduced.

If d_{ji}, $i = 1, ... , N$, are the continuous measurements over a very small sampling length (e.g. $10\mu m$) when the sensor is continuously moving, the mean M_j of the N output voltages can be accurately approximated with M_j^* as the average sample number N becomes large enough

$$M_j = \frac{1}{N}\sum_{i=1}^{N} d_{ji} \approx M_j^* = \frac{1}{N}\sum_{i=1}^{N} d_{0ji} + M_p \qquad\qquad j = 1, ... , K \qquad\qquad (3)$$

The mean of d_{0ji} is used to represent this small part of surface as a small step, and a $10*K\mu m$ surface can be represented with K steps. Fig. 2(b) is a sample surface profile measured this way.

It is possible to design a filter to remove the internal noise as shown in Fig. 1(b). The difficulty encountered is the determination of a reliable cut off frequency of the internal noise. A filter will also add in extra components into the signal (5).

2. RELATIONSHIP BETWEEN AVERAGE SAMPLE NUMBER N AND MEASUREMENT PRECISION.

Fig. 3 shows the relationship between the average sample number N and the measurement precision in terms of standard deviation σ based on the experimentation results of using the sensor to measure a fixed distance.

It is seen that when N increases from 1 to about 3000, σ decreases dramatically from $10mV$ to $1mV$. Even $N = 10$ will result in $\sigma = 6.076mV$. But in practice, the average sample number N is limited by the maximum sampling frequency of a sensor.

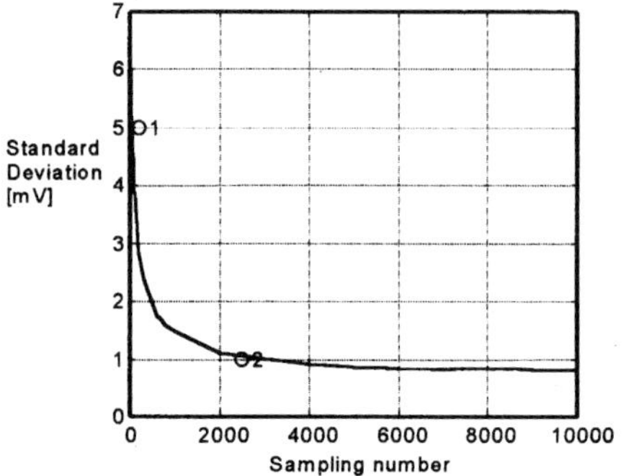

Figure 2. Relationship of Average Sample Number & Measurement Precision

It is necessary to determine the average sample number N for a predetermined precision σ. This is a typical non-linear problem which an ANN can solve. In Fig. 2, marks 'O1', 'O2' are the points *(32, 5)* and *(2321, 1)* obtained by using a trained ANN. An alternative would be the use of MATLAB Polyval command, which is able to compute coefficients of a polynomial equation.

3. CONCLUSION.

This paper proposes methods to achieve high-precision distance measurement with low-resolution laser sensors. The methods are suitable for in-process measurements of surface roughness or inspection of coated surfaces.

REFERENCES

(1) CyberOptics Corporation (1999), 2D Profiler Performs Non-Contact Measurement, USA

(2) Matsushita Automation Controls Ltd. (1989), Sensor Technology, England

(3) Precision Devices, Inc. (1999), Surface Profile Filtering, Italy

(4) Steven W. Smith (1997), The Scientist and Engineer's Guide to Digital Signal Processing, California Technical Publishing, USA, pp 11-34

(5) Emmanuel C. Ifeachor and Barrie W. Jervis (1993), Digital Signal Processing, A practical Approach, Addison-Wesley Publishers Ltd, UK, pp 252-490

Defects monitoring incidence on quality forecasting

I CRISTOFOLINI, G MANCUSO, and **G WOLF**
Department of Mechanical and Structural Engineering, University of Trento, Italy

Abstract

The present work aims defining the relation existing between kind and frequency of defects monitoring and effective defects trend, being it important developing reliable forecasting model. The data concerning the main type of defects encountered during two years of production are analysed, considering the case of a consolidated model of refrigerator by Whirlpool S.p.A.–Trento. FFT analysis of the data allows recognising for each kind of defect the same value for main frequency, related to defects monitoring frequency. By the use of dynamic data structuring a hypothesis for performance enhancement is finally proposed.

1. INTRODUCTION

Quality measurement is obviously critical in Quality System, to evaluate production process reliability, defectiveness trend, non-quality costs and so on (1). An accurate analysis of the "non quality data" allows, in fact, pointing out the defects, which most frequently verify, in order to look for causes and, consequently, solutions (2). In a previous work (3) we developed a data structuring method (the so called "dynamic data grouping") allowing to relate the different types of defects to the various production process steps.
Considering now forecasting model, we note that, as it often verifies, the quality level to expect is not critically evaluated, just imposed on the basis of a general request of defects reduction. On the contrary, we think it is important to analyse the effective production process trend, pointing out eventual critical steps, to define quality level reasonably expectable and consequently reliable forecasting.

The aim of this work is thus analysing the trend of the most frequently encountered defects, to evaluate if an "intrinsic defectiveness" can be found. The eventual relation to different steps of the production process, as derived by the data structuring previously defined, is also considered.

2. DATA ANALYSIS

We analysed the data concerning the main defects encountered during a period of two years, related to a class of refrigerators produced by the Whirlpool S.p.A.-Trento. In order to analyse an expectably "stable" product, it has been chosen a model, which has been produced since a long time.
The data information system actually used at Whirlpool S.p.A. collects the data related to the various types of defects taking them from claims and service calls. The data, adequately codified, are collected continuously and evaluated every three months (4).
For each type of the main defects, a curve simulating the trend during the two years has been elaborated, using the modeller Eureka Gold – CadLab. Figure 1 shows some example, related to the defectiveness trend of different components.

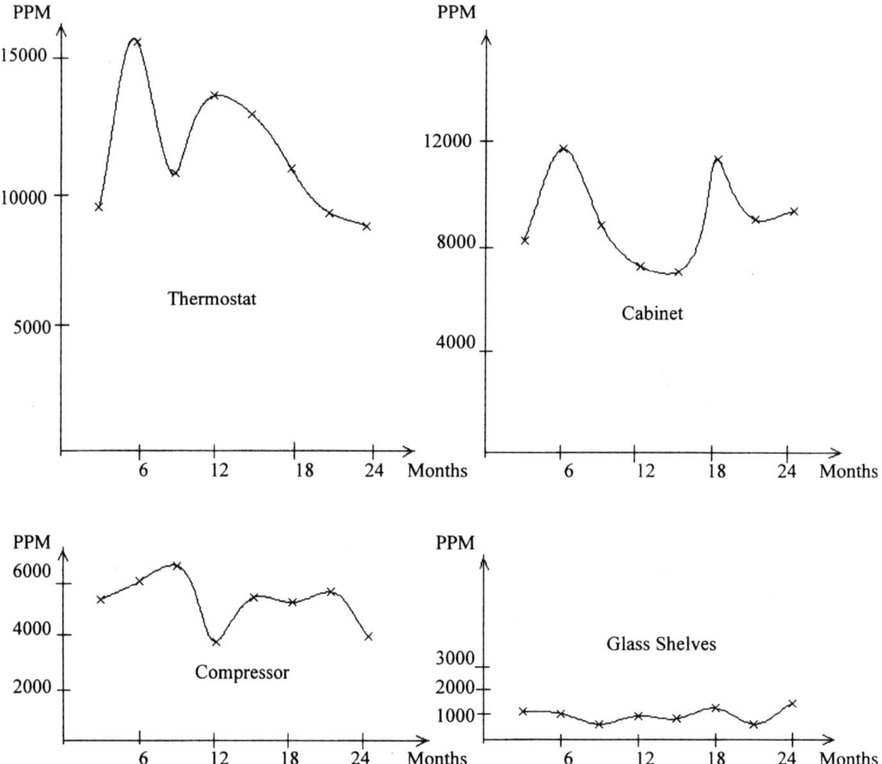

Fig. 1 – Defectiveness trend for different components (PPM vs. time)

The defectiveness trend look very different for the various components, but the subsequent FFT analysis of the data reveals that, on the contrary, some periodicity exists.

Let us firstly spend some words about the methodology used to analyse our data.

The use of Fourier transforms to analyse the defectiveness curves was suggested by the particular forms of these ones. They seemed to evidence some periodicity and, as known, any periodic function can be easily interpreted if we transform it in the frequency domain (5,6).

Let $d(t)$ a curve in the time domain. The Fourier transform identifies or distinguishes the different frequency sinusoids (and their respective amplitudes) which combine to form an arbitrary waveform. The mathematical relation is stated as:

$$D(f) = \int_{-\infty}^{\infty} d(t)e^{-j2\pi ft}\, dt$$

where $D(f)$ is the Fourier transform of $d(t)$.

The FFT analysis (7) requires sampling the function $d(t)$. Let the sampling interval T. The function $d(t)$ is continuous at $t = nT$, and a sample of $d(t)$ at time T is expressed as:

$$\hat{d}(t) = \sum_{n=0}^{N-1} d(nT)\delta(t - nT)$$

The discrete Fourier transform of the above equation is therefore:

$$\hat{D}\left(\frac{n}{NT}\right) = \sum_{k=0}^{N-1} d(kT)e^{-j2\pi nk/N} \qquad n = 0,1,\ldots,N-1$$

In the case at study we already have the availability of discrete values of defectiveness sampled at a time interval T of 1 month, being the total number of data $N = 22$. The data were computed by MicroCalc Origin.

Figure 2 shows the curves derived from FFT analysis related to the aforementioned components. In the following paragraph the interpretation of FFT results is explained.

3. CORRELATION QUALITY MONITORING-QUALITY TREND

The most interesting aspect revealed by the FFT analysis of the data (looking so different in the curves showing the number of defects versus time) is the presence of a frequency peak, localised in the same frequency interval for all the components, as evidenced in figure 2. Moreover, the presence of other, although lower, peaks, which are localised at frequency multiple of the main, allows recognising the main one as effectively characteristic of the system.

This "main frequency" reveals that every six months a "defectiveness cycle" tends to repeat. Being this trend confirmed for all the components analysed, we could reasonably relate it to the defectiveness monitoring frequency, as following explained.

We know being the defect data continuously monitored, but only every three months evaluated. Our analysis shows that such a monitoring frequency does not allow a sufficiently reliable control of the process. We can suppose, in fact, that when the analysis of the defect data shows an enhancement of the defects, related to the previous three months period, special attention is paid to the whole production process. By this way, an improvement of quality is revealed by the subsequent data monitoring, three months later, so that the attention is relaxed. As a consequence, the data revealed by the subsequent monitoring show again a defectiveness enhancement, which appears six months after the first one, and so on.

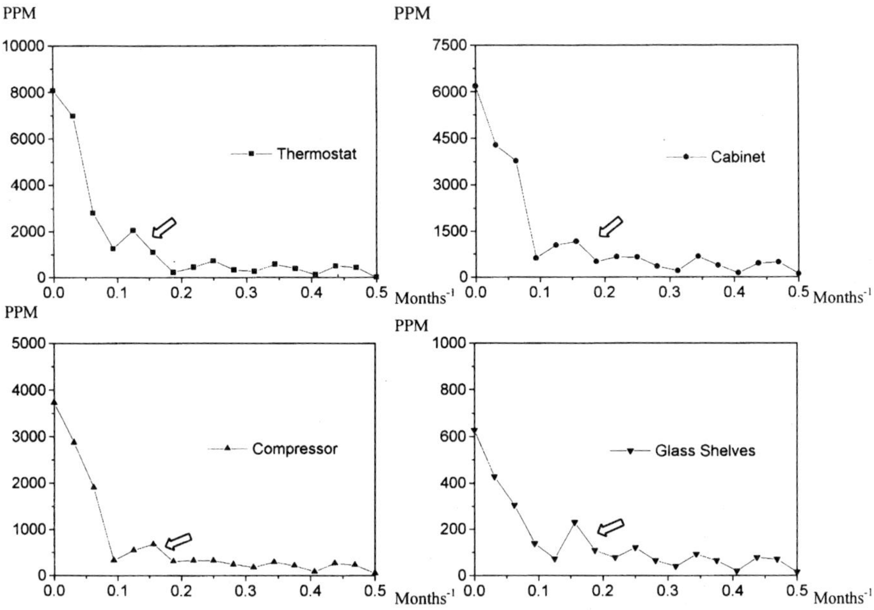

Fig.2 – FFT curves for different components (PPM vs. time^{-1})

Our analysis allows thus establishing a relation between data monitoring and quality trend, which can be extended to other cases. That means, such a relation can be reasonably hypothesised when the FFT analysis of the data shows the presence of a frequency peak, common to the various defectiveness curves analysed, which is localised in a zone corresponding to a multiple of the defects monitoring frequency.

Moreover, the analysis of the FFT diagrams allows us recognising a probable value for the "intrinsic defectiveness level" (IDL), which is characteristic of each kind of component analysed in the corresponding production process situation. This value can be identified in an interval close to the half value number of PPM corresponding to the main resonance frequency previously defined. Unfortunately the number of available data do not allow us showing the effective trend of the defectiveness curve around IDL.

4. QUALITY TREND ENHANCEMENT

The first hypothesis for quality trend improvement is thus obviously concerning an enhancement in defects monitoring frequency. As it always happens, it is necessary carefully evaluating the cost of this enhancement, comparing it to the "non quality costs" otherwise verified. Anyway, being the data collected continuously, it is probably possible enhancing the

allow them to be better prepared for the environment into which they will be exposed and will enable them to shape the destiny of their businesses rather than be shaped by the situations and traditional beliefs held by others.

Within Pedigree Masterfoods we have developed a simple model to highlight what we are trying to achieve. This model was taken from a paper written by John Moubray, Managing Director of Aladon Ltd on the subject of "The responsible custodianship of physical assets". It entails the requirements for building a house and focuses on the need for sound foundations as a starting point for any successful venture.

3.   RCM II AND THE OPERATING CONTEXT.

RCM II, as developed and copyrighted by John Moubray, is a methodology which enables the correct foundations to be defined and built thus enabling the walls (resources, training, skills, tools and spares) and the roof (systems needed to manage the resources) to be correctly identified and constructed.

This is done by defining the **"Operating Context"** for the equipment or system under analysis. The operating context enables the analysis group to identify **"the Wants and the Cans"** for the equipment or system within the Business Context.

The concept of the Operating Context is a simple one, after all, what is a machine specification other than a definition of what you want it to do within your business to enable you to achieve your goals ? But, from the many hours of discussions between myself, DataStream, Aladon and other F.M.C.G. companies and many of the challenges faced by me as a Reliability engineer, I have realised that:-

Defining the Operating Context is the FMCG industry's biggest challenge and also, currently, its greatest weakness.

Failure to specify the correct operational context for equipment and/ or reviewing it on regular basis is the primary raison d'être for all inefficiency, unreliability and risk.

Once the operating context is defined the RCM II methodology asks a further 6 questions, in a structured way, which enable the most cost effective reliability strategy to be derived and applied.

4.   ADVANTAGES OF THE METHODOLOGY.

The major advantages of the RCM II methodology over the preponderance of other asset maintenance strategy theologies are:-
In general:-
   It allows you to manage "risk" rather than take "risks".
In particular:-

a)   Safety and the environment - enables you to accurately assess the risks to your business without experiencing the failure and also, allows you to control your exposure.

b)   Hidden failures - Enables this type of failure, which has no impact during normal operation and only becomes critical as part of a multiple failure scenario e.g. Piper Alpha, Chernobyl, Bophal etc., to be identified, measured and managed.

c)   Its purpose is to preserve the functions of the assets within the operational context of your particular business in the most cost effective way.

None of these issues are addressed, in a structured way, by any other maintenance analysis tool which I have been involved with.

5.   DESIGN.

The issues detailed above could be dismissed as only applying to Pedigree Masterfoods, or to the F.M.C.G. environment, and only applicable to a small percentage of the engineers being produced by our academic establishments. To do this would be would be short sighted and would overlook one of the core concepts painfully learnt from applying the RCM II methodology. That is the relationship between the Operating Context and the designer.

Many companies rely on the design expertise of equipment suppliers to provide them with the correct "piece of kit" to do the job they want. Unfortunately few Suppliers understand the Operating Context and production dynamics of that company/ plant into which their equipment will be placed and, therefore, fail to fully interpret their requirements due to insufficient detailed information.

To reduce this impact to a minimum it is essential that the project specification is developed to compensate for this lack of understanding. It is also important that equipment suppliers are capable of asking the awkward, non standard, questions which help potential customers to more clearly understand what it is they actually want.

The equipment specification should be indistinguishable from the initial Operating Context i.e. it defines and compares all the "Wants" and the "Cans". It identifies where equipment will not satisfy the requirements of your business and what can be done to ensure that the impact of this is minimal and manageable.

6.   CASE STUDIES.

The three case studies which will be discussed during the conference will highlight many of the issues detailed above. Each shows how the Operating Context and Reliability were de-prioritised in favour of other business "needs" and how engineers failed to influence this course of action. They will also show how the application of RCM II methodology engineering retrieved each situation and actually achieved the original business goals more successfully. This, unfortunately, was done in hindsight rather than within a culture where "getting it right first time" actually means just that and not one where it actually meant for the "right" price, the "acceptable" numbers of man-hours or within the "required" time scale.

7.   **FUTURE ENGINEERS**.

Within many businesses RCM II is being used as the ultimate fire fighting tool and this has, in the past, been the case at Pedigree. Some senior managers have been surprised at what I can deliver, as an individual, and have therefore used my "skills" in critical situations. This perpetuates a vicious circle where the inference is that the abilities and skills are inherently mine and no one else's and, that any situation is retrievable. This simply isn't true.

The principles contained within RCM, if used correctly, will give any business the most cost effective design and reliability strategy for their Operating Context. This will be both sensible and defensible (especially where safety is concerned) and will also provide an auditable maintenance strategy. It would allow engineers to understand how they influence the success of their business both negatively and positively and, how they can contribute outside of the maintenance sphere. The biggest benefits from the application of this methodology are also the most intangible i.e. it has enabled all functions within the business to contribute together and therefore promoted ownership, team work, motivation and safety and environmental integrity.

The principles and methodology incorporated within RCM is only one tool in an engineers tool box although, my belief is that it is, a fundemental one. If our future engineers could be taught the theory and development of this methodology at an early stage of their career and links with industry were developed whereby this theory could be put into practice and the shown to be realistic, then all parties would benefit significantly. Businesses would be more competetive, engineers would grow in influence and respect and academic establishments could help grow the profile of engineering and their ability to offer services to businesses which help support their core role.

Flexible monitoring of quality with linguistic control charts

F FRANCESCHINI and **D ROMANO**
Department of Production Systems and Economics, Politecnico di Torino, Italy

ABSTRACT.

Quality attributes of products and services are frequently assessed using linguistic variables. A method to define innovative control charts for linguistic variables without using any arbitrary numerical encoding is proposed. It is based on the use of the OWA operator (Ordered Weighting Average) introduced by Yager to aggregate information given on a scale where only ordering property applies. A peculiarity of OWA is that the quality inspector is allowed to use a flexible decision logic to face the inherent uncertainty of linguistic assessments. An application to the control of a booking service in a local hospital is presented.

1. QUALITY ASSESSMENT BY LINGUISTICS

Quality attributes are frequently assessed on non-numeric scales using linguistic variables (1). This applies sometimes to products but nearly always to services. Typically final product conformance is assessed onto a two-state scale, e.g. acceptable/unacceptable, or a multiple-level scale, e.g. inadequate, poor, medium, good, excellent. In service quality, verbal scales with multiple levels are usually adopted. Whenever a person is asked a question about a product or a service we are measuring firstly the subject, secondly, and only indirectly, the object, see scheme in Fig. 1. What we really measure is the attitude of the subject towards the object and this attitude is expressed in linguistic form. Two sources of vagueness are present. Firstly, we do not know how the individual translates her attitude in words, whether in optimistic, neutral or pessimistic mood. Secondly, terms of human language have not a formal and unique definition, but rather a set of possible meanings, which the majority of people agree upon.

Nonetheless, subjectivity of measures fits very nicely with the semantic of the concept of quality as pure subject's satisfaction. Moreover, the language format, even if imprecise, is by far the more natural way to express a human attitude. The use of numeric scores would force the individual to an unfamiliar conversion of her psychical conditions resulting in loss of information and a deceptive sense of precision. We like to think of linguistic assessments just like the more adequate approximated rating where more accurate symbols are *de facto* impossible or unjustified.

Now we explicitly list two requirements for a consistent analysis on linguistic data.

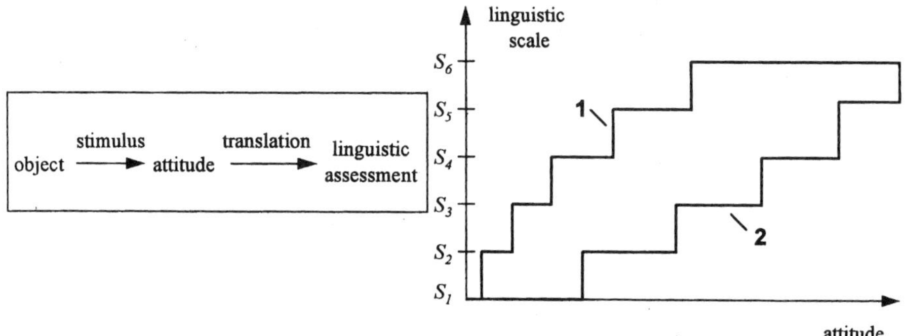

Fig. 1 – Left: steps involved in the generation of a linguistic assessment. Right: two different ways (1 is optimistic, 2 pessimistic) to convert attitude into linguistics.

1. Absence of numerical encoding of original data. Linguistic variables only satisfy an ordering property. Moving to numbers means introducing metrics which the original variable has not (for example, is it legitimate to assume that the difference between the 'poor' state and the 'medium' state is the same as between 'medium' and 'good'?). As conversion could be reasonably done in a manifold way reliability of results would be greatly questionable (a paradigmatic case is selection of ratings in QFD technique (2)). An important consequence of this requirement is that allowable computations solely involve operators that can handle ordinal data. Note that it is a current practice to adopt a numeric, but intrinsically ordinal scale and then perform standard arithmetic on it. An outstanding example is found in SERVQUAL (3), the most famous technique for assessing quality of services, where ordinary arithmetic average is applied to difference between numeric scores, ranging from 1 to 7.
2. Necessity of dealing with uncertainty deriving from lack of alignment of individual rating scales and vagueness of linguistics. A frequent danger of using numbers is the temptation to make exact calculations forgetting the uncertain nature of information in hand.

2. OWA-BASED CONTROL CHARTS

The problem of monitoring linguistic quality attributes has been addressed by a variety of proposals; a complete overview can be found in (4). Apart from charts where a straightforward and arbitrary numerical conversion is done, we signal the fuzzy control charts implemented in (5,6). However theses methods tend to resort eventually to numerics, after the assignment of a membership function to each linguistic level, thus violating requirement 1 above. Derived from fuzzy logic but with a minimal use of mathematics Yager (7) introduced a family of operators, called ordinal OWA (Ordered Weighting Average), able to aggregate a variety of information on a linguistic scale. Control charts based on ordinal OWA were already proposed by Franceschini and Romano in (8). Now we extend these charts allowing them to adequately dealing with the uncertainty in samples of linguistic assessments.

Let $S = \{S_1, S_2, \ldots, S_t\}$ the set of linguistic labels and $q_1, q_2, \ldots, q_n$ a sample of n evaluations. The ordinal OWA is defined as:

$$\overline{q} = \max_{k=1,2,\ldots,n} \min\left(B_k, Q(k)\right) \qquad [1]$$

where B_k is the kth largest of the q_1, q_2, ..., q_n and $Q(k)$, $k=1,2,...n$ is a non-decreasing function, called linguistic quantifier, with values in S, indicating how much the quality inspector (QI, the evaluations' evaluator) is satisfied having k positive evaluations out of n. Recalling that min and max in fuzzy set theory are akin to product and summation in ordinary numbers, definition [1] resembles a weighted average where Q weights the evaluations depending on their disposition in the descending ordered list. Top list evaluations are weighted less than bottom list. This is because an evaluation B_k in the sorted list represents itself plus all the preceding ones, namely all those who express satisfaction with a level at least B_k. Therefore B_1, representing none but itself, has the lowest weight, B_n, representing all the sample, the highest. Then the aggregate evaluation results from the interaction between the sample and the logic of the QI; it practically coincides with the element of S nearest to the crossover between descending list of evaluations and the ascending list of Q, where the degree of satisfaction of the QI, the weight Q, best balances the degree of satisfaction expressed by best k assessments, B_k (see Fig. 2). The possibility for the QI to change her attitude is given by taking a different quantifier. The most pessimistic and optimistic attitudes are expressed by quantifiers selecting the minimum and maximum assessment in the sample. All possible attitudes span between these two extremes and are accordingly expressed by a suitable quantifier. Yager has also suggested a way to measure the degree of optimism contained in a quantifier and different techniques to construct a quantifier expressing a given optimism. For the mathematical details on this topic see (7). Exploiting the flexibility of the QI's aggregation, we propose to monitor two averages, one optimistic and the other pessimistic, postulating that they copy the way humans express their own degree of satisfaction by linguistics. The rationale is that we aggregate consistently sample information by using a QI's logic as similar as possible to the logic of respondents. Unfortunately the latter is both unknown and discrepant among respondents. However QI can consider two extreme situations where all respondents are supposed to have the same optimistic/pessimistic attitude; thus the most suitable sample aggregation is unknown but lies

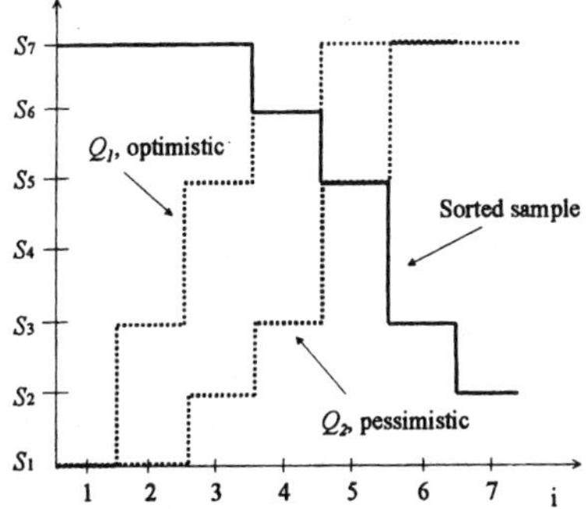

Fig. 2 - Graphical representation of ordinal OWA for samples with seven elements; the linguistic scale has seven levels too. The ascending stairs (dashed) are two linguistic quantifiers, the one optimistic and the other pessimistic; the descending stair is the sorted sample (solid). OWA aggregation yields S_6 for Q_1, S_5 for Q_2.

somewhere between lower and upper bounds established by the extreme aggregations. The line describing variations over time of quality becomes a strip of variable width. The chart completion needs to establish the control limits. They are duplicated as one pair holds for the optimistic aggregation, another for the pessimistic one. In practice, they are set up by a simulated sampling from the empirical distribution of individual observation obtained by a large set of stable data. Because of the discreteness of linguistic variables, it is impossible to realise a generic I-type error α, but we can only choose a limit whose α is closest to an established value (e.g. the standard figure of 0.0135). Note that: 1. graded out-of-control situations may occur, reflecting again uncertainty of the process; 2. the chart contains information also on the scattering inside the samples.

A quality survey in a local hospital was conducted adopting control charts as described before. An example of pattern is depicted by the $\bar{q}$ control chart for *service responsiveness* attribute of the booking facility. Fig. 3 shows the results obtained for 40 samples (eight samples per day) of n=10 questionnaires analysed on a hourly basis for five days of continuous service operation.

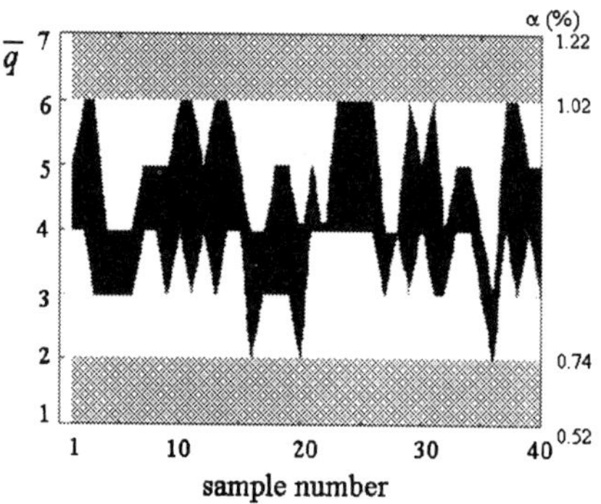

Fig. 3 - $\bar{q}$ control chart for the *service responsiveness* attribute.

REFERENCES

(1) Zadeh L.A., 1975, «The concept of a linguistic variable and its application to approximate reasoning: Part 1», *Information Sciences*, v.8, 199-249.

(2) Franceschini F., Rossetto S., 1995, "QFD: The problem of comparing technical/ engineering design requirements". *Research In Engineering Design*, v.7, 270-278.

(3) Parasuraman A., Berry L.L., Zeithaml V.A., 1991, «Refinement and Reassessment of the SERVQUAL Scale», *Journal of Retailing*, 67, 4, 420-450.

(4) Woodall W.H., 1997, «Control Charts Based on Attribute Data: Bibliography and Review», *Journal of Quality Technology*, v.29, n.2, 172-183.

(5) Raz T., Wang J.H, 1990, «Probabilistic and membership approaches in the construction of control charts for linguistic data», *Production Planning & Control*, v.1, n.3, 147-157.

(6) Kanagawa A., Tamaki F., Ohta H., 1993, «Control charts for process average and variability based on linguistic data», *Internat. Jour. of Prod. Research*, v.31, n.4, 913-922.

(7) Yager R.R. «Applications and extensions of OWA aggregations», *International Journal of Man-machines Studies*, v.37, 103-132.

(8) Franceschini F. and Romano D., 1999, «Control chart for linguistic variables: a method based on the use of linguistic quantifiers», *International Journal of Prod. Research*, v.37, n.16, 3791-3801.

Density profiles of wood composites as quality improvements

S HIZIROGLU
Department of Forestry, Oklahoma State University, Stillwater, USA

ABSTRACT

In this study, basic principles of density profile measurement is presented within the perspective of its application to various types of wood composites. Density profiles and surface roughness characteristics of laboratory manufactured medium density fiberboards were determined by using an X-ray density profiler and a fine stylus type of surface roughness profilometer, respectively. Based on the initial experimental results of this study, it was found that as density of surface layers of the specimens increases their surface characteristics could also be improved which would influence overall quality of the panels.

INTRODUCTION

Wood composites including particleboard and fiberboard are usually manufactured in the form of three layers where fine furnishes are used in the faces and coarse material is in the core layer. The mat faces are consolidated first and the resin in these layers is cured before core portion of the mat reaches to the final temperature and thickness of the board during the pressing. This process yields a panel with high density faces and a low density core layer developing a vertical density differentiation throughout the board thickness which is called the density profile (4). Non-uniform density profile of composite panels improves their many properties. For example a non-uniform density over cross section of medium density fiberboard is desirable due to its ability to smooth machining and easier application of finishing. A density gradient through the panel thickness is typically reflected by the presence of a higher density face layer and a lower density core layer in the board (7,8,9).

The density profile of wood composites is influenced by both raw material and manufacturing variables. Specie and particle geometry can be given as examples for raw material characteristics while press closing time and sanding are major manufacturing variables which should be considered to make boards with a better density profile. Density profiles are not a substitute for standard test. However, there is generally, a direct

relationship between the density profile and most board properties, namely static bending, internal bond strength, surface roughness, and thickness swelling. Therefore, it is important to analyze and monitor a density profile of the panels to deliver product with a desired quality for further manufacturing steps.

reducing quality of the final product. Several studies have been carried to evaluate surface roughness of fiberboard by using a stylus technique and it was found that this method is quite appropriate to determine surface roughness of various types of wood composites (1,2,3).

EXPERIMENTAL

Laboratory manufactured medium density fiberboard (MDF) panels made from a combination of softwood, hardwood, and recycled newsprint fibers were used for the density profiles and surface roughness tests. Various types of furnishes, namely recycled newsprint, softwood ,oak, hardwood fibers mixed with 9% urea formaldehyde resin were manually consolidated into a mat with a size of 40 cm by 40 cm in a forming-box prior to their pressing in a computer controlled press at the temperature of $160^{0}C$ and a pressure of 5100 kPa for 5 minutes. The samples with 5 cm by 5 cm cross section were conditioned in a computer controlled chamber until they reach to equilibrium moisture content of 5 % before any tests were carried out.

Density profiles were determined on a Quintek QDP-01X X-ray scanning system. The principle of this equipment is based on the relationship between the X ray attenuation and the density as can be expressed in the following equation :

$$I / I_0 = e^{-u1\,t} \quad (6)$$

where ;

      I : Intensity of the radiation beam after passing through the
          sample
      I_0 : Intensity of the radiation beam without passing through the
          sample
      u1 : Material linear attenuation coefficient (1/cm)
      t   : Material thickness
      e : Natural logarithm base

Material linear attenuation coefficient is related to material density and material mass attenuation coefficient which is a material property depending on the energy of the incident radiation and the material composition. Once I and I_0 are measured for each point in the profile then u1 is calculated for this point within the profile (6).

A fine stylus RC-4000 Hommel Unit was employed to determine surface roughness of the specimens. The device has three parts, namely pick-up equipped with a stylus, a drive unit, and data acquisition system. Stylus which moves along the surface at constant speed of 0.508 mm/s over 1.5 cm span. First, vertical displacement of the stylus is converted into an electrical signal and then amplified. Later, this signal is converted into a digital information which is used for the calculation of different roughness parameters (1). Details of the working principle of the stylus technique and definition of roughness parameters can be found in previous studies (1,2,5). A total of 5 measurements were taken from both sides of each sample for the roughness test.

Average of two maximum density points of each profile was calculated. Later these values were averaged for each type of sample and used as a criteria if there is any relationship between them and roughness measurements form the stylus profilometer. Board type C made a combination of 30% recycled newsprint and 70 % mixed hardwood fiber had the roughest surface having 6.0 um, 48.8 um, and 62.8 um for R_a, R_z, and R_{max}, respectively. An average maximum point density for the same type of board was also found to be the lowest one with 901 kg/m^3 among the others. It was also found that panel made from 100 % Oak fiber had the least rough surface with the highest maximum density point.

It was determined that a strong correlation existed between maximum point density and average roughness (R_a) parameter of the specimens as shown in Figure 1. Other two parameters considered in this study also followed relatively similar trend which is presented in Figure 1. This may suggest that as density of face layer of the board is increased surface quality of the board can also be improved and MDF panels may be overlaid without any major show-through problems due to the irregularities on the surface. In general combined data from density profiles and surface roughness measurements of the panels can be used to monitor products quality. It would be desirable to establish certain limit values for surface roughness and maximum density profiles for the

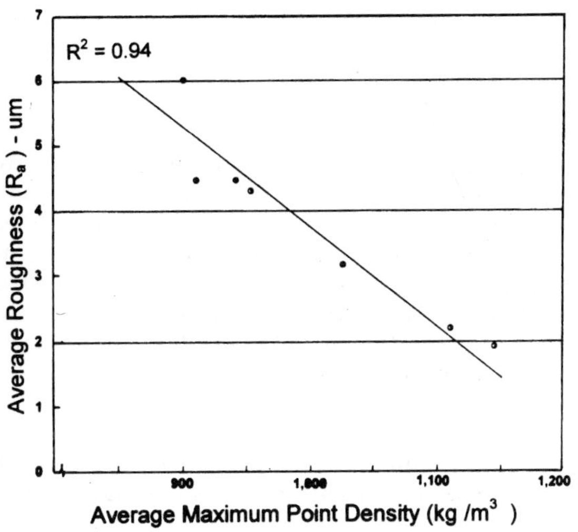

Figure 1 . Relationship between maximum point density and average roughness. (R^2 : Correlation coefficient)

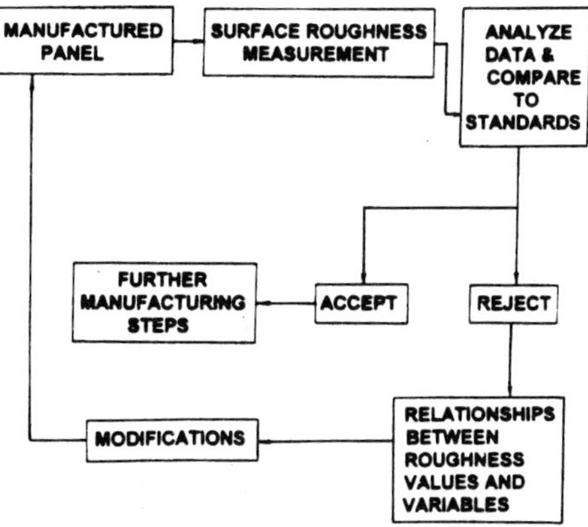

Figure 2. Quality control diagram to improve board properties.

MDF panels to improve their quality for value added products. If any deviations exist from those accepted values of the above test methods, production processes can be adjusted to have a better board quality by using   a quality control diagram as illustrated in Figure 2.

CONCLUSION

This paper discussed some of the preliminary findings of experimental data regarding surface roughness related density profiles of laboratory made MDF boards. It seems that maximum density points of the profiles and roughness parameters obtained from the board surface can be related to each other and analyzed together. Based on the results of this study, the data from a stylus profilometer and X-ray density profiler can be combined within the perspective of improving board quality.

REFERENCES

1- Hiziroglu,S. [1996] Surface roughness analysis of wood composites: A stylus method. Forest Products Journal. 46(7/8): 67-72.
2- Hiziroglu, S. and M. Graham. [1998] Effect of press closing time and target thickness on surface roughness of particleboard. Forest Products Journal. 48(3): 50-54.
3- Hoag, M. [1992] Surface testing of particleboard and medium density fiberboard for laminating application. Tappi Journal. 75(13): 111-121.
4- Maloney,T. [1977] Modern Particleboard and Dry-Process Fiberboard Manufacturing. Miller Freeman Publication Inc., San Francisco, California 94105.
5- Mumery,L.[1993] Surface texture analysis. The Handbook. Hommelwerke, Muhlhausen, Germany. 106 pp.
6- Quintek Measurement Systems Inc.[1998] Model QDP-01X. 201. Center Park Dr.,Suite 1140, Knoxville, Tennessee 37922.
7- Suchsland, O. and G. E.Woodson. [1987] Fiberboard Manufacturing Practices in the United States. Agriculture Handbook No. 640. USDA Forest Service, Washington, D.C.
8- Winistorfer, P.M., T. M. Young, and E. Walker. [1996] Modeling and comparing vertical density profiles. Wood and Fiber Science.28(1):133-141.
9- Young,T.,P.Winistofer, and S.Wang. [1999] . Multivariate control charts of MDF and OSB vertical density profiles attributes. Forest Products Journal. (49(5):79-86.

The responsible custodianship of plant and equipment

A J LANDI
Safety and Reliability, WS Atkins, Bristol, UK

Abstract
Developing a maintenance strategy is analogous to the building of a house. Success depends first and foremost on the laying of suitable foundations. These foundations are an organisation's maintenance needs.

Maintainers are the custodians of an organisation's physical assets and accountants are the custodians of an organisation's financial assets. If we are to improve the reliability of the assets under our control and meet the increased demands of safety and environmental legislation continually being introduced. The standards we utilise for the management of our assets must at least try to match the standards of custodianship expected of our financial assets.

There are many approaches that bombard industry on the right way to improve the performance of the maintenance function. Some address Maintenance Effectiveness some Maintenance Control activities and others Maintenance Strategies.

These are defined as follows:
Maintenance Effectiveness addresses data collection, problem solving, technical support etc.
Maintenance Control focuses on the work carried out by the department covering productivity, computer aids, administration etc.
Maintenance Strategy reviews what maintenance is required and when.

It is an organisation's Maintenance Strategy that should be the earliest step in any initiative.

Establishing a maintenance function is analogous to the building of a house. Success depends first and foremost on the laying of suitable foundations. These foundations can be described as the organisation's maintenance needs or strategy.

Only when the foundations are completed satisfactorily, can the walls of the house be started and it is the walls that represent the resources needed by an organisation to implement its maintenance needs. These resources are represented by people, tools, spares etc.

To complete the analogy, the roof of the house can be described as the management system required by any organisation for overall management control. It is not with the roof that we should start when developing any strategy; neither can we realistically consider the walls as an early project in our endeavours.

If we are to design and build a solid foundation, we must plan properly, prepare the ground, design our requirements and use the correct materials. To do this effectively we need to utilise people with appropriate knowledge and skills.

Appropriate people are those who have a thorough understanding of the assets whose maintenance needs are to be reviewed. They may include designers, owners or suppliers. They certainly include the engineering and operations staff associated with the assets and these two disciplines are key players in the process.

There are different pressures on the disciplines of Engineering and Operations with regard to their responsibilities for assets.

- Engineering specify, obtain and look after plant
- Operations use and manage the outputs of that plant
- Engineering tend to handle technical issues associated with the plant
- Operations tend to focus on the business pressures placed upon them

There is clearly good cause for conflict between engineering and operations because of the technical issues and business issues that need to be addressed. The approach exists to overcome these differing objectives and enable all disciplines to focus on the performance standards required of plant and equipment.

If maintenance is defined as the energy required by an organisation to preserve the functions of the assets under its control, then the first task is to define those functions and make it clear what maintenance is trying to achieve. By agreeing the performance standards required of the plant and equipment, and that the plant and equipment can meet the objectives of the business, failure, the likely causes of failure and the effects of each failed state can be identified and strategies developed to handle such failures.

By determining whether the failures identified are of importance to the organisation. It is possible to determine which of the five groups of failure management options are appropriate (predictive, preventive, function testing, run-to-failure or change to design or operation).

The processes available are very structured, enabling us to make decisions rationally, enabling us to document such decisions with confidence and justify such decisions to regulators.

Regulators expect companies to introduce effective controls. It is expected that risks are managed and controlled, and directors are now being held personally responsible for such issues as safety and pollution. None of the traditional financial controls in use by an organisation can be seen to control risk or impact on the issues of safety and environmental integrity.

The London Stock Exchange has recently published its requirements of business and in "The Principles of Good Governance and Code of Best Practice" companies have to include in their annual reports, statements on how they have met the requirements to actively manage risk. Processes do exist to ensure risk is managed and these processes meet their demands. It is the maintenance management team that is providing this input.

Because it is the maintenance function in any organisation that is given the responsibility to keep the assets in good condition. It can clearly be said that it is the custodian of those assets and parallels can be drawn between the custodianship of physical assets and the custodianship of financial assets.

In 1494, a Florentine named Pacioli invented double-entry bookkeeping, which can be described as the process at the heart of financial custodianship. The process continues into the twenty-first century with accountants, computers and an army of administrative staff still applying his principles on behalf of Directors, Shareholders and Regulators to balance a company's books to the nearest penny each and every month. It is the norm in all organisations in spite of the time and effort required.

Let us review an imaginary organisation:

Asset value	£10 000 000
Turnover	£ 3 000 000
Operating cost	£ 1 750 000
Maintenance cost	£ 1 000 000

Does the financial effort on the turnover of £3 000 000 per annum match the effort expended on the £10 000 000 worth of assets under the custodianship of the maintenance function? To exercise the same standards of custodianship to those rightly applied to the financial assets, organisations should be obliged to identify all the failure modes reasonably likely to affect the functions of any of the assets under maintenance care. They then have to understand the consequences of such failures, to select the most cost-effective failure management policies and match the resources and systems needed to manage such strategies.

It has been argued that the worst outcome from poor custodianship of financial assets is a bankrupt organisation, whereas irresponsible custodianship of physical assets could lead to the death of employees or even major environmental damage.

The physical and financial health of most organisations depends on the continued physical and functional integrity of their assets. Thus the pressures on maintainers is incredibly high.

The regulators are not only demanding greater precision and clarity in the manner in which assets are being managed but are asking for proof that such decisions are sensible and defensible. The punishments for failure are increasing and the idea of the crime "corporate manslaughter" being applied to senior executives of organisations who are responsible for fatalities that result from irresponsible custodianship of physical assets, might well become common place.

Maintainers need to raise their standards of custodianship to much higher levels than have been practised in the past. ARM, RCM and ILS techniques have only been around for twenty years or so, but slowly and surely, these tools are becoming fashionable. Asset management processes are no where near as widely accepted and rigorously enforced as those in the financial management world but then they have been in existence for some centuries.

Financial managers have always been able to submit their custodianship to exhaustive, mandatory external scrutiny at least once per year. Such pressures have not yet been placed on the maintenance manager but they increase each day and it is the new found ability, to include issues of risk and audit maintenance requirements, that has enabled the maintenance world to change so dramatically.

Applying quality through interactive design tools

J D LAW and **D HANDS**
Birmingham Institute of Art and Design, University of Central England, Birmingham, UK

ABSTRACT

This paper discusses current research at BIAD, aimed at providing Design Managers with an interactive design tool that will enable an organisation to 'design-in' quality attributes at both a strategic and project level. When complete, the design tool will draw upon a database of technical, consumer and market information, to provide an effective mechanism in assisting towards planning for New Product Development (NPD), where the goal is to satisfy and exceed the client's expectations.

INTRODUCTION

Doctoral Research conducted by one of the authors examined a range of issues pertinent to the formulation of business/design strategies aimed specifically at establishing and sustaining growth within EU Markets. However, a checklist of criteria does not constitute a strategy for successful independent learning or execution of management techniques. The construction of a model as an efficient delivery mechanism for the communication of such information is equally important if the gap between teaching and learning, experienced by many design managers within the industrial sector is to be bridged. In applying techniques at managerial level, recognition must be made of other factors influencing a design manager's ability to absorb and apply information communicated through a model. The research examined how strategies to encourage the 'designing-in' of quality attributes within a product or system, could feed and support the new product development process, linking with effective IT management to control the flow of information to the point of need, and importantly, at the time it is needed. In order to function effectively as a design management tool, the model has to fulfil the roles of 'information carrier' and 'pedagogic device.' In these contexts five key attributes were identified as desirable:

- Interaction between the user and the model
- Clarity in the portrayal of conceptual relationships
- Incorporation of a means of guiding the user
- The facility for concepts to be (fully) explored from a number of different perspectives
- Encouragement to explore relationships between cultural, technical and managerial issues from different starting points.

Despite Europe's current status as the largest trading area in the world, it is American and Japanese management models which have dominated attention. Published work on European management systems (Business International, 1992; The Economist, 1992; Group ESC Lyon Report, 1993; Calori & de Woot, 1994) suggest that while an emerging

European model may be no more successful in terms of guiding strategic intent, than US or Japanese models, it should at least be more adapted to European culture, traditions and Social values.

The European Union hosts a diversity of people and social traditions, reflecting centuries of cultural evolution. "Design" can be characterised as a 'people' business (designing artefacts and services for use by people) and for many companies aspiring to design pan-European products, it is of crucial importance to identify the 'type' of people who are likely to be targeted as customers. Europe's consumer population represents one of the main contradictions between the European Unions Aspirations of creating a single market and the present reality.

The significance of the model in this respect is in *guiding* the user of the model to recognise the importance of establishing an effective and efficient strategy that priorities targeting a market. This would allow a customer (and user, if they are different people) profile to be constructed to help analyse how particular consumers perceive product identity and quality attributes that represent 'value for money.' (Fig.1)

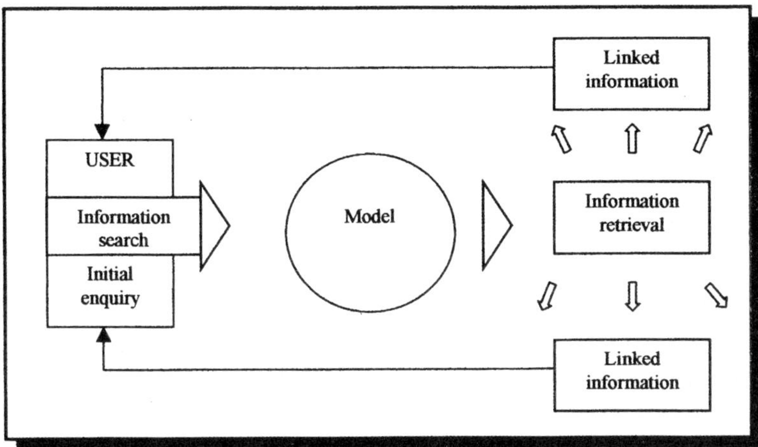

Fig.1 Model development – access to information

The Competitive Edge

With no shortage of published material to support the case for investing in design, why have many UK companies failed to capitalise on the potential benefits offered by such an approach. One factor may lie in the way in which design information is *communicated* throughout a corporate structure. Knowledge of a checklist of material helpful to European design managers does *not* constitute a strategy for successful learning or execution of management techniques. The construction of an interactive model as an efficient delivery mechanism for the communication of such information is equally important if the gap between acquisition and application, experienced by many design managers in industry is to be bridged.

In a paper by Svengren (1995) concerning Industrial Design methods and approaches for strategic development, a conclusion is drawn that in firms where design is integrated conceptually, there exists an attitude on an organisational level, that design is a means for strategic development. Svengren comments that *"Design is more than a method. Design is also a view of practice that is risk-orientated implying that actions are based on maximising experience rather than minimising risks."*

In applying techniques at a managerial level, recognition must be made of other factors influencing a design manager's ability to absorb and apply information communicated through an (interactive) model. These factors were investigated through a case study in the manufacturing sector, by one of the authors during doctoral research. The research explored the relevance on new legislation for manufacturers interested in export opportunities and implications for designers wishing to gain an insight into developing strategies for designing pan-European products. (Fig.2)

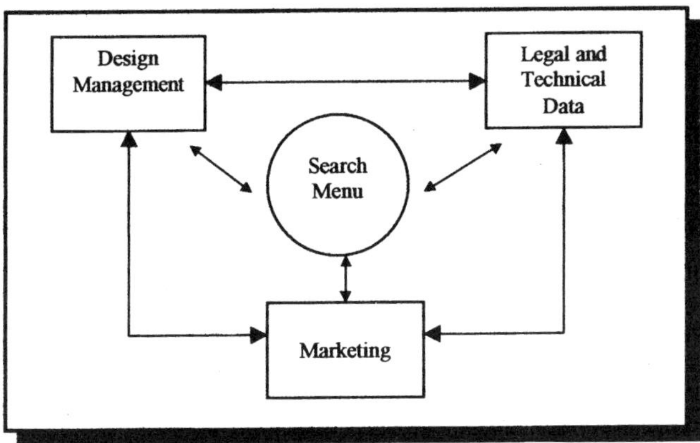

Fig.2 Design Management Model

The Quality Challenge

> *"Quality is everything that an organisation does, in the eyes of its customers, which will encourage them to regard that organisation as one of the best, if not the best, in its particular field of operation. In other words, quality is a measure of the achievement of customer Satisfaction."* (Hutchins, 1990)

This definition of quality argues the value to an organisation in striving to satisfy the client. Perceptions of products or services offered are fundamental to their very success. To develop and sustain a competitive edge in a post deregulated European marketplace, customer and user expectations need to be interpreted carefully and exceeded wherever possible. Every industry has its own version and metrics for measuring quality. For industrial designers, addressing client perceptions is as much a quality issues as those attributes which may be measured through analytical methodologies for assessing and controlling quality during the design process, such as Failure Mode and Effects Analysis, Quality Function Deployment and Design for Manufacture and Assembly.

To embrace the challenge of 'designing-in' quality throughout all facets of design activity within the organisation, the design management model provides an effective tool in identifying and satisfying customer needs. The model contains direct links between sections to enable the designer and (or) design manager to concurrently access information from different sections and different layers of information. Consequently, the model contains a high degree of flexibility to provide information at a strategic, project and product level. Jack Hagan continues this point, commenting that:

> *"...normally, launching involves a co-ordinated process to ensure that the entire marketing organisation, down to the last tier is properly prepared for the introduction. It must be armed with all the knowledge and support tools necessary to promote and to professionally support initial customer satisfaction. The quality of this preparation will often depend on how well marketing co-ordinates the inputs and involvement of other participating functions."* (Hagan, 1994)

By allowing the user easy access to specific information at the initial stages of the design project, the quality of information gained will inevitably significantly contribute towards the quality of the end product. As a result of having a greater awareness of customer needs and requirements, other potential benefits arise, firstly, owing to increased customer satisfaction, the customer or client is likely to remain loyal to the organisation. This will be critical when defending existing markets from competitors. With greater efficiency in understanding the clients needs, the potential for waste will be minimised, thus leading to cost savings in manufacture and production.

References:

Calorie, R., and de Woot, P., 1994: *A European Management Model: Beyond Diversity*, Hemel Hempstead, Prentice Hall.

Editors, Business International, 1992: *Management Europe: How companies are dealing with critical management issues*, London.

Editor, The Economist, 07/11/92: *The Elusive Euro Manager*.

Group ESC Lyon Report, January 1993: *The Characteristics of Management in Europe; Paths for the Development of Managers*, Brussels and Lyons.

Hagan, J. 1994: *Management of Quality – strategies to improve quality and the bottom line*. ASQC Quality Press, Milwaukee.

Hutchins, D. 1990: *In Pursuit of Quality*. Pitman, London.

Svengren, L. 1995: *A Study of Industrial Design Methods and Approaches for Companies' Strategic Development*, conference proceedings, The European Academy of Design, April, University of Salford.

Reliability and maintenance

A OUABDESSELAM
Ecole Nationale Polytechnique, Algiers, Algeria

Abstract

This paper relates succinctly the links between reliability and maintenance. We present a reliability function of equipments, of which we know only the renewal times. That is the case in many developing countries.

1 INTRODUCTION

Reliability is no doubt born during the second world war. It has been quickly established to become itself a science which applications participate in the quality increasing of equipments and mechanisms used in many fields of industry (transport, telecommunications...) (1).
Reliability measures the confidence that we may accord to a device working. It is characterised by the probability that an element accomplishes, without failure, a required function, in given conditions, during a specified time.
The maintenance could be defined as the whole of the actions which permit to maintain or to reinstate an equipment in a specified state, or to assure a determined service, at optimal cost.
In order to realise in best possible conditions an equipment maintenance, it is indispensable to know its reliability characteristics: MTBF, reliability function, failure rate, ...
For that purpose we present the different connections between reliability and maintenance. We will present an application of statistical methods to obtain the reliability function of equipments in the case when we do not have any information about the failures dates and the reparation duration.

2 MAINTENANCE-RELIABILITY LINKS

i- The reliability and the maintenance studies are down in parallel at different stages: project establishment, fabrication, reception, transport, exploitation and renewal; that studies being established in technical and economical point of view.

ii- The operations of as well as corrective or preventive maintenance are tied to the aleatory character of the element life duration, and then in consequence to the element reliability characteristics.

   a)- The frequency of corrective maintenance operations is function of the failure rate (or failure risk).

Let us be interested by an element in working from the instant zero and let us ask how many renewal could we have in the time range (0,t)?

Evidently, this question has not a strict answer. But if we know the reliability function of the element, we can calculate the probability to have one, or two, or three,..., renewals.

We may also calculate the average number of renewals in the range (0,t).

In the particular case of constant failure rate, the renewals number in (0,t) is distributed according to Poisson's law.

b)- The preventive maintenance operations are down when the failure risk is increasing. The renewal times are then determined from the reliability functions of the given elements.

If the failure rate is constant, it is evident that no preventive renewal must be done.

iii- Let us consider the following problem. We have a number $N(t)$ of identical equipments we would like to maintain in service at every moment; we wonder then how to realise this objective? With the help of reliability we may give an answer to this question. In the particular case of $N(t)=N_0$=constant and $\lambda(t)=\lambda_0$=constant, the number of replaced equipments since the instant zero to the instant θ is equal to $\lambda N_0 \theta$.

3 APPLICATION: DETERMINATION OF RELIABILITY FUNCTION

3.1 Reliability function determination

Let us suppose the following situation: N identical equipments are into function, in the same conditions, at a time t=0. The number of functioning equipments is continuously maintained equal to N during the time range (0,t). The renewal of a failing equipment is insured by a new equipment or a restored one, i.e. renewed. The renewal date is noted without any other information; in particular, that neither the substitute nor the old equipments are identified; the duration of the substitution operation is not mentioned and the repration duration is not indicated.

How in such conditions could we obtain information about the reliability of the considered equipments? In order to answer to this question we call the notions dealing with a renewal stochastic process (2).

Let us consider the discreet case, the time range (0,t) is cut into ranges of equal durations.

Given $h(\theta)$ the number of substituted equipments until a time $\theta<t$, and given $f(\theta)=h(\theta)-h(\theta-1)$, with $\theta>1$, the number of substituted equipments in the range $(\theta-1,\theta)$.

If we designate by R(t) the reliability function we look for, the number of equipments in service at the time t is:

$$N.R(t) + \sum_{\theta=1}^{\tau}[f(\theta).R(t-\theta)], \text{ equal to } N \text{ from hypothesis}$$

in consequence :

$$R(t) = 1 - \frac{1}{N} \sum_{\theta=1}^{\tau}[f(\theta).R(t-\theta)], \text{ with } R(0)=1$$

We know the f(t) values; thus, by recurrence, we subtract those of R(t).

3.2 Case study

The case study concerns the maintenance of an hospital hemodialyse unit (3). We present a numerical application of reliability function determination for thirteen observations. Each observation means an equipment renewal after a certain time use "t" (Table 1).

In table 1, R(t) represents the reliability function calculated as presented in the precedent paragraph, Fexp(t) is the experimental repartition function and Fest(t) the estimated repartition function.

Rank	T (in hours)	R(t)	Fexp (t)	Fest (t)	| Fest - Fexp |
1	124	1	0	0.017	0.017
2	247	1	0	0.055	0.055
3	371	1	0	0.108	0.108
4	494	0.667	0.333	0.172	0.161
5	618	0.667	0.333	0.244	0.089
6	741	0.667	0.333	0.319	0.014
7	865	0.667	0.333	0.396	0.063
8	988	0.444	0.556	0.470	0.086
9	1112	0.444	0.556	0.542	0.014
10	1235	0.444	0.556	0.609	0.053
11	1358	0.444	0.556	0.670	0.114
12	1482	0.296	0.704	0.725	0.021
13	1600	0	0	0.772	0.228

<u>Table 1:</u> Reliability and Repartition Functions

Using Weibull law as adjusted one, we find:

Parameters: $\alpha=0$; $\beta=1.75$; $\eta=1280$
MTBF=1140 hours

The Kolmogorov-Smirnov test gives:

Modified statistics of the test, for an error risk $\alpha=0.05$; $D_{n,\alpha}=0.361$
Critical value D= Max |Ftest(t)-Fexp(t)|=0.228
D is inferior to $D_{n,\alpha}$; thus, we accept the hypothesis considered, with $\alpha=0.05$

<u>Reliability determination</u>

The adjustment test confirms that the considered equipment life duration is distributed depending on Weibull law of parameters: $\gamma=0$, $\beta=1.75$ and $\eta=1280$.

In consequence, the reliability function could be written as follows:

$$R(t)= \exp (- (t/1280)^{1.75}) , \text{ error risk being } 0.05$$

4 CONCLUSION

Since many decades, in developed countries, reliability and maintenance notions are very well apprehended. But in the countries in developing industrialisation, these notions are not yet well felt, and moreover, the application conditions of maintenance introduction techniques exist only partially. Furthermore, the reliability teaching is not often assured. In this subject, if we admit that probabilities and statistics are learned in lyceum, the higher education programmes would must include the reliability. Particularly, in technology, students have to be familiarised with the different aspects of this science, because, whatever is his work, every engineer could be bound to apply the reliability principles and techniques.

REFERENCES

(1) Gnedenko B., Beliaev Y., Soloviev A., Méthodes mathématiques en théorie de la fiabilité, Edition Mir, Moscou 1972

(2) Carton D., Processus aléatoires utilisés en recherche opérationnell, Edition Masson, Paris 1975

(3) Ouabdesselam A., Etude de la maintenance des équipement d'hémodialyse d'un hôpital, rapport interne, ENP, Algiers 1996

Quality, reliability, and maintenance in semi-active control of vehicle vibration

C BOWLER, A J MEDLAND, and **C W STAMMERS**
Department of Mechanical Engineering, University of Bath, UK

SYNOPSIS

Three designs of semi-active dampers are discussed in terms of performance, control and reliability. A novel design is compared with two existing types.

INTRODUCTION

Semi-active devices for vibration control have in recent years attracted attention because they have a performance not too inferior to that of an active system while being significantly cheaper and more reliable. A simple form of semi-active control in a vehicle suspension is provided by the *switchable damper*. However this type of solution has drawbacks in terms of performance. The paper deals with the merits of controllable dampers as a function of complexity, reliability and maintenance.

SEMI-ACTIVE DAMPERS

Figure 1 portrays a variable orifice car damper believed to be similar in function to those being studied by a number of vehicle companies. The force in the damper is generated by oil flowing through sets of orifices. One set of small diameter holes (bleed holes) is permanently open. The force generated is modified by controlled opening of a secondary set, which are throttled by tapered pins. Pin position is controlled by a solenoid in response to a command signal. The simplest configuration is a bang-bang solution. In the 'off' configuration the orifice is open, the pins being held out by means of a spring. Energising the solenoid pulls the pins in and closes the orifices. The control logic requires sensor input, the acceleration of the quarter-car for instance.

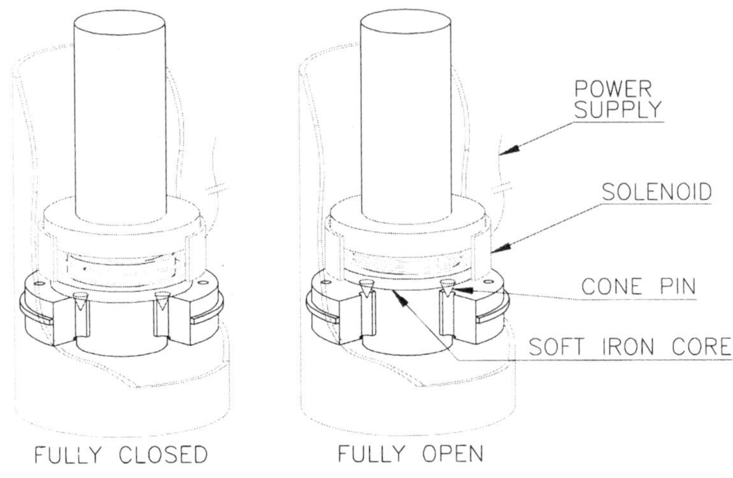

Figure 1 — Viscous damper A

An alternative solution (Figure 2) is to use a hollow spindle, the wall of which contains a hole overlapping one in the cylinder wall. The spindle is rotated by an electric motor so as to adjust the effective orifice area. This system can be operated in a bang – bang mode but has the merit that spindle position (and hence orifice area) can be monitored via an optical encoder on the motor shaft. Hence continuously variable damping is possible.

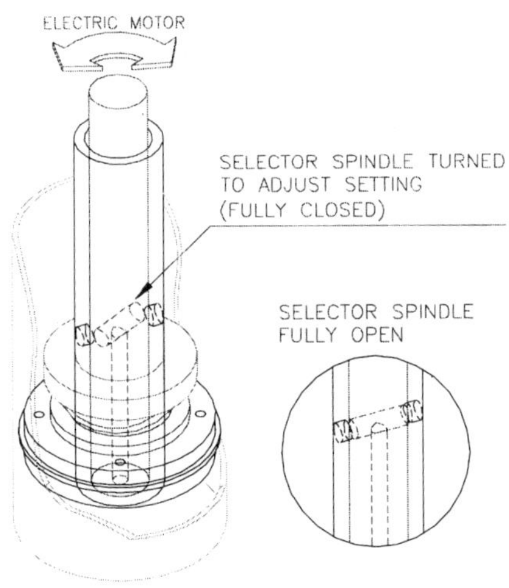

Figure 2 — Viscous damper B

Both types employ relatively simple well proven technology and should be reliable but the force generated, being dependent on relative velocity, may be inadequate at low velocity and too large at high velocity (since the damping ratio cannot be reduced below 0.1 or so)

An alternative design approach was sought at the University of Bath. This aimed at a superior variable vibration control device to the viscous damper which would be simple enough to ensure a cost low enough to be commercially acceptable in the high volume vehicle market.

In searching for a solution, the objectives were to use only existing engine mounted hydraulic supply systems as the power source and to design a semi-active that would simply replace the existing dampers. These new units should fit upon the same suspension mounting points and fit within the existing vehicle bodywork.

The establishment of these constraints led to a detailed investigation into means whereby this could be achieved. This concentrated upon evaluating existing damper designs and how they could be re-engineered to provide variable damping, under the control of a single variable pressure hydraulic supply.

A number of concepts were created and analysed. The attributes and engineering problems presented by each were identified and evaluated. This resulted in a preferred solution being developed.

The solution, which is presented here, is based on the control of friction forces generated within the cylinder assembly. The advantage of using friction is that forces are (more or less) independent of relative velocity.

Figure 3 indicates the type of system being developed. The theoretical and experimental performance is reported in references (1) and (2). The device can be fitted within the within the suspension spring like a conventional damper and has the same external appearance and dimensions as the passive damper it replaces. No changes to the vehicle design are required apart from providing a hydraulic line to the top of the 'damper.'

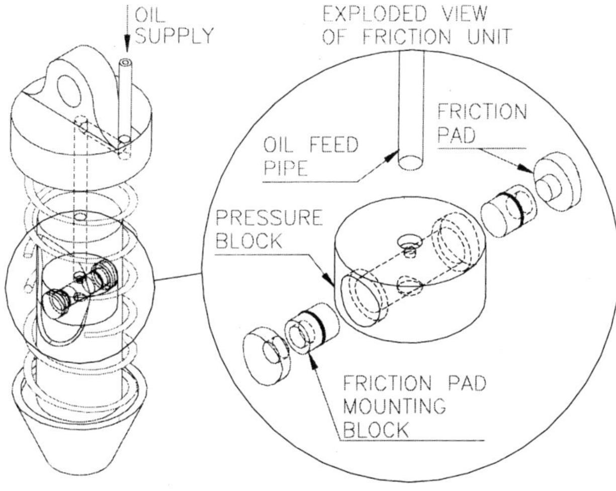

Figure 3 — Friction damper

Clearly the dynamic friction characteristics have to be ascertained. The system can mimic a viscous damper or can be used as an *anti-spring* device when the friction force acts in the direction opposite to the spring force in the suspension.

Oil, pressurised by the engine oil pump, is delivered to a central chamber via the central tube. There is virtually no flow, the control force being modulated by means of a pressure control valve. Two pistons carrying friction pads (vehicle brake material) are pushed by oil pressure against the cylinder wall. Pressure demand is set by the control algorithm employed.

Any control logic is possible (so long as the demanded force opposes the relative velocity). The device is currently being used in a predominantly anti-spring mode, although viscous friction can be emulated if desired. In the 'off' condition there is a small non zero force set by the return line pressure as the valve opens to tank, but this is not a practical problem.

A comparison of the simulated response of the friction damper with a continuously variable viscous damper indicates that that the former always produces a lower acceleration on the connected mass than does a continuously variable viscous damper.

An obvious concern in practice is that the friction coefficient may change with relative velocity, and temperature. Wear effects need to be studied. In the longer term a heuristic approach to identify the current coefficient of friction could be considered. This *might* be prohibitively expensive in the case of a passenger car, but should not be a problem for a large freight vehicle where the percentage cost of the device would be lower.

REFERENCES

1   Stammers,C W, and Sireteanu, T. Vibration control of machines by use of semi-active dry friction damping. *J. Sound and Vibration* 1998, 209 (4),pp 671-684
2   Stammers C W, and Sireteanu, T. A semi-active system to reduce machine vibration. *10th World Congress on the Theory of Machines and Mechanisms.* Oulu, Finland, June 1999, pp 2168-2173

Maintenance Management

Development of a knowledge-based production, planning, and control system

I LYALL and **Q PENG**
Wescol Limited and School of Engineering and the Built Environment, University of Wolverhampton, UK

1. ABSTRACT

This paper investigates and analyses the existing problems to integrate technology and manufacturing at Wescol, and provide an appropriate strategy to the solution. It also investigates how current developments in off-the-shelf software packages are providing SMEs with the tools and ability to cost-effectively manage quality, reliability and maintenance. A tool system, which has been developed based on VB, Access and AutoCAD, is represented. The system can support the production planning and control process with a good performance such as cost-effective, short lead-time and efficient use of resources.

2. INTRODUCTION

Current competitive markets require industries to be on the forefront of their business nature. The speed and efficiency of the companies' adaptation to markets determines their profitability and survival. The manufacturing environment is a complex system of processes and machines, using Complex algorithms to handle imprecise industrial performance. It is therefore logical to integrate information technology and manufacturing to enhance the

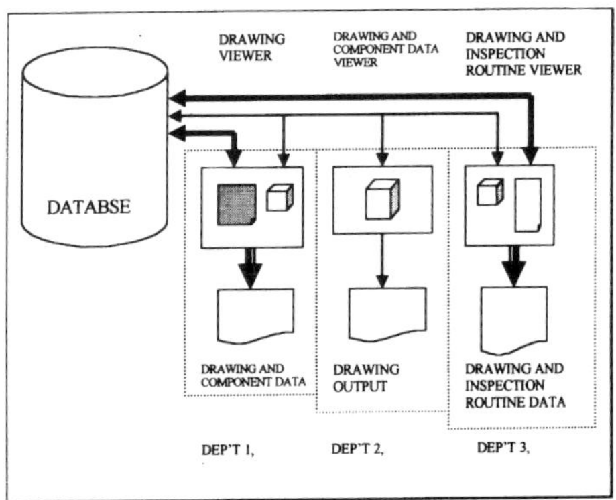

Figure 1. Departmental Data usage.

production process. Wescol is a leading independent manufacturer of gas welding, cutting and safety equipment. It manufactures for its own assembly departments and sells through distributors both in UK and world-wide. In recent years, Wescol faces the move from a very heavily labour intensive manufacturing system to one that can employ low-level computer-based automation techniques along with extensive product redevelopment and reduced labour requirement. It is strongly believed that the key to future success is through co-ordination, integration and the computerisation of many manual operations. Wescol prides itself on the exemplary relationships with its supply chain, and aims ultimately to develop a virtual enterprise. Part of this aim is to automate many of the intermediate processes, involved in processing customer's requirements.

Wescol has developed a knowledge-based production planning and control system using readily available software. It can reduce loss by providing discrete tools based on a central archive of company knowledge, creating a visualisation link between the customer's requirements and shop floors capabilities. In the paper by G. Campbell, D. Webster [1], the inadequacies of the SME's computer system and implementation are analysed. Namely, the use of the system for estimating and sales orders processing and the work involved in adapting the system to perform adequately. It was mentioned that "The software firm had implemented the system with minimum involvement from within the company's estimating department". In addition, "Due to this the company had experienced many of the hallmark problems, which are synonymous with computer system implementations". Problems listed included, pour support from the vendor, a lack of support from senior management and scepticism from low level users. The system described was a Non Windows based interface (Unix). It is difficult to visualise, the user becomes disabled by the interface rather than empowered. This was the dominant system at this time (early nineties) although since then, development of software packages has enabled SME's to develop their own systems. Fast, effective user interfaces based on the Windows operating system are now possible. In this environment, people can develop useful tools by linking common software. The result is an application based around a relational database, with tailored viewing tools, such as the systems used by the authors which is shown in figure 1.

3. ANALYSIS OF THE CURRENT PLANNING SEQUENCE AND THE MANUFACTURING PROBLEMS.

Using CNC (computer numerically controlled) machines, Wescol is now able to produce 80% of components in one operation. It is possible to simplify manufacturing further by re-designing components [2]. These advances along with others adopted have enabled Wescol to dramatically reduce lead times and have made manufacturing easier. However, it is still difficult to bridge the gap between the customer's requirements and an integrated co-ordinated effort to produce. What it perceives as being the nature of the manufacturing environment is clearly far from reality. These problems delay production and ultimately require that substantial additional work be carried out to meet the customer's requirements on time. Figure 2 shows the stages undertaken to determine the production requirements. Only a small amount of production is scheduled, whilst the majority is triggered by low stock levels. This has proved effective however, the cycle's triggers are not clearly defined or understood and the effects that external problems have on this system are difficult to visualise. Islands of both technology and knowledge exist within the organisation, whilst other areas lack the information to support the decision making processes. By linking in all contributing factors,

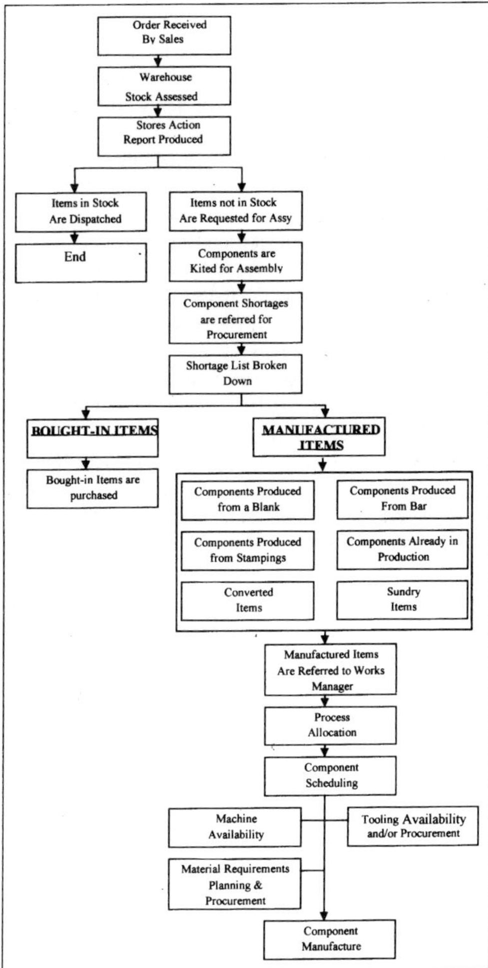

Figure 2. Current Production Planning Sequence.

figure 2 becomes highly complex. However, it becomes a detailed road map of how to get from customer order to dispatch of goods. The obvious goal would be to design a system that would enable the user to visualise their location along this road map, demonstrating what was required to proceed further. It is believed that automation alone only provides information quicker, it becomes necessary to view that information in the best possible way to truly improve productivity. Furthermore, once solutions have been established, the knowledge should be made available for future use.

4. THE SOLUTION TO THE PROBLEM

A tool that would better organise the data within the enterprise has been developed. It keeps information such as drawings and work instructions up-to-date easterly. This has been achieved through the integration of many areas of the enterprise by designing and implementing a collection desecrate software tools. It provides the user with instant access to information concerning the enterprise and gives the user the methods of visualising possible outcomes before making decisions. The software comprises a tabbed user interface providing both tools and information in twelve subject areas. Figure 3 shows the hypothesis area providing a tool that dynamically demonstrates the effects on production from alterations in process parameters. The user has the ability to model hypothetical production characteristics by adjusting sliding controls. An example to operate the hypothesis tool is shown as follows.

Input variables include; Batch quantity, Set-up time, Component, Bar length, Component length, Bars available, Cycle time (Max, Min, Avg. or User), Bar-end length, Start time, Start date, Bar clean-up length, End time, End date, Face-off amount, Part-off width. Output values include; Batch quantity, Bars required, Start time, Start date, End time, End date.

Output values are depending on the information known. To calculate the end time of production, The user will first locate the part number from a drop-down list. This list is generated from an Access database that provides the system with the component length and the cycle time (max, min and average). By clicking the start calendar and moving the start time slider, the start time and date are set. The set-up time and quantity required sliders are than set to the required values causing the bars required and end time sliders to dynamically move to the calculated positions. If for example, this time is inappropriate, the end time slider

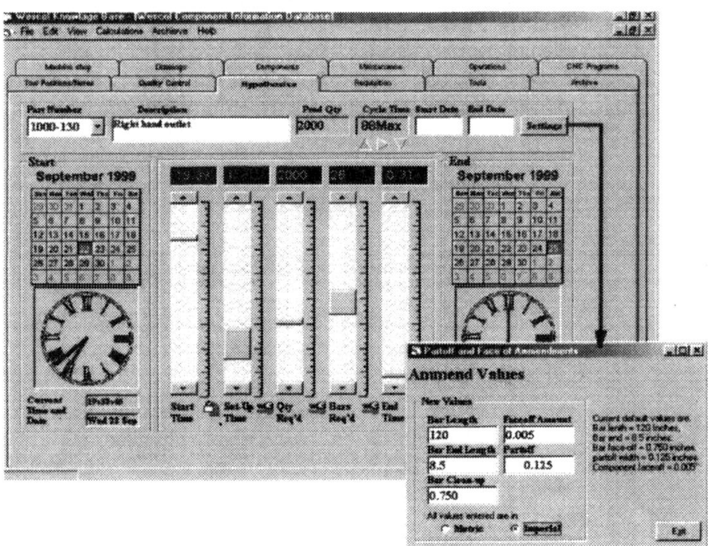

Figure 3. Hypothesis area of software.

can then be moved to indicate an earlier or later end time. The bars required and quantity required sliders will move to show the alternative requirements.

5. SUMMARY

In the management of a world class enterprise, it is important to balance the requirements of all departments, understanding their functions and designing systems that accommodate all areas of the enterprise. If this is achieved, integration becomes possible, leading to the customer receiving their product on time, with minimum expense. By developing their own software, using off-the-shelf building blocks, Wescol has achieved an improvement in productivity (less scrap, downtime and management problems). Effective discussions can only be made if the cause and effect relationships within all areas of the enterprise are clearly understood. To implement systems, companies often employ the services of consultants and software vendors who have difficulty in appreciating the true nature of an enterprise [3]. Systems should be developed from within, using company knowledge. Further development of the system will include knowledge from the complete supply chain, linking directly with suppliers stock inventory and dispatch systems, effectively ordering on-line.

REFERENCES

[1] G. Campbell, D. Webster, Implementation of an integrated manufacturing system into a SME. Proceedings of the 15Th international conference on computer aided production engineering, CAPE'99 (1999).
[2] GC. MacKerron, G. Campbell, Operational impact: competitive advantage through improved information. Proceedings of the 15Th international conference on computer aided production engineering, CAPE'99 (1999).
[3] E. Mazharsolook, D. Robinson, IT-based problem solving: helping small manufacturing companies. Proceedings of the conference on integration in manufacturing, Göteborg, Sweden, (1998).

Manufacturing maintenance organizations – evolution and development

K M PHIPPEN
PFH Total Maintenance Limited, Sheffield, UK

Abstract

This paper examines the development of maintenance activity within organisations. It draws upon the evolution of manufacturing and its increased reliance upon maintenance to support the industrial process. Insodoing it emphasises the need to measure and cultivate people in addition to providing pure technological support. In addition to the lessons that can be learnt from the past it argues the need for more positive, pro-active maintenance to support manufacturing industry in the future.

1.0 Background

Manufacturing industry, as we know it today began largely within the British textile industry in the nineteenth century. Prior to the automation or mechanisation of the process textiles were produced from the home either on a piecework basis for a manufacturer or by self-employed tailors. In home based textile manufacture the machines that were used were small & hand powered having evolved little since the middle ages. With the invention of the steam engine came the ability to replace humans as the source of power to run machines and the ability to collect together many smaller machines to form an industrial process as we know it today. With this shift in focus came the move from a highly skilled series of process workers toward the utilisation of an unskilled workforce. Consequently the vacuum created by the expansion of industry drew unskilled labour from the previous agricultural society into the towns & cities of the newly created industrial society. The British success created a model for industry that came to be applied throughout developing countries.

Milestone No 1. – From the early nineteenth century onwards the need existed for an effective manufacturing maintenance service.

Moving on from the industrial revolution the turn of the century industry saw the emergence of its first effective management techniques when F.W.Taylor devised his system for the management of factories. With the publication of "Scientific Management" in 1911 Taylor created probably the most influential and wide ranging series of elements and principles ever applied to manufacturing. Most manufacturing industry today owes its success and to a lesser degree its future upon the visionary yet practical thinking of Taylor. Whilst criticised by many latter day visionaries, mainly for the translation of his principles by others and the contemporary terminology used in his publication, there can be no doubt that Taylor's fundamental approach to the principles of measured performance hold good. In fact the translation of Taylor into "Taylorism" created an association with "time & motion" study which did little to reinforce his progressive views on the potential within people. The core content of his principles "the development of each man to his greatest efficiency and potential"[24] perhaps being given second priority to the measurement of individual task elements. Taylor recognised the potential to translate his methods in a negative sense when he wrote "the knowledge obtained from accurate time study, for example, is a powerful implement, and can be used, in one case to promote harmony between workmen and the management, by gradually educating, training, and leading the workmen into doing a larger days work for approximately the same pay that they received in the past." [17] It is interesting to note that Taylor developed his approach from the difficulties he observed whilst working as a foreman in the Steel industry. His observations arose from the problems he experienced when an influx of unskilled, often migrant, workers failed to contribute as greatly as their regular counterparts. This may in part explain his tendency to use words and phrases that today appear somewhat barbaric and hence cloud his otherwise sound philosophy.

Milestone No 2 – From the early years of the Twentieth Century manufacturing industry recognised the need for measurement in order to standardise and control productivity.

2.0 What of the wider implications?

As we enter the 21st Century, effectively 150 years since the birth of manufacturing maintenance, it is disappointing to note that the lessons learnt from Taylor in manufacturing productivity have been built upon by Fayol, the application and humanisation of the classic theories have been enhanced by such notables as Maslow, Mc Gregor & Herzberg, their collective work has in turn been broadened and applied to greater effect by Drucker, Waterman & Peters whereas generally maintenance appears from the evidence to have moved along without parallel development.

Techniques have been, and continue to be applied which have demonstrated improvement. A small minority of manufacturing maintenance operations can and do demonstrate excellence. We have seen the re-invention of pre –industrial revolution practices such as TPM. We have seen the birth and decline of pure planned maintenance in favour of a more systematic RCM approach. We have seen the use of more sophisticated equipment to analyse & predict failure. We have seen tremendous advances in the technology associated with equipment leading to higher outputs and lower incidents of failure. We have seen the availability of sophisticated Maintenance Management Systems at realistic cost.

What we have not yet seen is a degree of simultaneous, widespread development within the sphere of manufacturing maintenance that develops its potential to the fullest extent.

Studies within the UK Maintenance sector clearly reinforce the lack of progressive, sustainable betterment currently taking place. The last major national survey commissioned by the Department of Trade & Industry for example highlighted the deficit in progress when it concluded "factories could work plant 2.5 times harder with a mix of better working patterns & better maintenance"

The lessons learnt in all areas of manufacturing about the humanisation of process and the benefits of involvement seem to have missed the fact that maintenance is primarily a people based function. Hence its success relies as much upon involved, competent & committed people as it does upon the application of technology. It also requires a high degree of diverse measurement if the people within it are to demonstrate value.

If we return to the experience of Taylor he recognised right at the beginning that "soldiering" or organised under utilisation of potential in people regularly occurred. This he termed "systematic soldiering." Though worded in a derogatory way by modern standards the essence of his observations are valid today

a) People believed that increased output would lead to less workers
b) Inefficiencies in the management & control systems led to poorly designed incentives and no positive link to productivity
c) The work itself often had "poor performance of design"

It is unlikely from his conclusions that Taylor recognised the subtlety of his observations. His somewhat prejudicial assumptions that unskilled workers tended to "loaf around" when allowed freedom whereas skilled workers did not is perhaps more a reflection on the social & economic conditions of the day than a scientific observation. In maintenance terms of course the skilled nature of the role meant that it fell in Taylor's view into the skilled / management category. In fact Taylor states when citing the case of his conversation with a worker called Schmidt " this seems rather rough talk. And indeed it would be if applied to an educated mechanic, or even an intelligent labourer"

Perhaps we can see from this example the beginning of the un-paralel evolution in industry. Management initially became separated from workers forming two distinct groups and as mechanisation continued to dilute the requirement for manual process skill so a third group of technical workers were formed. Developments that occurred through the study of management created evolution in the management group and conversely exploitation generated the need for representative bodies who shaped the thinking of the worker group. In terms of its evolution the third, disenfranchised group became more introverted and concentrated on technical advancements allied to their original skills.

Milestone No 3 – The formation of centralised maintenance departments with a strong technical emphasis.

With concentration on the technical and disenfranchisement from the social & management evolution is it really any wonder that so few maintenance organisations recognise the need to measure people performance, to measure effectiveness and to become more integrated into mainstream operations. In essence after 150 years of un parallel evolution the distance between the typical maintenance organisation & the ideal maintenance organisation for today's manufacturing industry is huge.

3.0 Conclusion

Where will evolution now take place?
We can clearly see that in all areas of manufacturing a significant shift is taking place. From the post-industrial process where bigger & bigger operational platforms meant more output, and in turn more output led to greater return we have shifted to saturation of many markets. Manufacturers in the global economy have recognised that to remain competitive their process has to be multi- faceted, has to be capable of adapting to short life cycle products and has to operate much more efficiently to protect and enhance return. The era of hierarchical management and bulk process has already been left behind as smaller adaptable organisations emerge.

What we are seeing is the growth and creation of what has become known as the "virtual business". Large multi-national companies retreating from layers of management and large process mainland's toward flat reporting structures and small, detached process islands. The whole process of outsourcing, concentrating on core business, has created the realisation that large corporations can in fact exist without necessarily having any visible assets.

As in the majority of business cyclic trends create the reinvention of previously redundant methods and structures. Is the "virtual business" really any different from the textile industry in pre-industrial revolution times? Is the horizontally integrated maintenance/process technician who is trained in the art of TPM really any different to the homeworker who operated and maintained his or her own machine?

Within the manufacturing maintenance industry there is a need to learn from the past and adapt to the present. In order to support the revised needs of the process correctly and effectively maintenance needs to develop into small flexible organisational units. In this way the available technology can be applied consistently and progressively. Maintenance Technicians need to learn & emphasise the people skills regularly applied elsewhere in manufacturing without losing sight of the contribution to be gained from technical input. The modern maintenance organisation needs to become an agent of change and not claw towards the old centralised and detached structures.

Taylor studied the immediate post-industrial revolution and established many of the guiding principles of process measurement. As we move back into the pre-industrial style of manufacturing we must not lose sight of the contribution required by maintenance organisations and of the need to continually define, measure & evolve effectiveness. Once this can be achieved then maintenance will begin to enhance and demonstrate its existence

Ultimately the recognition and reward equally emphasised by Taylor but perhaps forgotten by many will allow the perpetuation and development of this vital service to the manufacturing sector.

Developing an Internet-based intelligent process monitoring and management centre

P PRICKETT, R I GROSVENOR, A JENNINGS, V KENNEDY, D NOWATSCHEK, and J TURNER
Cardiff School of Engineering, UK

Abstract

This paper outlines the current stage in the development of an Internet based Intelligent Process Monitoring and Management Centre. The aim of the Centre is to support the optimisation of manufacturing systems. This will be achieved via the real time remote monitoring of production processes using Internet based communication protocols. The work is based upon the development of advanced computer based process modelling, data acquisition and analysis tools. The resulting data will be used as the basis of intelligent process monitoring systems that can aid the development of optimised process management.

1. INTRODUCTION

This paper considers the initial stages in the establishment of an Intelligent Process Monitoring and Management Centre. The Centre is the focus of developments aimed at improving the competitive position of collaborating SMEs. It will do so by supporting process optimisation strategies that will save process time and reduce levels of rejected products, wasted resources and energy. The project will develop and deploy advanced process management engineering tools and provide support for this activity by developing and managing a dedicated computer network. Researchers located within the Cardiff School of Engineering will directly support activities undertaken within these SMEs using Internet links. This will link the SMEs to the established Centre and, equally importantly, to each other. By developing this "cluster" of collaborating companies examples of best practice will be shared and experiences pooled to aid innovation and communication.

The Centre deploys a range of process monitoring tools including an established Petri-net based modelling and analysis system, (1). This system has been specifically developed to monitor elements of machine tools and Flexible Manufacturing Systems based upon clear indications obtained from collaborating industrial partners relating to their perceptions of why such systems were failing to meet anticipated performance criteria, (2,3). The Petri-net modeller and analyser allows users to develop a graphical model of the process or system under consideration. Inputs into the system, such as signals from switches and sensors are identified and associated with the action that they control. This model can then be incorporated into a real time analyser, which follows the process by monitoring each input as it occurs. The aim of the work currently being undertaken is then to make this information available via the Internet to provide real time process management information about the system under consideration, from which performance data and diagnostic information can be extracted.

1.1 Petri-net process monitoring

A simple representation of part of a typical model produced is shown in Figure 1. Operational conditions and associated machine states are represented as Places in the net and are indicated by circles such as P1. Events or actions that must proceed to allow the completion of a machine or system cycle are represented as Transitions and are indicated by bars, such as T4. These Places and Transitions are linked together in the model using Arcs which are represented by arrowhead lines. Arcs define the path through the Petri-net, and connect areas of the Petri-net to other Petri-nets and external functions. If it is a required condition for one event that a certain other event should not have occurred then an Inhibitor Arc, represented as a line terminated with a small circle, is used.

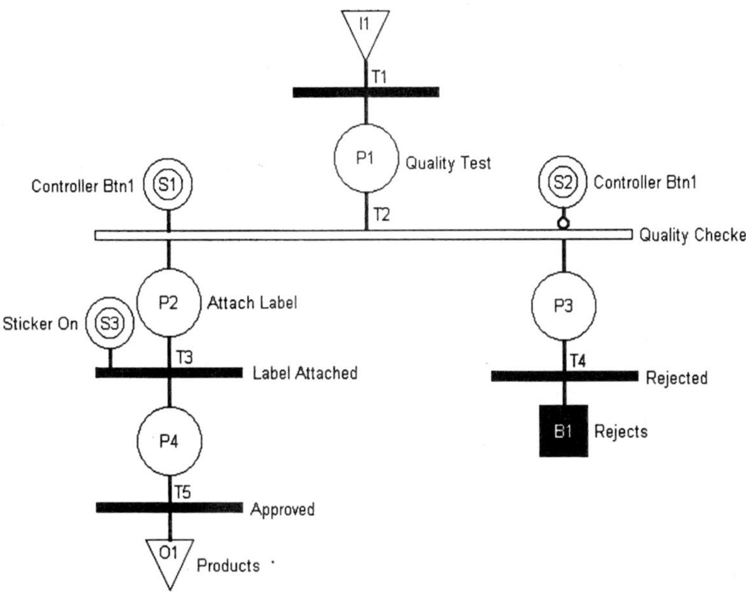

Figure 1. Example Petri-net based process model.

As events proceed tokens are fired through the associated transition, which normally allows an associated condition to be met. In this way the current status of every element within the system being modelled is represented on the Petri-net. Process modelling and understanding is made more effective with the addition of comments and descriptions of each stage in the process. In this way the current status of elements within the machine tool can be represented and monitored. The resulting information can then be deployed using Internet based protocols as part of an integrated maintenance management system to monitor aspects of the process, and provide diagnostic routines upon machine or process failure.

1.2 Condition Monitoring

The established Intelligent Process Monitoring and Management Centre will also incorporate previously developed Condition Monitoring research capabilities. These capabilities cover most aspects of Condition Monitoring, including current research aimed at in-process cutting tool monitoring (4,5) and in the continuing development of flexible and cheap data acquisition systems (6). The new aspects of this work relate to the acquisition and management of data using Internet protocols. This will allow for the automatic transfer of data and process information between the established Centre and collaborating SMEs.

2. DATA MANAGEMENT

The latest developments in software technology as applied to object-oriented data and databases have evolved system architectures in the form of a so-called "n-tier" model as shown in Figure 2. This model forms the basis of the approach being developed within the Centre. The first tier is the client application, such as a Web browser. This will be the means of access for all companies

First Tier:
    user Interface
    application presentation

Second Tier:
    Appplication logic control and service:
        Web Service
        Remote Client Service
        Distributed Transaction Management
        Data Translation
        Machine Diagnosis
        Process Monitoring...

Third Tier:
    Data service:
        Data Validation and Access

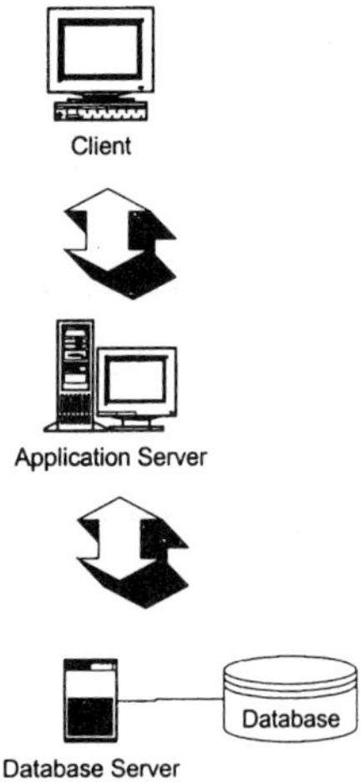

Figure 2 The N-Tier Data System Architecture

using the established Centre. The second tier is dedicated to providing application logic control and services, and is the most complex part in the system. The third tier is the data service tier which is primarily concerned about how to improve the performance of data access for multiple users and in a distributed concurrent environment how to prevent invalid data from entering the system.

The main functions provided by this approach are to provide secure data access in a web application environment, to support basic data manipulation and to allow dynamic data monitoring. The key technologies used to allow this include the use of a common data access interface to achieve application independence and data source independence. This will be developed to allow users to insert, update, delete and search data in the database through Web pages.

3. SYSTEM DEVELOPMENT

The deployment of flexible data acquisition and processing techniques is key element in the continuing development of the capabilities of the established Centre. The techniques include the utilisation of relational databases to marry acquired data to previously defined processing methods to achieve data reduction whilst maintaining data integrity. The approach taken will be to develop the facilities, including intelligent process monitoring tools, within the established Centre and then deploy them to collaborating SMEs initially in a very controlled manner via an Intranet. As expertise is developed access will then be widened to other users.

The work is expected to lead to a greater level of the utilisation of process monitoring and optimisation techniques. This will result in an associated improvement in process management, and hence to higher levels of productivity. Although the benefits of these developments will initially be confined to directly collaborating SMEs the lessons learnt and approaches taken will be made continuously available to a wider population via the established Internet protocols.

The Centre is supported by the European Regional Development Fund.

REFERENCES

1. Prickett P and Grosvenor R.I. (1995) A Petri-net based machine tool failure diagnosis system. Journal of Quality in Maintenance Engineering. 1 (3) pp 47-57.
2. Prickett P. & Grosvenor R.(1996) Integrating monitoring and maintenance in FMS's.
Condition Monitoring and Diagnostic Engineering Management '96 pp 733-742.
3. Davey A, Grosvenor R., Morgan P. & Prickett P. (1996) Petri-net based Machine Tool failure diagnostics. Condition Monitoring and Diagnostic Engineering Management '96 pp 723-732.
4. Johns C. and Prickett P. (1997) Machine Tool axis signals for tool breakage monitoring.
Condition Monitoring and Diagnostic Engineering Management '97 Vol 2 pp 450-459.
5. Prickett P. and Johns C. (1999) An overview of approaches to end milling tool monitoring.
Int.J.of Machine Tools and Manufacture. Vol. 39, No. 1, pp 105-123.
6. Jennings A.D. and Drake P.R. (1997) Machine tool condition monitoring using statistical quality control charts. Int.J.Machine Tools and Manufacture, V37, No. 9 pp 1243-1249

Innovation in water treatment technology – the CFFA process

B A QUAYE
Consultant, Sheffield, UK
R Y G ANDOH
Hydro International plc, Clevedon, UK

Synopsis

The paper describes the contact flocculation-filtration and adsorption (CFFA) process and details design and operational features of its components. Results of more recent work investigating the efficacy of the process in treating ferruginous discharges from disused coalmines into the River Don in the UK are presented. The characteristics of an effective adsorption media (Bone Charcoal) enabling tri-media rapid gravity filtration without chemical addition to achieve removals of colour, turbidity and metal cations such as iron, manganese and aluminium, are described. The CFFA process is shown to be a modular cost-effective water treatment process.

1.0   INTRODUCTION AND BACKGROUND

The contact flocculation-filtration and adsorption (CFFA) process is a physico-chemical water treatment process comprising a number of optimised unit processes described in detail by Quaye and Andoh (1). The CFFA process uses an inclined polyzonal upflow clarifier to remove the bulk of suspended solids from raw water prior to filtration. Filtration is achieved in a multi-media rapid gravity filter with an optimised **water only** backwashing regime (2).

More recent work involving the replacement of anthracite with natural charcoal in the multi-media filter has resulted in the enhancement of the adsorption of dissolved materials and removal of micro-pollutants. The high density of natural charcoal results in its intermixing with the sand during the backwash fluidisation stages with a resultant reduction in the expanded height of the mixed media. This in turn has lead to a number of operational benefits such as a lower height of filter weir.

The CFFA process can be used in a staged approach to cost-effectively upgrade existing Water Treatment Works, as described elsewhere (1). A schematic of the Contact Flocculation-Filtration System (CFF) – predecessor to CFFA, is shown in Figure 1.

A packaged water treatment based on the CFF system was tested at the Morehall Filter Station, Sheffield (Yorkshire Water Services) and produced results which met with World Health Organization (WHO) standards and 'EC Guideline Values' for potable water. Optimal upflow wash rates at various temperatures for summer, spring and autumn, and winter respectively derived from earlier work (3)(4) were used to backwash the dual-media filter bed with **water only**.

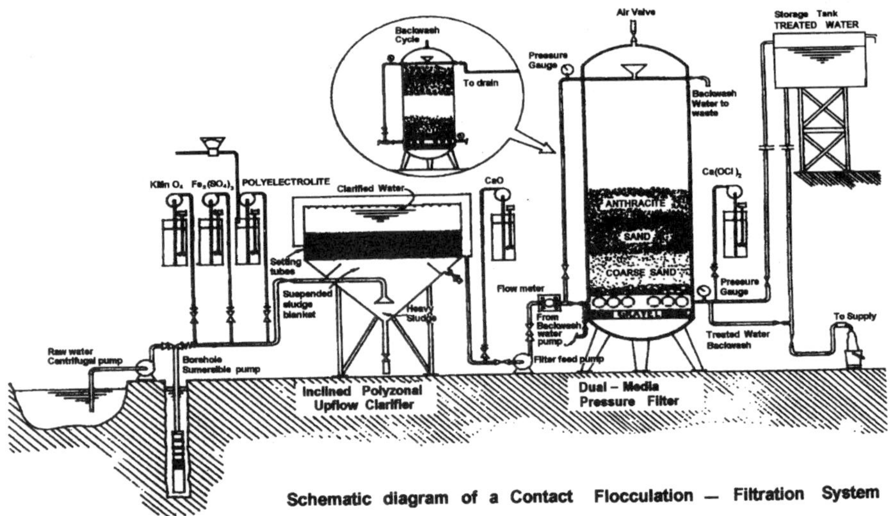

Figure 1: Schematic of Contact Flocculation – Filtration System

This paper focuses on an investigation into the feasibility of extending the filtration component to incorporate a new adsorption media – Bone Charcoal and the use of this configuration to treat ferruginous discharges from a disused coal mine into the River Don in the UK without chemical addition.

2.0   CFFA PROCESS COMPONENTS

2.1   INCLINED POLYZONAL UPFLOW CLARIFIER

As shown in Figure 1, the inclined polyzonal upflow clarifier is pyramidal in shape. Lying across the top of the clarifier prior to the clarification zone is a bank of steeply inclined settling tubes that increase the rate of clarification and settle out the suspended solids at a rapid rate. The operating characteristics of the polyzonal upflow clarifiers units are such that they can accommodate wide fluctuations in throughput whilst absorbing instantaneous increases in flow rate.

Retrofitting the polyzonal upflow clarifiers with steeply inclined settling tube modules enables a staged increase in throughput with minimal additional expenditure and without the need of extensive capital works. Above the settling tubes is the clarification zone, where the water still rising upwards flows over the castellated weir plate into a collecting channel and is piped away to the filter feed pumps.

In the normal physico-chemical treatment mode, three chemicals are added to the raw water prior to the inlet zone of the inclined polyzonal upflow clarifier. These are namely: ferric sulphate to aid flocculation in the removal of iron and aluminium at a pH of 5-6;

polyelectrolyte which being cationic aids rapid settling of the flocs formed; and potassium permanganate facilitates the oxidation of organic and inorganic material (iron) in the raw water. Concentrated sludge is periodically removed through the sludge discharge pipe.

2.2 MULTI-MEDIA PRESSURE FILTER

At the filter feed pump, lime is added to raise the pH above 9 so that manganese is removed in the pressure filter bed. The multi-media pressure filter is designed to operate at three times the conventional filtration velocity and is manufactured (to BS 5500 Class 3), as a pressure vessel to withstand the pressure required. The inlet to the vessel is via a bell mouth designed to give an even distribution of the flocculated water over the surface of the multi-media filter bed.

The particle size ranges of the filter media for the new configuration is as follows:

- Natural charcoal (1.2 - 2.4mm)
- Fine sand (0.5 - 1.0mm)
- Coarse sand (1.0 - 1.18mm)

The underdrain system is modified into a manifold (header) and perforated laterals with orifices drilled in two rows as pairs directed downwards at angles of 45^0 to the vertical. This results in an even collection and distribution of wash-water and minimises the risks of blockage.

The pressure filter contains all necessary access holes and covers, lever butterfly valves, air release valves, gauges and flowmeters. The CFFA plant is fitted with its own electrical control panel which incorporates a number of safety devices so that the system will not start in the event of either the source (e.g. borehole) being short of water or the elevated water tank being full. The electrical control panel also operates the filter feed pumps, the chemical dosing pumps, the chemical mixers and all visual and audible alarms. The filtrate from the pressure vessel is treated with calcium hypochlorite to disinfect the water prior to flowing into the elevated storage tank.

3.0 TREATMENT OF DISUSED COAL MINE DISCHARGE

3.1 COAL MINE DISCHARGES

Acid mine drainage poses a major environmental threat in mining regions throughout the world. In the UK, the problem is severe with an estimated two hundred kilometres of rivers and streams affected by discharges from abandoned coalmines (5). The quality of mine drainage varies from mine to mine. Discharges from disused mines into the River Don in the Yorkshire region of the UK have been found to be rich in iron and sulphates and usually have a low pH and are thus acidic. These ferruginous discharges give rise to significant discoloration of receiving waters. The colour results from the deposition of iron hydro-oxides on the bed of the watercourse. This precipitate has a smothering effect depleting the oxygen content and causing other adverse environmental impacts.

This paper reports on results of an initial study investigating the efficacy of a revised configuration of the multi-media filter incorporating Natural Bone Charcoal (trade name Brimac 216) as a new filter medium instead of anthracite. Results of both downward flow and upward flow filtration configurations without the use of chemicals are presented in Table 1.

Table 1: Comparison between the Water Properties before and after Filtration

Parameter	Downward Flow Filtration		Upward Flow Filtration	
	Raw Water	*Treated Water*	*Raw Water*	*Treated Water*
Colour (Hazen)	250 – 350	5	250 – 350	10
Turbidity (NTU)	29.0	0.47	51	0.38
pH	6.42	6.38	6.61	6.60
Iron (mg/l)	7.9	0.14	7.2	0.05
Manganese (mg/l)	3.35	1.8	3.2	2.3
Aluminium mg/l	0.14	< 0	0.14	0.01

The results show dramatic reductions in colour, turbidity and iron concentrations without chemical addition. The pH levels remained largely unchanged. Chemical addition mainly for pH adjustment would be required to improve on the observed removal levels for manganese.

3.2 BACKWASHING REGIME

After a period of filter run, the multi-media filter bed begins to clog up and this is detected by reference to the pressure gauges on the inlet and outlet piping of the pressure filter. At this point the filter requires backwashing by an upflow water flush for 5min at velocity barely sufficient to fluidize the bed, followed by a full expansion of the filter bed for 5min at the incipient fluidization.

The conventional backwash rates of $23mh^{-1}$ ($6.4mms^{-1}$) to $46mh^{-1}$ ($12.8mms^{-1}$) ($15"$ to $30"$ per min) for 30% to 50% expansion of surface of a sand bed of a size commonly used in a rapid sand filters are not sufficient to expand the entire bed. Theoretical and practical investigations by Quaye (3) showed that the optimum expanded porosity for anthracite and sand is 0.75 and 0.78 respectively. The procedure also described elsewhere (4), was used as the criteria for calculating the optimal wash rate and for predicting the optimal expansion of the tri-media filter bed using the measured media physical properties shown in Table 2.

Table 2: Physical Properties of Filter Media

Parameter	Sand	Bone Charcoal
Settling Velocity (m/s)	0.101	0.0907
Uniformity Coefficient	1.44	1.50
Density (kg/m^3)	2,654	2,590
Sphericity	1.137	0.418
Porosity	0.402	0.725

Table 2 shows that the uniformity coefficient (d_{60}/d_{10} – the ratio of size at which 10% by weight is finer to the size at which 60% by weight is finer), is within the recommended range of 1.3 ~ 1.7 for rapid gravity filters.

The total expansion of the multi-media filter bed incorporating the bone charcoal was 0.993m compared with a total expansion of 1.28m for a multimedia filter bed incorporating anthracite. This suggests that use of bone charcoal instead of anthracite should result in minimal loss of filter media during backwashing and savings in the construction costs relating to the height of the filter weir.

4.0   CONCLUDING REMARKS

The CFFA process is a flexible and modular physico-chemical process for treating water sources to potable standards.

Recent advancements have included the incorporation of natural bone charcoal as a filter media to facilitate and improve on the adsorption processes.

The CFFA system has proven to be capable of treating ferruginous discharges from disused coalmines without the need for chemical addition.

Other benefits from using bone charcoal include its effective intermixing with sand resulting in significantly reduced media expansion heights during backwashing thus saving on construction costs.

Acknowledgement

The Authors would like to acknowledge the contribution of Safa Z. Nasser, a former student of Sheffield Hallam University in the UK.

References

1.   Quaye, B. A., and Andoh, R.Y.G. "Upgrading and Uprating of Water Treatment Plants Using the CFFA Process" In 2nd International Conference on Advances in Water and Effluent Treatment, Cumbria, UK, 8-10 December, BHR Group Conference Series Pub. No. 8., pp 181- 197, 1993.

2.   Quaye, B. A., and Isaias, N. P. "Contact Flocculation-Filtration of Low-Turbidity, Highly Coloured Acid Moorland Water", Journal of the Institution of Water Engineers and Scientists, Vol. 39, No. 4, pp 325-340, August, 1985.

3.   Quaye, B. A. "Contact Filtration of Reservoir Water", PhD Thesis, University of London, 1976.

4.   Quaye, B. A. "Predicting Optimum Backwash Rates and Expansion of Multi-media Filters", Wat. Res. Vol 21., No. 9, 1987.

5.   National Rivers Authority "Abandoned Mines and the Water Environment", HMSO Publications, 1994.

Just-in-time production system and its application in a furniture factory

E TANRITANIR
Department of Forest Products Industry, University of Istanbul, Turkey

Abstract
Just in time (JIT) production techniques is reported for a furniture factory. The production system is uses Siman simulation language. The application carried out with the aid of computer showed that performance values are suitable for the utilization of JIT.

Keywords: Just-In-Time Production System, Kanban System and Pull System.

1. INTRODUCTION

The firms during their activity have to keep the production factors at the level which provides the highest productivity. The use of these sources productively is up to the effectively of the production system. The most important problems which decrease the productivity of the production systems of the firms in Turkey are insufficiency of production planning and control, unpredictability the demands, the excess quantity of stocks; the length of setup times, the LENGTH of total operation periods, the highness of transportation and scrap costs.

There are two different systems to decrease the stocks which constitute an important percentage of the costs: One of them is Material Requirement Planning (MRP) or Manufacturing Resource Planning (MRP II). The second is Just-In-Time (JIT) Production System which has been developed in Japan and found a wide application in the manufacturing companies of many countries.
Toyota Production System also known as JIT, has been developed by Toyota Motor Company which today has a distinguished place, in the world car market. In addition, the terms Zero Inventory Production (ZIP), Material As Needed (MAN) and Kanban System are also used to refer JIT (1).

Just-In-Time Production is the production of the necessary parts in the necessary quantity, quality and time. The approach of this system is to find permanent solutions to the problems instead of keeping high level inventory and buffer stock and of working with long lead times.

The most important target of JIT is to keep out everything seen as waste from the system. Waste is the activities as handling, storage, counting, ordering, scheduling which do not increase the value of the product, but rather increase the cost, and everything which is above the required level.

JIT can be succesfully applied in batch production. The specialty of this production is to apply a repetitive manufacturing system and to have a simple work flow. In order to increase the output quantity in repetitive manufacturing system, standard parts are being processed in small parties and to simplify the work flow, cellular layout is being preferred instead of process-oriented layout.

JIT avoids the accumulation existence of stocks and recognize it as river bed. If the amount of the materials in the production is taken as water in the river, as the level of the water -that is the stocks- decreases the rocks in the bottom of the river -that is the problems- will be seen easier and it will be possible to overcome them. Therefore, raw material, work-in-process and product stocks are being decreased as much as possible in the JIT System. The decreasing in stocks are related to the decreasing of lead and total processing time and total cost (2).

The results of a survey which made by APICS (American Production and Inventory Control Society) on benefits of JIT, was given as below (Table 1):

Table 1. The benefits of JIT.

	Mean Improvement (%)	**Standard Deviation (%)**
-Reduced inventories	41	19
-Reduced manufacturing costs	17	8
-Reduced lead time	40	23
-Improved product quality	26	19
-Reduced production and warehouse space requirements	30	15
-Improved competitive position	15	9
-Increased profit margin	54	130
-Improved worker efficiency	25	27
-Reduced labour requirements	12	7
-Reduced paperwork	30	17
-Improved equipment efficiency	16	8
-Improved worker motivation	33	35

The goals of JIT are as follow (3): To

- decrease the raw material, product and work-in-process stock levels as much as possible.
- simplify the control of the stocks by decreasing the changes at the stock level as much as possible.
- minimize the uncertainty in the production as much as possible.
- minimize the mistakes in the manufacturing and to increase the quality level.
- simplify the shop-floor control by decentralization and connecting to the last stage.
- increase the productivity by using the right combination of labour and equipment rate.

In order to achieve the objectives mentioned above, Japanese developed and used two concepts These are **"Elimination of Waste"** and **"Respect for People"** . The basic elements of "Elimination of Waste" concept can be ordered as below: Focused Factory, Reduced Setup Times, Group Technology, Pull System, Total Preventive Maintenance, Total Quality Control and Automation, Leveled Production, Keeping the Parts on Moving, Just-In-Time Delivery of Purchased Parts, Kanban. On the other hand, Respect for People concept includes the following elements: Lifetime Employment, Company Unions, Attitude Toward Workers, Automation and Robotics, Subcontractor Networks and Quality Circles

In this study, a furniture factory which has chosen an application place. The products of this factory are standard furniture and occupies small places in houses. This furniture are twelve different types (wardrobe, table, bed,, book shelf, etc.). The manufacturing area of the factory is 1435 m² and the number of workers is 123.

There are the product-oriented layout of the machines (M1, M2, M3, M4, M5, M6, M7, M8, M9, M10, M11, M12, M13, M14, M15, M16, M17, M19, M20) and process-oriented layout of the machines (M21, M22, M23, M24, M25) in this factory. As a result, while the work flow is simple at the product-oriented layout; it is complex at the process-oriented layout. Since the set up times of M5, M6 and M7, are longer, units being processed have been accumulated in front of these machines. For this reason the distances between them (M1, M2, M3, M5, M6 and M7) are kept longer.

3. METHODOLOGY AND APPLICATION

First of all apply JIT system studies which would organise database have done. These are work analysis, the determination of standard times and demands.

Then to study on cellular manufacturing system four different cells have been determined by the help of group technology. Because of being heuristic of King's algorithm having the cells which have been found examined part moving has been simplified more (4,5,6).

To apply the kanban system; plant layout has been put into a new order according to cellular layout. The numbers of withdrawal kanbans (WK) which move between cells and production-ordering kanbans (POK) which move in cells, have been found out (6,7,8,9,10). Later, in order to organize the kanban system the flowing movements of kanbans have been determined

Having this new production system by Siman simulation language the following results have been gained. To compare the actual production system with JIT System five different performance criteria have been determined. These are; Time in System, Average Number of the Parts in the System, the Number of the Parts Completed, Total Machine Utilization and Total Quelength (1,6).

4. RESULTS

When Siman results have been evaluated, it has appeared that the performance values of JIT are suitable to apply JIT System. On the other hand by the help of cellular manufacturing system not only % 36 space, but also % 44,3 from the distance for the transportation between machines have been saved as well (6).

5.0 CONCLUSION

This work has demonstrated that the JIT principle can be applied to small and medium size production units, in this case a furniture The paper shows that the Siman simulation language is suitable in te application of JIT in the small production unit. The long lead times and high level inventory and buffer stock can with this application be reduced.

6.0 REFERENCES

1. KOSAL, H., Productivity Improvement in a Multistage Manufacturing System: A Simulation Approach to the Just-In Time Production System, Ph.D. Thesis, University of Pittsburgh,1988.

2. CRAWFORD, M.K., BLACKSTONE, J.H, and COX, J.F., A study of JIT Implementation and Operating Problems, Int. J. Prod. Res., Vol. 26, No. 9, pp. 1561-1568, 1988.

3. MONDEN, Y., Toyota Production System, Industrial Engineering and Management Press, Norcross, Georgia, USA, 1983.

4. DURMUSOGLU, B., Comparison of Push and Pull Systems in a Cellular Manufacturing System, Proceedings of the International Conference on Just-In-Time Manufacturing Systems, Edited by Ahmet SATIR, pp. 115-132, Concordia University, Canada, 1991

5. DURMUSOGLU, S., Simulation Analysis of Just-In-Time Production and Its Applicability Ph.D. Thesis, Istanbul Technical University, Istanbul, 1989.

6. TANRITANIR, E., Just-In-Time Production System And Its Application In Woodworking Industry (Ph. D. Thesis), University of Istanbul, Faculty of Business Administration, Department of Production Management, pp. 333, 1993, Istanbul.

7. DAVIS, J.W., and STUBITZ, S. J. Configuring a Kanban System Using a Discrete Optimization of Multiple Stochastic Responses, Int. J. Prod. Res., Vol. 25, pp. 721-740, 1987

8. MIYAZAKI, S., OHTA, H., and NISHIYAMA, N., The Optimal Operation Planning of Kanban to Minimize the Total Operation Cost, Int. Prod. Res., Vol. 26, No.10, pp. 1605-1611, 1988.

9. PHILIPOOM, P.R., REES, L.P., TAYLOR III, B.W., and HUANG, P.Y., An Investigation of the Factors Influencing the Number of Kanbans Required in the Implementation of the JIT Technique with Kanbans, Int. J. Prod. Res., Vol. 25. No. 3, pp. 457-472, 1987.

10. WANG, H. and WANG, H.P., Determining the Number of Kanbans: a Step Toward Non-Stock-Production, Int. J. Prod. Res., Vol. 28, No. 11, pp.2101-2115, 1990.

Identifying and minimizing risks of discharges of toxic materials to water courses from water treatment works

R VAS
Yorkshire Water Services Limited, UK

1. Abstract

The historic location of water treatment works to sensitive water courses poses a continuous risk of accidental discharges from the handling, storage and usage of ecotoxic materials used in the water treatment process. This paper briefly describes the process undertaken to identify such risks and the measures put in place to reduce these risks.

2. Introduction

The continuous drive to reduce the risk of accidental discharges of ecotoxic materials into a water course and avoid the risk of prosecution by the Environmental Agency has resulted in the development of a simple generic risk study framework which defines the objectives, the scope, the risk assessment and the evaluation using and adaptation of the HAZOP method. This framework can be applied to all water treatment sites operated by the company.

2.1 The objectives and scope of the study were stated as follows:
- to review the entire facility to identify any potential sources of polluting discharges to nearby water courses.
- to evaluate the likelihood and consequences of such potential discharges.
- to identify and prioritise those risks that are in need of remedial action, taking into account the nature of the chemicals, the sensitivity of the watercourse, and existing protection against inadvertent pollution.
- to record possible remedial measures identified in respect of the significant risks (without any cost - effectiveness evaluation - which should be done before the measures are finally committed for implementation).

The scope would include all hazardous chemicals that are present within the site boundary, whatever their origin. In practice, this was interpreted as meaning from their entry into the works site, their storage and subsequent usage in the water treatment process.

2.2 Risk assessment and evaluation can take many forms, from the very qualitative to the completely numerical. However, all risk assessments involve the evaluation of the same essential matters:
- Comprehensive identification of potential hazards.
- For each individual hazard, an assessment of the causal mechanisms and the preventative measures, to arrive at an assessment of the likelihood of occurrence (frequency analysis).
- For each hazard, an assessment of the damage that would be caused if it were to occur (consequence analysis).

- For all hazards, the generation of a "risk profile" which expresses the whole range of risks and the relative importance of different risks, and from which recommendations for action can be derived.

In general the studies focused on hazard identification and consequence analysis. The technique used for hazard identification utilises the knowledge and experience of the company's employees and external consultants working in "workshop" format under the guidelines of a chairman. The workshop format was an adaptation of "Hazard and Operability Study", HAZOP, rather than the full method, which is repetitive and time-consuming for the type of process found in water treatment works.

The consequence aspect was dealt with by assessing the ecotoxicity of the chemicals involved, judging the quantities that might be discharged, detection times and the sensitivity of the watercourses to the particular chemical spilled.

Detection was addressed by identifying which instruments would respond to the event, what the indication would be, and how long it would take for the indication to be noticed and for corrective action to be taken.

The likelihood aspect was addressed by considering the frequency of operations, and applying knowledge of the relative likelihood of breaches of containment in different types of equipment, then identifying the number and type of the protective devices or procedures that would have to fail for the discharge to reach the watercourse.

2.3 The adapted HAZOP procedure first follows the normal path of each chemical through the works noting at each point along that path the potential for leakage and the particular drainage system or watercourse that would receive that discharge. As a preliminary step, the drainage systems were first described in terms of their layout, areas served, and the capacities of any interceptors into which they discharge.

Second, a segment of the normal pathway was selected and the team was asked to identify possible mechanisms for leakage or overflow. Consideration was given to:
- Hardware failure leading to leakage
- Wrong procedure leading to leakage
- Wrong procedure leading to material in wrong location, or contamination
- Mechanical impact or other external interference leading to leakage
- Process abnormality such as out-of-spec fluid conditions or overflow
- Defective maintenance

Third, when a hazard was identified, this was recorded, along with its causation mechanisms and consequences.

Finally, if the hazard appeared significant, the team was asked to suggest potential improvements that would address either the prevention or the mitigation of the incident. In practice, some of the discussions were taken as read because of similarities between the risks due to different chemicals, which would otherwise have led to repetition and waste of time

Each hazard was individually assessed and a weight was given to reflect such factors as:

- The likelihood of the leak arising in the first place
- The likelihood of the leak reaching a water course
- The quantity (rate of each chemical that could reach the water course)
- The vulnerability of the ecology to each specific chemical

These will be recognised as the classic "frequency/consequence" dimension of any risk assessment. They are conveniently expressed on a "Risk matrix" which is a table having a number of categories horizontally, representing severity of consequences, and a number vertically representing severity of the likelihood. Each individual hazard can be placed in one of the grid squares, thus expressing both aspects of the risk it presents in a convenient form.

In presenting the conclusions from each study, the following definitions were adopted for frequency, (Table 1) and consequence (Table 2), used in the risk matrix.

Table 1 - Frequency Categories

Frequency Category	Qualitative Definition	Quantitative Definition (approximate)
A	likely once in the next year	1 per plant - year
B	possible but not likely	1 per 10 plant - year.
C	unlikely	~ 40% chance in one plant lifetime (1 per 100 plant - years).
D	very unlikely	~ 4% chance in one plant lifetime (1 per 1000 plant - years).
E	remote	~ 4% chance in ten plant lifetimes (1 per 10000 plant - years).

Note: For the purpose of this study 1 plant lifetime is taken as 40 years.

Table 2 - Consequence Categories

Severity Category (consequence)	Qualitative Description	Definition (semi-quantitative)
1	Catastrophic	very serious, widespread and permanent damage causing loss of more than £10 million, national level of concern.
2	Major	permanent or long term damage to a large fraction of a habitat, clean up costs of more than £1 million, tightening of statutory controls
3	Very serious	local permanent or long term damage, substantial clean up costs, prosecution by regulator
4	Serious	significant effects, fully restorable after some expense and delay, local regulatory concern
5	Minor	minor effects, easily restorable, no loss of resources, no complaints from the public and no regulatory action.

The risk associated with a particular hazard is composed of its severity category in both consequence and frequency, and is often expressed as a pair of characters, for example B4 which means a hazard with a frequency 'B' and a consequence '4'.

Hazards with very high assessment, such as A1 falling in the black squares are thought of as being very severe and require immediate action. Hazards with low assessments falling in the white squares are considered to require no further action. Hazards in the grey squares are considered worthy of some improvement if a cost effective solution can be found. The closer the result is to the black area , the more urgent is the need for improvement.

Hazard Matrix

Frequency Category	Consequence Severity Category				
	5	4	3	2	1
A					
B					
C					
D					
E					

2.4 Typical risk reduction measures identified during the HAZOP studies are as follows:
- Improve segregation of drains for all chemical delivery areas
- Provide local indication of high level alarms at filling points
- Test ducts to ensure integrity
- Provide chemical/diesel interceptor tanks
- Review tanker delivery procedures
- Install double walled chemical dosing pipes in high risk areas

It is sometimes difficult to evaluate the cost effectiveness of these measures. This should be done by examining which of the hazards each measure addresses, and estimating how much of the total risk is removed by each measure and relating this to the corresponding costs, the law of diminishing returns applies. Ultimately, improvements should be selected as a compatible, effective set of measures which comply with the legal duty of "BATNEEC" - "best available technology not entailing excessive cost," which applies to environmental protection measures.

3. Conclusions

A hazard identification and risk assessment exercise using an adaptation of the HAZOP method is an easy and powerful tool that can be used in the identification and ranking of many potential incidents which are capable of polluting nearby watercourses.
The ranking scheme makes use of approximate ecotoxicity data and spill size estimates to express the consequences of chemical spills, and takes into account preventative systems equipment integrity and historical experience in the evaluation of spill frequencies. The overall results are expressed by plotting all the hazards on a "Risk Matrix"
The successful study is very much dependant on engaging operational staff from within the organisation with the right experience and knowledge and for the company to provide a level of funding to implement adequate environmental protective measures.

The use of neural networks in the control of water quality

G WATERWORTH
Faculty of Information and Engineering Systems, Leeds Metropolitan University, UK

ABSTRACT:

In the treatment of raw water, to make it suitable for use as drinking-water, it is necessary to monitor and control colour as a key parameter of concern to the customer[1]. The ability to predict raw-water colour is a desirable feature in the optimisation of the treatment process. In order to remove this discoloration, a coagulant is added. The required dosage is affected by a combination of the six parameters: raw water colour, raw water pH, raw water turbidity, raw water conductivity, dosed water pH and raw water temperature. These six variables are continuously monitored, and may be used to predict the necessary setting for the coagulant flow valve. Raw-water colour is inherently non-linear as a time series relationship, and it is difficult to achieve accurate modeling with conventional time series modeling techniques. This paper considers a number of possible methods including multiple regression, subtractive fuzzy clustering and artificial neural networks (ANNs). From this prediction it is possible to calculate the required dose of coagulant to eliminate the discoloration. Results show that the ANN modeling scheme is as good as the other methods in forecasting the raw-water colour in river water, and lends itself to the design of an ANN controller for dispensing the appropriate dose of coagulant[2].

1. INTRODUCTION

Water taken from rivers and reservoirs in upland areas is often contaminated with discoloration due to the peatland through which it passes[3]. In addition to the usual disinfection with chlorine to make the water safe to drink, it is necessary to carry out further treatments. This usually involves a number of processes both physical and chemical, which will vary depending on the quality of the raw water to be treated. In order to remove this harmless discoloration, the water is passed through tanks where a coagulant such as aluminium sulphate is added. This process is called flocculation, and produces a dense layer called the 'floc' to which the colour impurities adhere, and can thus be removed as a sludge. This chemical approach to removing the brown colour from raw water is expensive and only economical for large treatment works. It is appropriate to attempt to minimise the quantity of coagulant required, so as to reduce the costs of this process. River flow is affected by underlying water table factors and short term rainfall events. The former is affected by both soil parameters and meteorological conditions. The time constants of the variables are very disparate ranging from say rainfall rate variation over a few hours to the effects of soil conditions over several years. The approach to modeling needs to address these diverse time constants [4][5][6][7].

2. COMPARISON OF PREDICTION METHODS

In order to investigate the relative advantages of the various methods of modeling and predicting, a data set is used which describes a model identification problem often used as a benchmark, using 100 data points. In this problem we are interested in estimating the number of car journeys generated from an area based on the area demographics. Five factors are considered: population, number of houses, vehicle ownership, income and employment.

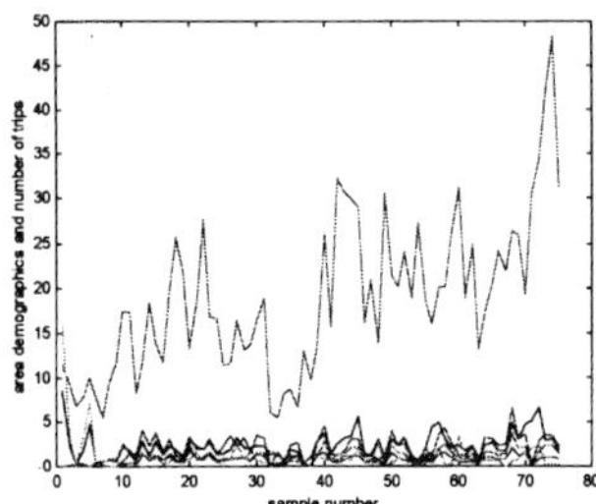

Figure 1 Plot of input data for auto trips generated from an area based on area demographics

Figure 1 shows a plot of the input data used for the model identification problem. The figure shows 75 data points for the five input variables which are used to generate the model, and the corresponding car journeys generated. The problem is to generate a model which allows accurate prediction of car journeys from different areas for which the area demographics are provided. In order to test the model 25 of the original 100 data points were set aside as checking data. Since we did not use this data to create our model, it is a useful measure of how good the model is.

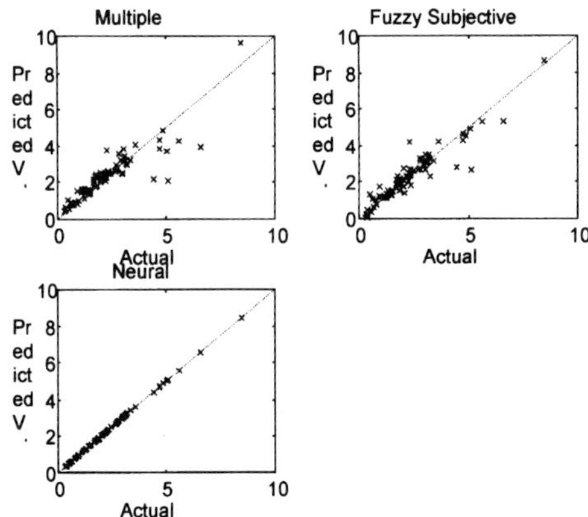

Figure 2 Correlation between actual and predicted values for the three modeling methods for the 75 sets of training data

Ultrasonic sensor for steam quality monitoring in power plants

W BOLEK, E SLIFIRSKA, and **T WISNIEWSKI**
Wroclaw University of Technology, Poland

Abstract

Industrial measure of steam quality is difficult and usually indirect. The sound speed a_d in a homogenous mixture of steam and water depends on pressure p and steam quality x. The inverse formula for x was derived as a positive root of a second order algebraic equation. This proposal of measurement system consists of pressure and ultrasonic distance sensors, and analogue-digital converter board and PC. The main advantages of this method are continuous measurement, broad range and sufficient accuracy.

1. NOMENCLATURE

x [-] – steam quality coefficient
m_p, m_w [kg] – mass of dry saturated steam and mass of the saturated water respectively
p [Pa] – pressure of wet steam
T [K] – temperature of wet steam
$p_k = 22,12 \cdot 10^6$ $[Pa]$ – critical pressure for water
a_d $\left[\frac{m}{s}\right]$ - sound speed in homogenous wet steam
v, v', v'' $\left[\frac{m^3}{kg}\right]$ - specific volume of wet steam, saturated water and dry steam respectively
s, s', s'' $\left[\frac{kJ}{kg}\right]$ - specific entropy of wet steam, saturated water and dry steam respectively
d [m] – distance, on which sound speed is measured
t [s] – time for sound speed measurement

2. INTRODUCTION

The measurement of steam quality is important in power plants (the nuclear ones in particular) and in industrial processes, where power steam is used.

Steam generators in nuclear power plants produce a saturated steam. Therefore, the expansion in turbine proceeds in wet steam area. Water drops are dangerous for turbine blades, because of erosion. It is assumed that the steam quality at the turbine outlet must not be lower than 0,9. The safety of turbine depends on the steam quality control in the last stages.

An accurate measurement of steam quality is also required to compute the rate of mass flow for wet steam, which in industrial processes is transferred from supplier to recipient .

Methods for x measurement, known so far, have several inconveniences, e.g. low accuracy, narrow range and steam leakage.

The dependence of sound speed a_d on pressure p and steam quality x is known for the homogenous mixture of water and saturated steam [1,4]. An inverse relation $x(a_d, p)$ can be derived in explicit form. It is possible to evaluate steam quality, when sound speed and pressure are measured. However this relation is quite complex and requires computations.

The hardware necessary for steam quality measurement consists of a pressure sensor, an ultrasonic distance sensor for sound speed measurement, an analogue – digital converter (ADC) and a computer (PC).

Monitoring of steam quality in turbine is an important element of reliability of nuclear power plants, because excessive moisture in the last turbine stages may destroy the blades, which rotate at a very high tangential velocity.

3. THERMODYNAMIC RELATIONS

Steam quality coefficient is by definition equal to the ratio of dry steam mass to wet steam mass.

$$x = \frac{m_p}{m_w + m_p} \tag{1}$$

Specific volume and entropy of wet steam is a function of temperature and steam quality.

$$v = v(T, x) = v'(T) + x\,[v''(T) - v'(T)] \tag{2}$$
$$s = s(T, x) = s'(T) + x\,[s''(T) - s'(T)] \tag{3}$$

It is assumed that wet steam is a homogenous mixture of dry steam and saturated water. The sound speed a_d in such a mixture is given by the relationship:

$$a_d = \frac{-v^2}{\left(\dfrac{\partial v}{\partial p}\right)_s} = \frac{-v^2}{\left(\dfrac{\partial v}{\partial T}\dfrac{\partial T}{\partial p}\right)_s} \tag{4}$$

The derivative in denominator in (4) can be evaluated by using (2), (3) and the Clausius-Clapeyron equation (5):

$$\frac{\partial T}{\partial p} = \frac{v''(T) - v'(T)}{s''(T) - s'(T)} = F(T) \tag{5}$$

Eventually, after some manipulations the derivative has the form:

$$\frac{\partial v(T,x)}{\partial T} = \left(\frac{dv'(T)}{dT} + x\left(\frac{dv''(T)}{dT} - \frac{dv'(T)}{dT} \right) - F(T) \cdot \left[\frac{ds'(T)}{dT} + x\left(\frac{ds''(T)}{dT} - \frac{ds'(T)}{dT} \right) \right] \right) \cdot F(T) \tag{6}$$

As it can be easily seen, the relation between a_d, T, x is a quadratic equation of x.

$$a_d^{\,2} \cdot \frac{\partial v(T,x)}{\partial T} + \left(v'(T) + x(v''(T) - v'(T)) \right)^2 = 0 \tag{7}$$

Steam quality can be evaluated as a positive root (8) of equation (7).

$$x = x(T, a_d) = \frac{\sqrt{4z_3(v'z_2 - v''z_1) + z_4^2} - (2v' + z_4)}{2(v'' - v')} \tag{8}$$

where:

$$z_1 = \frac{dv'(T)}{dT} - \frac{ds'(T)}{dT} F(T); \quad z_2 = \frac{dv''(T)}{dT} - \frac{ds''(T)}{dT} F(T); \quad z_3 = \frac{a_d^2}{s''(T) - s'(T)}; \quad z_4 = z_3(z_2 - z_1).$$

The relation (8) allows to evaluate steam quality, when temperature and sound speed have been measured for a wet steam. Specific volumes and entropies together with their derivatives are known functions of temperature. They can be approximated by 9th order polynomials [3], which gives accuracy of six significant digits. However, from practical reasons, it occurs that it is better to measure pressure of wet steam instead of temperature. In this case a relation (9) between temperature and pressure for wet steam has to be applied.

$$T(p) = \sum_{i=0}^{11} b_i \left[-\ln\left(\frac{p}{p_k}\right) \right]^i, \tag{9}$$

where b_i are polynomial coefficients [3].

The relation (8) is shown on Fig.2. It occurs that steam quality x depends mainly on sound speed a_d. This relation is almost of a parabolic type and could be approximated with a sufficient accuracy by polynomial of third order. Then tedious computations of (8) could be omitted.

4. MEASUREMENT SYSTEM

The steam quality measurement system consists of the piezo-resistant pressure sensor, an ultrasonic distance sensor [2] and a PC with an ADC board (Fig.1). Due to the fixed and known distance (which is traversed by a sound wave), the ultrasonic sensor works as sound speed sensor. The PC evaluates steam quality using formula (8).

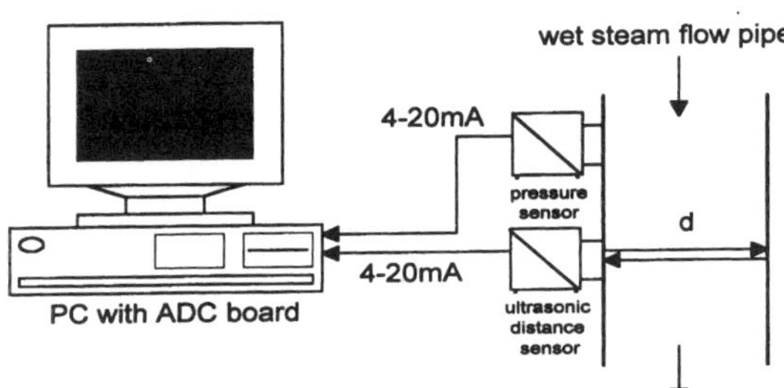

Fig.1. The steam quality measurement system

5. ACCURACY ANALYSIS

The relative error was evaluated based on the total derivative method. The class of a piezo-resistant pressure sensor is assumed to be δp = 1%. The absolute error of pressure measurement is equal to $\Delta p = \delta p \cdot p = 0{,}01\,p$. Sound speed is measured as time t, during which a sound wave has to traverse the known distance e.g. $2d = 0{,}4$ m. The absolute error of the sound speed measurement can be evaluated by a total derivative method. The distance $2d$ is known with accuracy $\Delta d = 0{,}001$ m. The accuracy of time measurement depends on the sampling rate of the sensor, e.g. the accuracy is equal to $\Delta t = 10^{-5}$ s, for frequency *100 kHz*.

The relative error δx was evaluated numerically, the partial derivatives were computed by a difference quotient. The result of the computations is shown on Fig. 3.

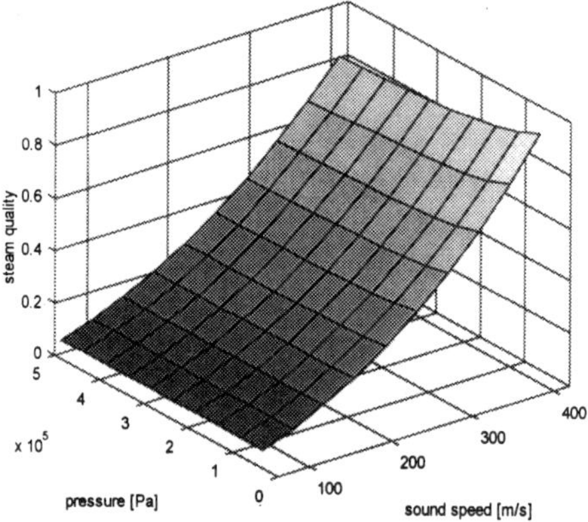

Fig.2. The surface of the function $x(p,a_d)$; the pressure range is 0 - 500 kPa, the sound speed range is 60 - 420 m/s

Fig. 3. The surface of the relative error δx of steam quality measurement; the pressure range is 0 - 500 kPa, the sound speed range is 60 - 420 m/s

6. CONCLUSIONS

The method of ultrasonic steam quality measurement has been described. The value of x depends mainly on sound speed. The estimated relative error is less than 3%. It is the highest for highest sound speed values, because higher speed requires better accuracy of time measurement. If sampling rate of *1 MHz* can be implemented in an ultrasonic distance sensor, then the relative error will be less than 2%.

The main advantage of this method is a continuous measurement of x in the whole range with the sufficient accuracy.

7. REFERENCES

1. Chorowski B., Wisniewski T.: *Fluidikwandler fur Dampfgehaltmessung.* Jablonna, Fluidic Conference. Dresden 1978.
2. Ebel F., Nestel S.: *Proximity sensors, textbook.* FESTO KG, Esslingen 1992
3. Matla R.: *New approximation formulas, representing interdependences of thermodynamic quantities of water and saturated steam.* Archiwum Energetyki, **1**, pp 83 - 115, Warszawa 1972.
4. Wisniewski T.: *Steam quality sensor.* Patent No 107165, Wroclaw University of Technology, 1980.

Neural networks application in the field of high-voltage cable insulation thermal ageing

A BOUBAKEUR, A KHELIFI, and **A L NEMMOUR**
Laboratoire de Haute Tension, Ecole Nationale Polytechnique, Algiers, Algeria
L MOKHNACHE
Institut d'Electrotechnique, Université de Batna, Algeria

ABSTRACT

Artificial neural networks have been used to predict the life duration of high voltage electrical insulation after thermal ageing. In this paper we present the obtained results in the case of Polyvinyl Chloride (PVC) . The theoretical and the experimental results are concordant. As a neural network application, we propose a new method based on Radial Basis Function Gaussian network (RBFG) trained by Random Optimisation Method (ROM).

1 INTRODUCTION

In practice, the high voltage solid and liquid insulation systems, electrical and mechanical strengths could be reduced by relatively high temperature application. When thermal stress is applied to the insulation during a long time, many complicated physical and chemical phenomena could appear leading to the degradation of the used material. This degradation may accelerate the electrical insulation breakdown, reducing in consequence equipment and apparatus life duration.

In the High Voltage Laboratory of Ecole Nationale Polytechnique, some investigations were carried out together with the ENICAB Cable Manufactories, on the influence of thermal ageing on the properties of polymer material probes and industrial cables (1)-(6). The used ageing temperatures are 80°C, 100°C, 120°C and 140°C and the ageing maximum time is 5000 hours. This test time is sufficient to obtain degradation for relatively high applied temperatures, but not for lower temperatures which need longer times. The experiments were carried out respecting IEC recommendations. In this paper, we have used experimental results obtained in the case of Polyvinyl Chloride (PVC) probes.

To reduce the ageing experiment time, but with obtaining same results as in the case of long duration tests, we have used artificial neural networks. For this purpose we would like to present a new method to predict the insulation electrical and mechanical properties after long duration thermal ageing stress. Using learning times lesser than 2000 hours, we have obtained the same non-linear characteristics as given by laboratory experiments in the case of 5000 hours.

After testing different kinds of artificial neural networks, we have chosen a Radial Basis Function Gaussian network (RBFG) (7) trained by Random Optimisation Method (ROM) (8) to predict the variations of PVC properties (1), (2).

2 NETWORK ARCHITECTURE

The architecture of the RBGF used network is shown in figure 1. It consists of three ordered layers : an initial input layer with n units, a hidden layer with m units and an output layer with a single node. Each node in the hidden layer has a radial symmetric response around a node parameter vector called centre. The output layer is a linear combiner with connection weights(7)-(9). Giving a set of input and output data x_i, y_i i=1,2...N, the state of the hidden unit j will be denoted by :

$$\rho_j(x_i) = \exp\left(-\frac{1}{2}\sum_{i=1}^{n}\frac{(x_i - c_{ij})^2}{\sigma^2_{ij}}\right) \tag{1}$$

where c_{ij} : i=1,...n. and j=1,...,m are the RBFG centres and σ_{ij} define the width of the Gaussians.

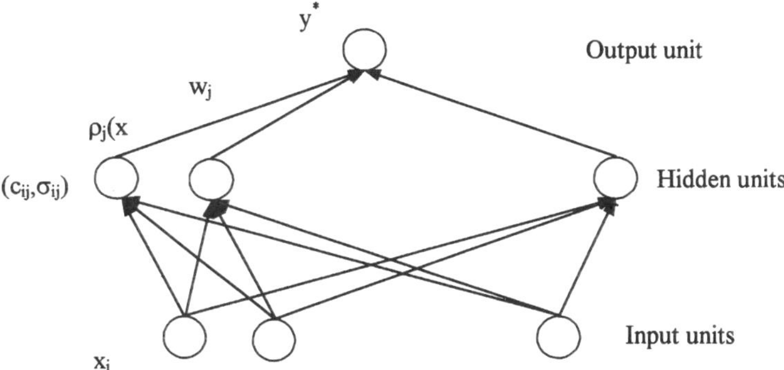

Figure 1: The feedforward Network with a single hidden layer and a single output unit.

The RBFG centres are vectors of n dimensions and they can be selected from the training data by some mechanisms cited in (7). In our case, we have used a simple technique which consists of an arrangement of these centres in a regular trellis in order to cover uniformly the data space input(10).
The covering rate is given by:

$$\tau = \exp\left(-\frac{1}{2}\frac{\left(\frac{\delta}{2}\right)^2}{\sigma^2}\right) \tag{2}$$

where δ denotes the distance between two neighbour centres.

The network response y* is given by:

$$y^* = \frac{\sum_{j=1}^{n} w_j \rho_j(u)}{\sum_{j=1}^{n} \rho_j(u)} \tag{3}$$

where $\left(\sum_{j=1}^{n} \rho_j(u)\right)$ is a normalising factor, and w_j is the weight connection between the hidden unit j and the output unit.

3 LEARNING PREDICTIVE METHOD

To predict a future value y_{n+1} of a set of measured data y_t, t=1,...,n, the algorithm will be trained on a set of examples having the form (y_i, y_{i+1}), i=1,...n-1. After this training, the weights of the net are updated so as when the network receives the value y_i, it response will be y_{i+1}. Now, we have a new set of data with n+1 values y_t, t=1,...,n+1. The training is repeated from the beginning in order to predict the value y_{n+2}, the new set of data will contain n+2 values y_t, t=1,...,n+2, and we repeat this procedure until the prediction of all values which constitute the remainder of the studied curve property is done.

In order to improve the prediction quality, after each future value prediction, we omit the first value from the set of data. For example, we omit y_1 after the prediction of y_{n+1}, and we omit y_2 after the prediction of y_{n+2} etc. In this way, the network is trained by the same number of values. More details on the learning predictive algorithm and the training method could be found in A.Khelifi and A.L.Nemmour's senior project (11).

4 NONLINEAR PVC PROPERTIES PREDICTION

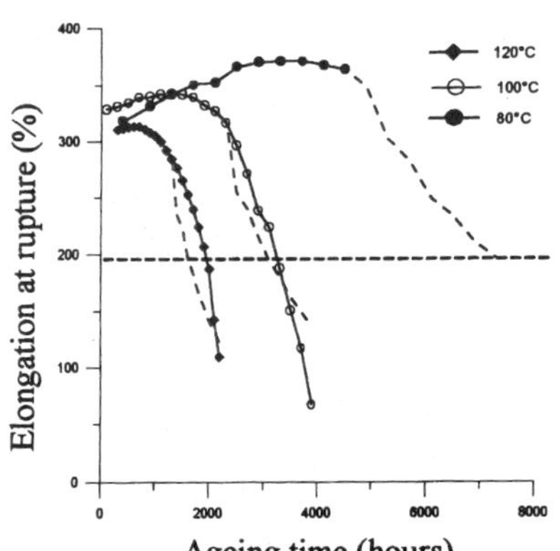

Figure 2: Life duration determination from the elongation at rupture
---- network output.
——experimental curve.

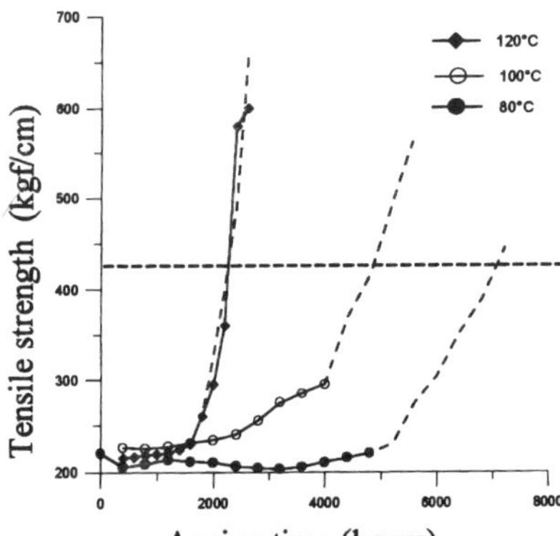

Figure 3: Life duration determination from the tensile strength.
----network output.
——experimental curve.

For illustration purpose, we present results of few numerical examples where the RBFG neural net trained by ROM is applied to predict some non-linear mechanical properties of Polyvinyl

Chloride (1), (2). All these results are obtained with a relative error of 5%.The figures 2 and 3 show the measured and predicted values of the elongation at rupture and the tensile strength in function of the ageing time. We observe that the network predicts very well the experimental curves, and for 80°C, we need the network predictions to reach the criterion of life duration defined by 50% of degradation in each property.

5 CONCLUSION

The Radial Basis function Gaussian network (RBFG) trained by Random Optimisation Method (ROM) constitutes a powerful tool to predict a non-linear function, and in particular, the non-linear variation of PVC properties versus the ageing time.

In practice, it would be very economical to use artificial neural networks in the investigations on high voltage insulation thermal ageing. In fact, we may reduce the laboratory test ageing time and let the network to predict the remained properties values at longer times leading to life duration determination.

We may propose the application of this method to other kinds of ageing (electrical, electrochemical…) and particularly, in the cases where we need to determine the extrapolation of non-linear functions giving the variation of a property versus a given parameter at different constraint levels.

REFERENCES

(1) A.Boubakeur, M.Nedjar, R.Khaili, "Influence of thermal ageing on the properties of PVC", Middle East Power System Conf., MEPCON-92, Assiut, Egypt, 1992, pp.124-127.
(2) M.Nedjar: "Influence du vieillissement thermique des polymères utilisés dans les câbles H.T.", Magister thesis, University of Tizi-Ouzou, Algeria, 1991.
(3) A.Medjdoub: "Influence du vieillissement sur l'apparition des D.P. dans les câbles isolés au PRC",Magister thesis, University of Bejaia, Algeria, 1997.
(4) Y.Mecheri: "Influence du vieillissement thermique continu, sur les propriétés des câbles isolés au PRC",Magister thesis, Ecole Nationale Polytechnique, Algiers, Algeria, 1998.
(5) A.Boubakeur, M.Medjdoub, IM.Boumerzoug "Influence of thermal aging on the properties of XLPE used as electrical insulation in M.V. cables", 10th International Symposium on High Voltage Engineering, ISH'97, Montreal, Canada, August 1997, paper 3054
(6) A.Boubakeur, M.Mecheri, M.Boumerzoug, "Influence of continuous thermal ageing on the properties of XLPE used in medium voltage cables",11th International Symposium on High Voltage Engineering, ISH'99, London, United Kingdom, August 1999, paper 280.
(7) S. Khemaissia, A.S. Morris, " Review of Networks and Choice of Radial Basis Function Networks for System Identification", Technologies Avancées N°6, 1994, pp 55-85.
(8)F. J. Sollis, R. J. B. Wetts, "Minimisation by Random Search Techniques", Mathematics of Operations Research, The Institute of Management Sc., Vol.6., N°1, Feb.1981, pp 19-29.
(9) Poli and R. D. Jones, "A Neural Net model for Prediction", Journal of the American Statistical Association", March 1994, Vol. 89, N°. 425.
(10) J.M. Renders, "Algorithmes Génétiques et Réseaux de Neurones", Application à la Commande des Processus, Editions HERMES, 1995.
(11) A.Khelifi and A.L.Nemmour, " Application des réseaux de neurones artificiels dans l'étude du vieillissement thermique du PVC utilisé dans les câbles de Haute Tension", Projet de Fin d'Etudes (Senior Project), E.N.P., Algiers, Algeria, 1999.

Investigation into the use of wavelet transforms to analyse acoustic emission (AE) signals

C BRASHAW
CB Services, Sheffield, UK
T J HOLROYD
Holroyd Instruments Limited, Bonsall, UK

ABSTRACT

Use of AE sensors for condition monitoring is increasing due to the enhanced signal to noise ratio achievable and hence potential for early detection of defects. However for a deterministic analysis the AE signals are not necessarily best served by the direct application of a Fast Fourier Transform (FFT). A possible alternative is the Wavelet Transform which decomposes the signal in a similar way to the FFT but uses a range of analysing sub component waveforms that more closely resemble the characteristics of the AE signals. This paper demonstrates an application of the Wavelet Transform to AE signals emitted from a bearing with seeded defects.

1. INTRODUCTION

The rolling element bearing is one of the most ubiquitous key components used in industrial machinery and is often monitored for signs of impending failure. Therefore identification of bearing signal components have been used as the subject for study in this paper. The examples used are extracts from a collaborative research project carried out on behalf of Holroyd Instruments Ltd and Sheffield Hallam University. For bearing analysis within a condition monitoring regime vibration analysis using the Fast Fourier Transform (FFT) has been the accepted method for signal processing of defect signals. The FFT has been applied to mainly harmonic signals generated by vibration transducers sited on vibrating structures such as industrial machinery. However there is an increasing use of AE sensors for condition monitoring due to the enhanced signal to noise ratio achievable and hence potential for early defect detection. AE signals are characteristically non harmonic and more impulsive because they are generated principally by interactive micro contacts between mating faces. By design the AE transducer responds to frequencies well above the structural resonance of machine components. AE measurements are therefore insensitive to, and effectively filter out, low frequency vibration. To enable low frequency analysis the high frequency signals are enveloped to produce a rectified dc type waveform representing low frequency repetitions of high frequency defect events. The selective nature of the measured signals and method of processing produce a waveform mixture of spikes from impacts and more prolonged bursts of activity from rubs as independent or combined events. Vibration signals measured at the machines surface will also contain impulsive components from impacts mixed with the ringing responses of the structure. Therefore the sinusoid as the basis for the decomposition of these waveforms may not be the optimum choice for interpreting the impulsive patterns in such signals. It is considered that machine defect signals generally would be analysed more

"

effectively using a method which is specifically designed to identify impulse patterns, however AE signals would benefit most from such a method. A relatively recent development in signal processing is The Wavelet Transform (WT) which decomposes the analysed signal in a similar way to the FFT but is based on a range of analysing sub component waveforms, known as wavelets rather than just the sinusoid. A dedicated analysis package, *The Wavelet Packet Laboratory ©,* acquired from Digital Diagnostic Corp of Yale University, has been used in this investigation to analyse AE signals. There are a number of pre-generated wavelets provided in the package to give a choice of wave shapes The wavelet shapes range from sinusoidal to impulsive offering a range of basic analysing waveforms some of which more closely resemble the impulsive content of the AE time signals. The function of the WT and generation of the analysing wavelets is a complicated exercise using mathematical series subject to strict constraints and is a subject for discussion in its own right, and is beyond the scope of this paper, the reader is directed to further reading on the subject (1).

2. DISCUSSION

See figure 1 for an example of a sample AE event extracted from a measured time trace taken during testing of a damaged bearing. Figure 2 is the selected wavelet used, by dilation and scaling, inside the programme, to analyse the full time trace. The Wavelet transform was applied to the time signal and a best basis selected being the most economical configuration of the smallest number of coefficients required to reconstruct the signal. The reconstructed signal is regarded as the Coherent component. The remaining signal component or the residual is reconstructed and may be considered to be noise. Examples of the separation of signal components can be seen in the example in figure 3 of a sine wave embedded in noise. The (S) trace is the source signal the (C) trace is the coherent signal and the (R) trace is the residual signal or noise. Only two coefficients were required to reconstruct the waveform when the most similar wavelet is used i.e. a sinusoid. The sinusoidal waveform was also identifiable when using the impulse like wavelet C06 which correlated with the noise component when compressed at higher levels. However the transformed signal required twelve coefficients to reconstruct a representative signal. The example is a simplistic demonstration of the importance of selecting the correct wavelet for the type of signal being analysed. In this case the sinusoid is the obvious choice given the prior construction of the signal. The small number of required coefficients is an indication of the effectiveness of the analysis. The same WT process using wavelet D06 was applied to signals measured from

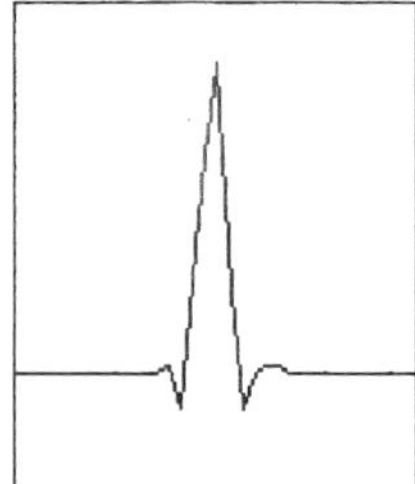

Figure 1 AE Event

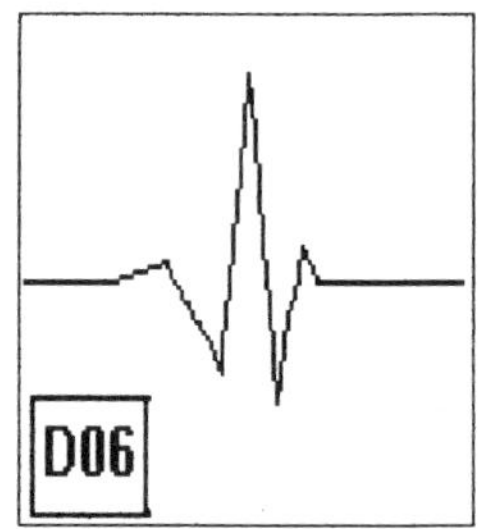

Figure 2 Sample Wavelet

defects seeded on elements of a standard SKF209 roller bearing run at 4kN load, and speed of 20Hz see figure 4. A distribution curve in window B displays the sorted coefficient amplitudes in descending order the shading under the curve represents the coefficients selected for the coherent reconstruction. The reconstructed signal was then analysed using the FFT to determine the repetitive elements in the reconstructed signal and hence carry out a deterministic analysis. The selected Wavelet analysis page was printed on the same page as the spectrum see figure 4. To enable discrimination of the spectral peaks the fault frequencies for the bearing defects were over plotted on the spectrum. The size of the defects were deliberately significant to demonstrate the method and the spectrum should contain defect peaks at each calculated fault frequency. It was notable that the only significant peak ,130Hz, which could be regarded as being at a fault frequency, with a margin of error of 5Hz, was the BPFO (Ball Passing Frequency Outer) relating to the stationary outer race. It was also observed that many of the distinctive impulses in the source signal were missed in the reconstructed signal. It was concluded that the WT decomposition was not effective if the defect signal events were distorted by sliding or amplitude modulation eg. due to the dynamics and geometry of the moving elements. Defects on mobile bearing components produce impulse magnitudes at varying levels as a function of the relative position of the rolling elements within the loaded region of the bearing. Also considerable sliding takes place which modulates the period of the repetitive impacts and hence signal events.

The Residual (R) signals referred to as child signals after processing can be continuously re-analysed with different wavelets to extract variations in the original pattern. However the possible combinations of distortion, in amplitude and duration of an event due to dynamic effects, exceeds the choice of wavelets available.

3. SUMMARY

The success of a particular analysis is dependent on identifying the correct sub-waveforms in the raw time trace, or source signal, which represent the actual defect activity, and a wavelet being available which most closely represents the defect characteristic waveform.

The effectiveness of the WT is very dependent on the consistent shape of the measured waveform components, slight differences in the selected analysing wavelet, and the source signal, have a significant effect on the resulting analysis.

Defects on moving bearing components produce modulations in impulse magnitudes at varying levels as a function of the position within the loaded region of the bearing. Also sliding modifies the shape of the signal and modulates the period between events which distorts the characteristics of the sub-waveform and impairs the analysis.

For bearing analysis the WT is therefore considered only effective in certain ideal circumstances and could not be relied upon to produce a representative analysis in all cases.

REFERENCES

1. D.E. Newland *An Introduction to Random Vibrations, Spectral & Wavelet Analysis* Third Edition Longman Scientific & Technical ISBN 0582 21584 6

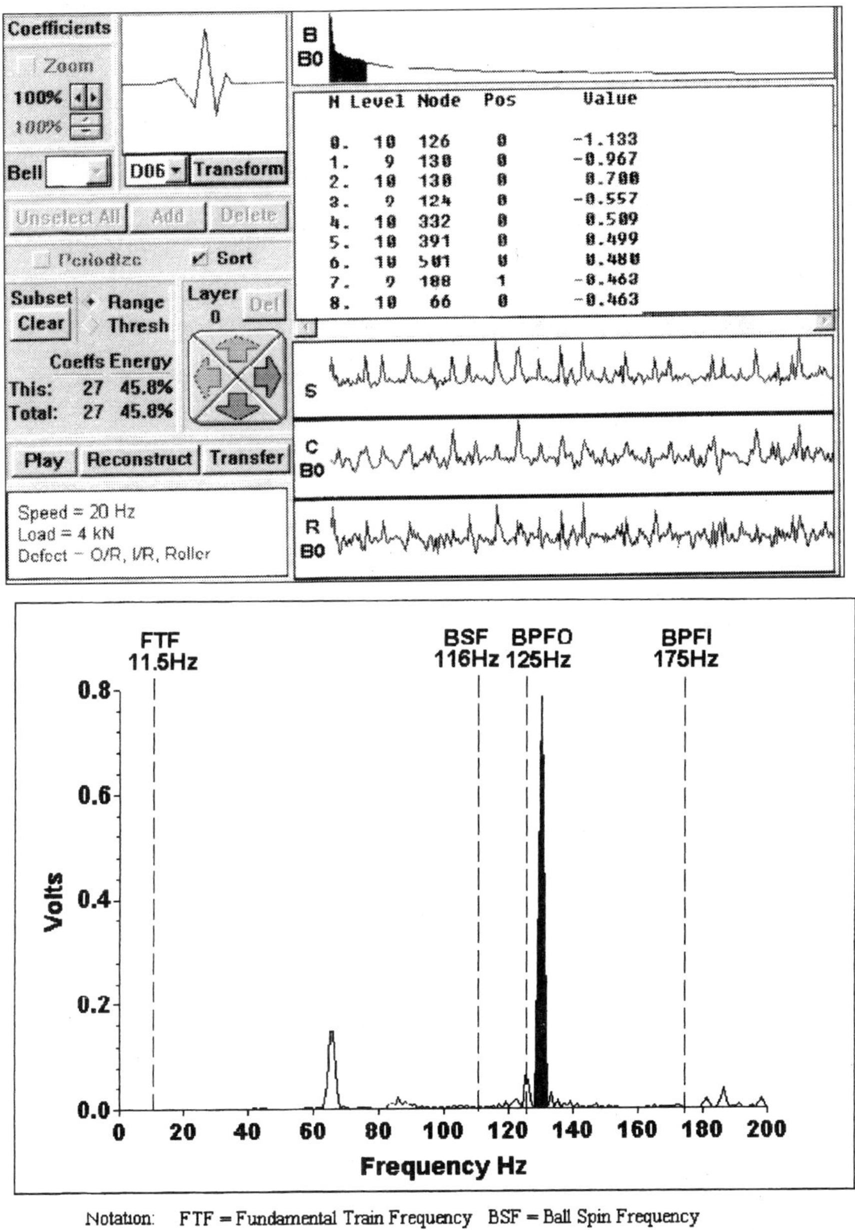

Figure 4 Wavelet Transform of time signal from seeded bearing defects and spectrum.

Electrical metrology of dynamic features of machine tools

I DUDAS and **T CSERMELY**
Department of Production Engineering, University of Miskolc, Hungary

Abstract: The Hungarian products must satisfy the increasing quality demands being in Hungary and in abroad. Developing the quality of the machine production is essential requirement of competitiveness of the Hungarian industry. The Department of Production Engineering applies a wide range of the Computer Aided Industrial Metrology and Quality Assurance Software's in education and in research. The paper introduces some methods for solving of the quality assurance tasks.

1.0 INTRODUCTION

The static and dynamic behaviour of plastic forming and cutting machine tools determines the quality of the workpiece produced. The accuracy of manufacturing is effected on the deviation of the main forming movements occurring between the workpiece and the tool.

The aim of the examination was to decrease the geometric and kinematic errors of the machine tools, furthermore to decrease the displacements, which are caused by the static and dynamic load as well. By the use of the method, developed by us, the dynamic stiffness of the press could be increased in that extent, so they satisfied the requirements prescribed to the precision press as well.

2 APPLICATION OF ELECTRIC METROLOGY

Some factors affecting the dynamic behavioius are: direction, extent and feature of the static and dynamic load; the displacements of the machine tool elements (frames, slides, spindles, rams, spherical shells, etc.) The inaccuracy of the dimensions of the workpieces produced results non-appropriate stiffness of the machine tool. When the dynamic stiffness is smaller than the required valueresulting in unwanted vibrations.

The result are the error of the surface roughness of the workpiece; the reduction of tool life, and even the failure of the tool.

Measuring techniuques are used followed by theoretical considerations.

There will be shown the examination of two concrete equipment's in the following sub-chapters. One of them is an eccentric press family type DKS having welded rigid frame. We made our examination on one member of this press family called 25 GH.

The other equipment was a copying lathe type KDM-A-18 produced by GF.

The other equipment was a copying lathe type KDM-A-18 produced by GF.
We will show different examining methods for these equipment's which are different in there construction and operating principle. The methods, which will be shown, can be applied for other type machine tools as well.

2.1 Measuring of the Elastic Deformations of Machine Tools

Specific elastic deformation measurement of the machine tools was performed in case of static and dynamic load. We consider the static deformation measurement of eccentric presses as the case of elastic deformation related to $F = 0$ Hz, so that we defined automatically the magnitude with the measuring of the dynamic elastic deformation. The specific elastic deformation or reciprocal stiffness means the displacement brought about by unit force.
The machine tool is interpreted as transfer circuit according to the concepts of control technique. So the transfer characteristic of the machine tool is characterised on the basis of the frequency transfer function of the elastic deformation. In accordance with it the qualification of the plastic forming machine tool is performed in the view of the stability of the deformation processes and possibilities of forming of induced vibration.
In order to determine the elastic deformation and its reciprocal value – the dynamic stiffness – of the machine tool, it was induced with variable force of increased frequency of sinusoidal type but with the maximum amplitude and we measured the displacement amplitudes brought about by the induction in the direction of the force and perpendicularly to it. The measuring configuration shown in Fig. 1. was applied for the elastic deformation measurement of the eccentric press.

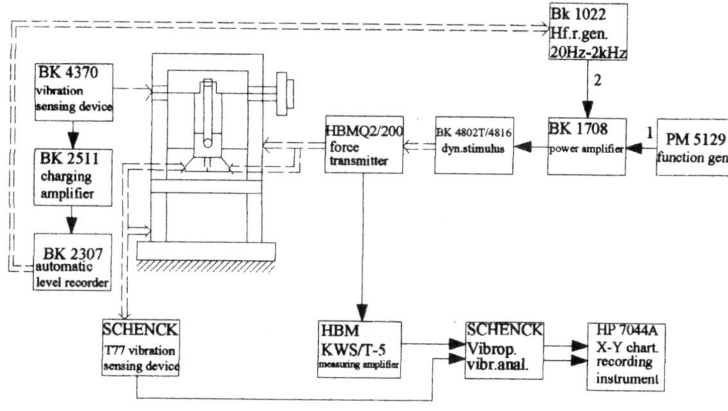

Fig. 1 Vibration analysis of eccentric press type DKS

In the generating eigendirections, and in two directions perpendicular to it the vibration amplitude were measured by electrodynamics acceleration sensing device type Scheck T23 and the registration was made by the piezoelectric acceleration sensing device type Brüel-Kjaer. The output signals. The output signals of the charging and the measuring amplifiers were recorded by automatic level recorders and measuring magnetophons. Interposing of the Fourier vibration analyser type Scheck Vibroport we had the direct possibility to draw the amplitude-phase curves (Fig. 2) by the x-y chart-recording instrument type HP 7044.

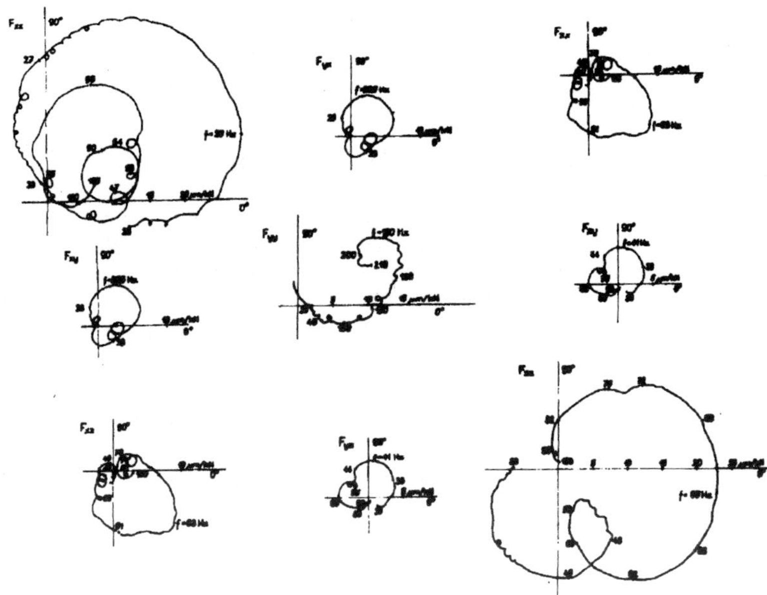

Fig. 2 Dynamic stiffness matrix

Simultaneous registration of the answer response vibration amplitude related to the generating force of sinusoidal amplitude made the transferring function possible to the determined with direct locus. By the use of the applied electric measuring technique methods the displacements and the dynamic stiffness matrix of the eccentric press type DKS-25 GH. The general equations, which describe the relative elastic deformation, can be determined in every position of the parts to be examined of the machine tool.

As the simulation of the different loading (e.g. pressing force, cutting force, punching force, bending force, striking force, striking force, shearing force) at the tool clamping place of the plastic forming machine tools the elastic displacement brought about with sinusoidal loading effect of changing direction and magnitude is determined by

$$\vec{t} = \underline{\underline{D}} \cdot \vec{F}$$

In details:

$$[u; v; w] = \begin{bmatrix} D_{xu} & D_{yu} & D_{zu} \\ D_{xv} & D_{yv} & D_{zv} \\ D_{xw} & D_{yw} & D_{zw} \end{bmatrix} \cdot \begin{bmatrix} F_x \\ F_y \\ F_z \end{bmatrix}$$

where:

$\vec{t}$: displacement vector. Its components are the displacements in the 3D orthogonal co-ordinate system matched to the machine in the direction of u, v, w.

$\underline{\underline{D}}$: 3 x 3 matrix of the changing characteristic of displacement and force.

From its 9 eigenfrequency 3 in the main diagonal contains the direct, and the other six the cross frequency characteristics.

$\vec{F}$: the vector of stimulus/generating force.

2.2 Examination of Vibration Pattern of Machine Tools

The measuring set up is shown on Fig. 3. which was used for examination of vibration pattern – of main spindle of coping lathe type KDM-A-18 made by GF – by use of sinusoid stimulus force. From the vibration pattern of the tubular shaft of the main spindle we determined the resonance modal points and amplitudes related to bucking half-wave. The measured critical frequencies were compared to those, which were calculated from the kinematic chain of the machine tool.

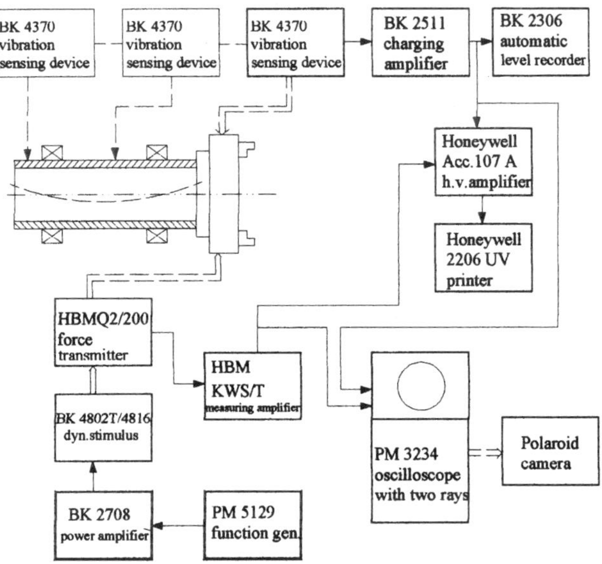

Fig.3 Examination of GF copying lathe

On the base of the measuring experience we made the change of the main part and executed the other important changes, which improved the dynamic relations.

3 SUMMARY

Related, that on the base of our previous measurements we could make more perfect the construction which improved the dynamic stiffness – perpendicular to the forming main direction – of the machine tool which has fundamental effect on the accuracy of guiding of the tool. The recordings show that the showed eccentric press type DKS-25, having welded rigid frame, satisfies the requirements prescribed to the precision press as well. By the measuring method showed in this paper – we think – we managed to present some new results of the machine tool development by the use of up-to-date electronic measuring technique.

4 REFERENCES

[1] Weck, M. - Klaus T.: Dynamisches Verhalten spanender Werzeugmaschinen. Einflussgrössen. Beurteilungverfahren, Messtechnik, Berlin, Springer- Verlag 1977.
[2] Schenk GmbH: Vibroport operating manual
[3] Hottinger Baldwin Messtechnik: Messtechnische Briefe, 1/89
[4] Brüel-Kjaer: Technical Review, 3/88

Acoustic emissions – a CM technology finally living up to expectations

T J HOLROYD
Holroyd Instruments Limited, Matlock, UK

ABSTRACT

The highly specialised non-destructive technique of AE has been continually developed as a means of structural integrity assessment since the pioneering work of Josej Kaiser which was published in his thesis in 1950. Parallel developments have been aimed at applying the technique to monitoring the condition of industrial machinery. Despite numerous research studies this second application field was slow to develop and reasons for this are identified. However it is shown that a systematic approach to applying AE has overcome the obstacles to its successful implementation across a broad front.

1 INTRODUCTION

Josef Kaiser is credited with being the father of the modern day Acoustic Emission (AE) technique. Using electronic amplification he discovered that a wide range of materials produced broadband sounds when stressed (1). The potential of this phenomenon as a means of materials investigation and defect detection in structures was recognised and initially developed in USA government laboratories in the 1960's. AE activity was observed to be random in nature and sensitive techniques for detecting and processing AE signals were developed and refined. Today commercially available AE instrumentation for structural integrity assessment is highly specialised. Alongside these developments other researchers investigated the use of AE techniques for monitoring the operation and condition of industrial machinery. In these applications the source of high frequency activity was not incremental defect growth but processes such as friction, impacts, turbulence and cavitation. Some of the early observations were :

- The response at the resonant frequency of the transducer was a good indicator of incipient failure (2).
- Signals at the high detection frequencies monitored (up to 90 kHz) had energies much higher than background noise (2).
- The spectrum of the envelope of the resonant high frequency signals enabled defect types to be diagnosed from the defect repetition frequencies (3).
- AE is more effective in detecting very early mechanical failure than low frequency vibration techniques (4).

These early observations have since been supported by many hundreds of published papers on the use of AE for machinery condition monitoring which, almost without exception, have reaffirmed the benefits that AE offers in terms of sensitivity to machinery problems (ie improved Signal to Noise Ratio or SNR). It is therefore surprising that the AE technique has taken such a long time to become established as a mainstream Condition Monitoring (CM) technique. With the benefit of hindsight a number of reasons for this can be identified :

- All too often the reported results referred to one-off or limited number tests performed using the same AE transducer permanently mounted in a laboratory environment.
- There was little commonality in the experimental, instrumentation and signal processing set-ups between the work of different researchers.
- The practicality of applying AE as a CM technique in the industrial field was inadequately addressed.
- Those instances relating to real industrial problems were often concerned with specialised rather than general items of machinery.
- The AE instrumentation used was too expensive, skill intensive and impractical for general Condition Monitoring use in industry.

The above factors tended to obscure the benefits of AE as a CM technique and made tenable the widespread belief that AE was a quirky and non-reproducible technique.

2   A SYSTEMATIC APPROACH

The obstacles to acceptance listed in the previous section can only be overcome by the systematic application of appropriate techniques and instrumentation to a wide range of machinery in the industrial environment. Addressing these areas has been the underlying task of Holroyd Instruments throughout the 1990's. Experience gained prior to this in applying AE to a wide range of components, structures, machines and processes had already led to the conclusion that enveloping techniques offered significant benefits (5). Compared with traditional 'rf' and threshold based AE techniques enveloping produced more consistent readings and removed the need for operator dependent settings.

The transducer is also a vitally important part of any measurement system and new developments led to improved consistency in the response from AE sensor to AE sensor. A design was chosen with a nominally 100 kHz resonance.

These developments have been incorporated in commercially available AE based CM instruments since 1993 and considerable experience has been gained in their use across a very wide range of machines in all industrial sectors. The main benefits of this approach have been found to be :

a)   Easy detection of machine faults
At high frequency machinery degradation (eg through deterioration of bearing and gear teeth surfaces) is readily detectable compared to normal running. During the earliest stages of a fault condition discrete transients are typically observed on the AE signal whereas more advanced damage results in an overall increase in the mean signal level. Inadequate

constrained to $1 \leq x \leq S$. This means that any mutation that occurs will be bounded within the limits set at the definition of the genome. The classification performance of the trained network using the whole dataset was returned to the GA as the value of the fitness function.

In this case, the GA uses a population size of 10 individuals, starting with randomly generated genomes. The probability of mutation was set to 0.2, whilst the probability of crossover was set to 0.75. An elitist population model is used, meaning that the best individual in the previous population is kept in the next population, and preventing the performance of the GA worsening as the number of generations increase.

6   TRAINING AND SIMULATION

Training was carried out using three data sets. As a comparison, a neural network was trained using each feature set. These were trained for a total of 350 epochs, and allowed to choose the best size of hidden layer between 2 and 15 neurons. Also using the genetic algorithm running for a total of 40 generations, each containing 10 members (meaning the training of 400 neural networks), eight separate cases were tested using various numbers of inputs, varying from five to twelve.

7   RESULTS

Tables 1 and 2 show a summary of results for all three feature sets used. The "No. Neurons." quoted in the second column is the number of neurons used in the hidden layer of the best network in each training run. "Classification Success %" represents the percentage success rate using both training and test data.

7.1   Performance of ANNs only

As can be seen from Table 1, all the feature sets have a performance greater than 88%, with two cases in excess of 91%; the spectral feature set gives the best performance, at 97% for the overall data set.

7.2   Performance of ANN with GA

Table 2 shows the performance of the different feature sets after running under the GA for 40 generations. All of the datasets have their best performance in excess of 97.5%. The feature set using all the available training data has managed to achieve an accuracy of 100%, indicating accurate classification. This is achieved using only six inputs out of the possible 156. Using 9 neurons in the hidden layer, a relatively small network has been created that is successful. A network of this size would be ideal for a realtime implementation on a small chip or micro-controller.

8   CONCLUSIONS

It has been shown that the Genetic algorithm is capable of selecting a subset of only 6 inputs from a set of 156 features that allow the ANN to perform with 100% accuracy. The performance of networks trained using the GA feature selection in an automatic manner was consistently higher than those trained without feature selection. This technique offers great potential for use in a condition monitoring environment, where there are often hundreds and even thousands of different measurements available to a monitoring system, and selection of the most relevant features is often difficult.

9 ACKNOWLEDGEMENTS

We are indebted to Weir Pumps for the loan of the machine set used in the experiments. Financial support was provided by Weir Pumps, Solatron Instruments, and the University of Liverpool. Thanks must also go to Dr. A. C. McCormick, for his invaluable help and advice.

REFERENCES

[1] McCormick A. C. and Nandi A. K., Classification of Rotating Machine Condition using Artificial Neural Networks, Proceedings of the Institute of Mechanical Engineers, Part C, Vol 11, No. 6, pp 439-450, 1997.

[2] Sorsa, T. Koivo H. and Koivisto H., Neural Networks in Process Fault Diagnosis, IEEE Transactions on Systems, Man, and Cybernetics, Vol 21, No. 4, 1991.

[3] Liu, T. I. and Mengel, J. M., Intelligent Monitoring of Ball Bearing Conditions, Mechanical Systems and Signal Processing, Vol. 6, No. 5, pp 419-431, 1992.

[4] McCormick A. C. and Nandi A. K., Real Time Classification of Rotating Shaft Loading Conditions using Artificial Neural Networks, IEEE Transactions on Neural Networks, vol 8. no. 3, pp 748-757, 1997.

[5] McCormick A. C., Nandi A. K., and Jack L. B., Digital Signal Processing Algorithms in Condition Monitoring, International Journal of COMADEM, vol 1., no. 3, pp 5-14, 1998.

[6] Nandi A. K. *ed.*, Blind Estimation Using Higher Order Statistics, Kluwer Academic Publishers, Boston, 1999

[7] Nikias C. L. and Mendel J. M., Signal Processing with Higher Order Spectra, IEEE Signal Processing Magazine, pp 10-37, July 1993.

[8] Goldberg G. E., Genetic Algorithms in Search, Optimisation and Machine Learning, Addison Wesley, New York, 1989.

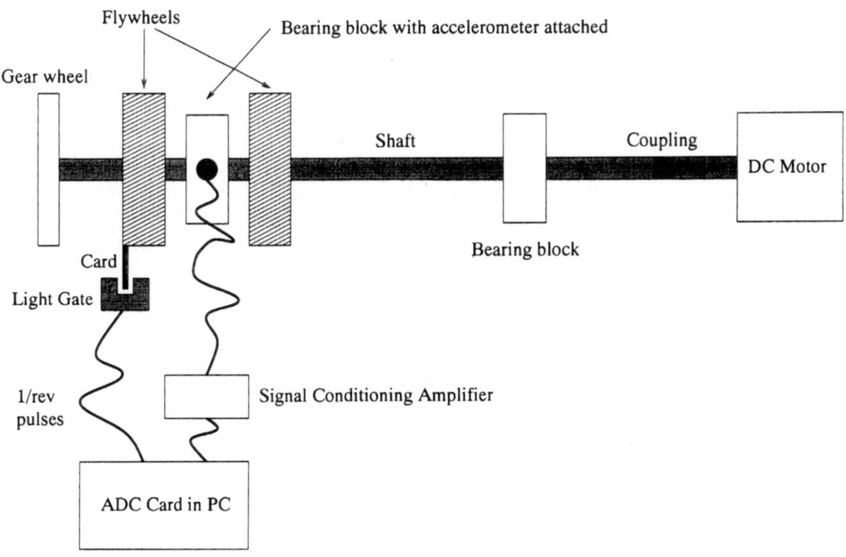

Figure 1: The machine test rig used in experiments.

Data Set	No. Inputs	No. Neurons	Classification Success (%)	False Alarm Rate (%)	Fault Not Recognised (%)
All Statistics	90	7	88.1	0.2	7.4
Spectral Data	66	8	97.0	0.2	2.2
All Data	156	6	91.1	0.1	10

Table 1: Performance of straight ANNs using different feature sets.

Data Set	GA with Best ANN			GA with ANN	
	No. Inputs	No. Neurons	Classification Success (%)	Mean Performance (%)	Performance Range
All Statistics	7	14	97.7	97.1	96.0 - 97.7
Spectral Data	6	11	99.8	99.2	97.3 - 99.8
All Data	6	9	100	98.3	95.0 - 100

Table 2: Perfromance of ANN with GA (after 40 generations) for all three data sets.

Performance monitoring of a small drilling machine using air–particle acceleration signals

M J M NOR and **M DIRHAMSYAH**
Department of Mechanical and Materials Engineering, Universiti Kebangsaan Malaysia, Selangor, Malaysia

In this paper the effectiveness of utilising air-particle acceleration signals to monitor the performance of a small drilling machine is studied. Vibration and sound signals are also measured for comparison purposes. This study focused on the ability of air-particle acceleration signals to detect faulty condition in drilling operations such as tool wear and breakage. Several types of material are used as the test-specimen including aluminum, copper, and low alloy steel. Different type of signal processing methods are also incorporated in the study such as statistical, spectral, and wavelet transform methods. The non-contact and non-intrusive nature of an air-particle acceleration transducer combine with a high signal-to-noise ratio shows that this is a very attractive technique to be used for monitoring the performance of small drilling operations.

1.0   INTRODUCTION

Drilling is one of the most popular activities in machining operation which constitute about 45% from a total number of 1161 machine tool facilities which were surveyed by Byrne (1) The other machining equipment indicated were 8% for milling operation, 38% for turning, 3% for grinding and 6% for other facilities 6%. Future generation of machine tools, working performance of the tool's operation should be able to be monitored effectively where any faulty condition such as tool wear and breakage can be detected and diagnosed at an early stage. For such cases, any potential indication of an impending catastrophic failure can be avoided by shutting off the machine tool's operation immediately.

There are few research works available today that are related to monitoring of drilling operations. For instance, Lee at al (2) presented a paper on monitoring of tool failure in an end milling operations using the spindle current signals and wavelet transform analysis method. Konig et al. (3) showed that acoustic emission signals could be utilised for monitoring small drills operations. There are some problems regarding the utilisation of contact-type transducers for monitoring small drill operations because the weight of the transducer itself will have direct effect on the performance of the whole system. In addition, finding the best position for installing the transducer is also a challenge for the researchers. Therefore, this study is carried out to propose a new method for monitoring small drilling machines in operation using air-particle acceleration signals. The proposed method is expected to give better results due to the non-contact and non-intrusive nature of the transducer. Moreover, since it has a vector property the signal-to-noise ratio is better compared to the conventional sound and vibration signals. However, sound and vibration signals are also recorded in this study for comparison.

Ren et al. (4) indicated that drill performance is significantly affected by its cutting angles and some researchers over the past few decades have tried to optimised cutting angles for better drill performance by modifying drill geometry. Different shapes and orientations with respect to drill axis greatly influence the performance of a drill designed. In this study artificial wear is fabricated which refers only to the point angle constraint. The measurement for quantifying the quality of the defect is not within the scope of this paper. However, the study on this particular topic is mentioned in the paper presented by Aoyama et al. (5) Several kinds of sample materials were used in this experiment such as aluminum, copper and low alloy. It is a well-known fact that large diameter tool bit will naturally have large contact zone between the drill bit and the workpiece materials. It has been shown that low friction coefficient between chip materials and the tool bit can be more effective at extending the tool life compared to using a harder tool bit material (Dautzenberg et al.,(6)

In this paper, studies were carried out to monitor the performance of different types of workpiece materials and several sizes of drill-bit-diameter with the same torque and spindle speed. The air-particle acceleration signals were derived from a two-microphone transducer and an accelerometer was used for detecting the vibration signals. M. J. Mohd Nor (7) introduced initial application of the air-particle acceleration signal, and it was successfully used for monitoring the performance of a rolling element bearing component. Subsequently, air-particle acceleration signals were also used for monitoring machine tool condition and a technical paper on the utilisation of these signals for monitoring a CNC-Lathe machine in operation has also been published recently (M.J. Mohd. Nor et al.,(8)

2.0   EXPERIMENTAL SET-UP FOR SMALL DRILL OPERATION

A small drilling machine was used in this study and the specifications are listed below: (i) Downfeed space of spindle = 1.25″, (ii) Spindle speeds = 8000 and 12000 rpm, (iii) Table dimension = 6.5″ FR x 6.5″LR, (iv) Drill bit material - high-speed-steel twist drill (standard ISO 235 / BS 328) with 118° point angle, (v) Drill bit diameter = 2mm and 3mm.

The artificial wear of the drill bit were initiated with a 175° point angle for the 2mm diameter tool bit and 150° point angle for the 3 mm diameter tool bit. Three types of sample material were used in this experiment including copper (HDHC-BS2874/C101-1986), aluminum (Al-Zi-Mg-CoAlloy), and low-alloy steel (KRUPP 2510 / AISI 01). Scientific Atlanta™ four-channel dynamic signal analyzer with SPS390 software were used for analysing the signals. An SKF CMSS9952™ accelerometer was mounted on the workpiece clamping, in which the distance from accelerometer to the tool bit is 7 cm.

The two-microphone transducer were utilised for capturing air-particle-acceleration signals and the distance for the nearest center microphone to the tool bit was 6 cm and from center to the center microphone is 1.5 cm.

3.0 DISCUSSION

From the results obtained there are indications of different spectrum signatures in the case of defective drill bit when compared to the spectrum signatures obtained from a drill bit with normal condition. Samples of the results are shown in Figures 1(a) and (b). It is interesting to see that spectrum from sound pressure signals are consistently similar to the spectrum obtained from vibration signals. Further research work will have to be carried out in order to explain this phenomena.

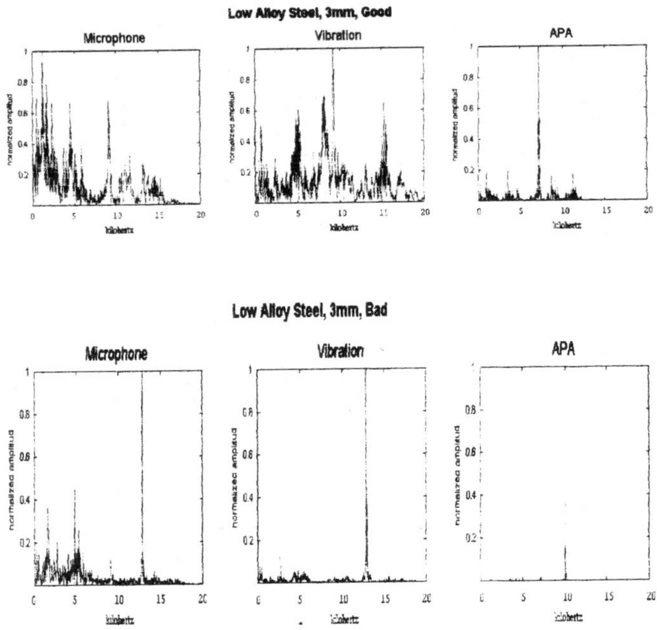

4.0   CONCLUSION

Initial results show that both sound and air-particle acceleration signals are capable of indicating the presence of an abnormal drill bit. However, further study need to be carried out in order to explain why they indicate high peaks at different frequency values.

REFERENCES

1   Byrne, G. et al., (1995), "Tool Condition Monitoring (TCM) – The Status of Research and Industrial Application," Annals of the CIRP Volume 44/2/1995, p. 541-567

2   Lee, B.Y. and Tarng Y.S., (1999), "Application of the Discrete Wavelet Transform to the Monitoring of Tool Failure in End Milling Using the Spindle Current,"International Journal of Advanced Manufacturing Technology, volume 15, p. 238-243.

3   Konig, W., Kutzner, K., Schehl, U., (1992), "Tool Monitoring of Small Drills with Acoustic Emission," International Journal of machine Tool, Vol 32, No. 4. P.487-493

4   Ren, K and Ni, J., (1999), Analyses of Drill Flute and Cutting Angles, ernational Journal of Advanced Manufacturing Technology, volume 15, p.546-   553.

5   Aoyama, E., Nobe, H., Hirogaki, T., (1999), "Drilled Hole Damage of Small Diameter Drilling in Printed Wiring Board," Proceedings of the International Conference on Advances in Materials and Processing Technologies and Annual Conference of Irish Manufacturing Committee, p. 531-538.

6   Dautzenberg, J.H., Jaspers, S.P.F.C. and Taminiau D.A. , (1999), "The Workpiece Material in Machining," International Journal of Advanced Manufacturing Technology, volume 15, p. 383-386.

7   Mohd Jailani Mohd Nor and Muhammad Dirhamsyah, (1999), "Development of Machine Tool Monitoring System Using Acoustic and Vibration Signals," Procs. Of the World Engineering Congress 1999 – Mechanical and Manufacturing Engineering, Universiti Putra Malaysia, pp. 45-50.

8.   Mohd Jailani Mohd Nor, (1996), "Development of a Machine Condition onitoring System Using Audio Acoustic Signals", Ph.D. Thesis, Sheffield   Hallam University, U.K.

Model-based predictive maintenance of naval turbochargers

N G PANTELELIS, A E KANARACHOS, and **N GOTZIAS**
National Technical University of Athens, Greece

1. ABSTRACT

The present work deals with the development of an automatic fault diagnosis for naval turbochargers. The development of this system is based on four sequential steps: the development of simple but realistic Finite Element models based on dynamic simulations of the complete system, the monitoring of the real turbocharger, the accurate modelling of the foundations and the excitation from the main engine. In the final step possible faults of the machine are identified using the Artificial Neural Networks taking advantage of their learning capabilities for the real time diagnosis of potential faults. The application of the proposed system to a real naval turbocharger with vibration data obtained on working conditions show promising results.

2. INTRODUCTION

The condition monitoring of naval turbochargers is a challenge as apart from the low rotational speed of the flexible rotor and the journal bearings' non-linear behaviour [1], the rotor foundation has large excitations from the large diesel engines of the ship. The ultimate target of the proposed approach is the automatic and accurate condition monitoring of turbochargers using only a series of acceleration data from the bearings of the machine. This will be possible by integrating the measured data together with simple Finite Element models and Artificial Neural Networks (ANN) techniques, incorporating only the advantages of each method. The concept of the proposed method is that by feeding the system with acceleration data, the ANN will identify several parameters of the FE model in order to match the simulated output to the measured data. These tunable parameters of the FE model refer to certain potential faults of the machine, mass imbalance at the present case but also cracks, oil whirl etc.

For the simulation of the general case of a flexible rotor with disks which is supported in journal bearings, the classic finite element, the linear beam element, has been employed [2]. Assuming that the journal bearings have a linear behaviour, the mass unbalance at the disks would allow the journals to rotate in an elliptical orbit around the equilibrium position (no mass unbalance) at synchronous speed. The bearings characteristics are calculated based on the modified Sommerfeld number [3]. Except from the simulation of the mass unbalance fault several other potential faults such as oil whirl, cracked shaft, can be introduced in our model. As the turbochargers are always founded on the main engine low frequency vibrations are transferred from the main engine to the rotor. However, the possibility to measure only the bearings' vibrations and not the journal, and consequently the impeller, itself complicates further the problem. Thanks to the significant distance between the two basic frequencies filters can be introduced to separate the lower and the higher frequency regions. Thus, using a lowpass and a highpass filters with cut-off frequency at 6 times the running speed of the main engine the bearings and the journal vibrations can be identified.

3. THE FAULT DIAGNOSIS SYSTEM

The proposed fault diagnosis procedure can be depicted in the figure below (fig. 1).

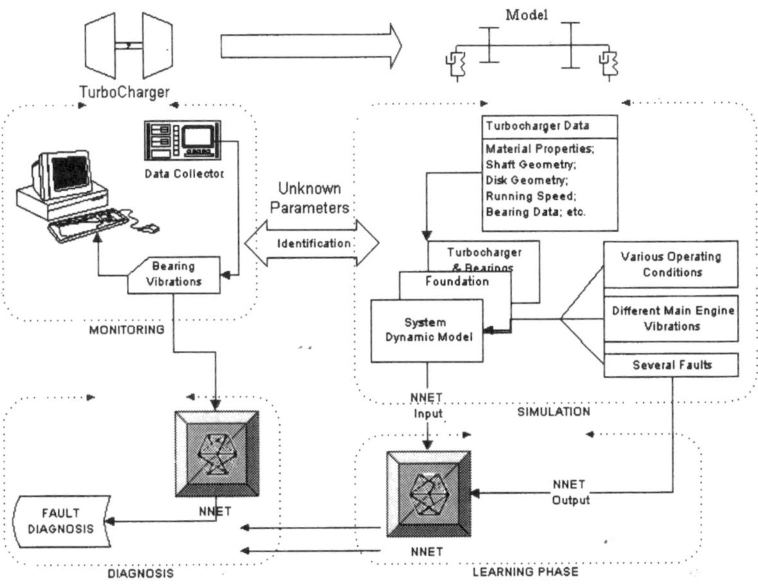

Figure 1. The flow diagram of the diagnostic system.

The development of the fault identification procedure consists of the neural networks training phase and the validation phase. To make things easier for the Neural Networks firstly the low frequency vibrations from the main engine are not considered at all because at this low frequency region faults of the turbocharger cannot be revealed. Secondly, the harmonics of the input vector are scaled with respect to the running speed of the turbocharger, as the running speed fully characterises the faults behaviour. For the output vector the unbalanced masses at the disks have been used. For the learning procedure the back propagation algorithm with optimal learning rate has been employed. The input vector consists of the first 30 harmonics and semiharmonics of the vertical velocity at the bearing that corresponds to circular speeds from half to 15 times the running speed of the rotor. At this point further adaptation of the input vector to the potential faults of the machine could add more efficiency to the system considering that faults are not present at all the range of harmonics but only in specific frequencies. For example in the present case where only mass unbalance fault is studied only the half and the running speed of the rotor could be used in the input vector. One hidden layer was introduced with 20 neurons and for the output vector the mass unbalance was considered. The logistic sigmoid transfer function has been employed both for the input vector to the hidden layer as well as from the hidden layer to the output vector. Using the feedforward algorithm and the optimal weights and biases found from the learning step unbalanced masses can be identified from an unknown deflection signal with acceptable accuracy.

4. RESULTS

For the identification of these low frequency vibrations of the casing of the turbocharger from the main engine of the ship the Hooke-Jeeves first order optimisation method has been introduced. The idea is to identify the mass (M) and the moment of inertia (J) of the foundation as well as the first four sinusoidal forces from the diesel engine so that the response from the model matches the real behaviour of the machine. To achieve this the first few low harmonic peaks of the spectrum of the measured data at the foundation near the journal bearings of the rotor should match to the harmonics reproduced by the model at the same positions of the foundation (X_1 and X_2 in fig. 2).

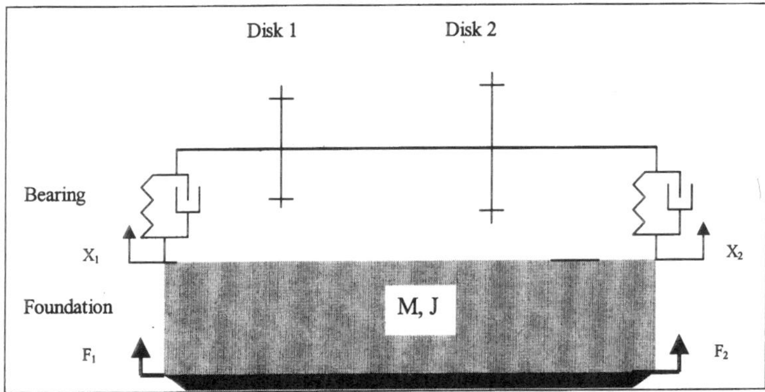

Figure 2. The full model of the rotor, the journal bearings, the foundation (X_1 and X_2) movement and the excitations (sinusoidal forces F_1 and F_2) from the large diesel engine.

With respect to the implementation of the Neural Networks after several trials the optimal performance was achieved employing one hidden layer with 20 neurons. For the input vector the magnitude of the peak velocity at synchronous speed has been used whereas for the output vector the magnitude of the unbalanced mass is considered.

The test case is with both disks unbalanced. Only three unbalanced masses for each disk (9 cases) were fed into the ANN for the learning procedure and the converged accuracy is shown in fig. 3 (left). The performance of the ANN with respect to «unknown» mass unbalances can be depicted in fig. 3 (right). The accuracy is very satisfactory considering that from the power spectral density the system does not find only the type of fault but also the severity of the fault in some adequate accuracy.

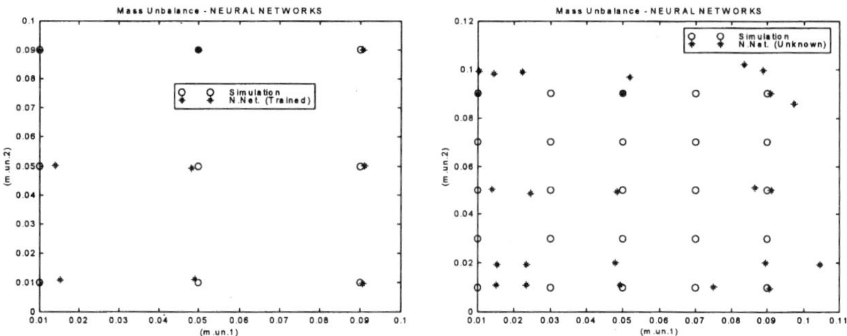

Figure 3. Learning performance of the ANN: On the left the correct outputs (circles) are compared with converged outputs (stars), on the right the «unknown» outputs (stars) are compared to the correct ones (circles).

5. CONCLUSIONS

A method for the automatic condition monitoring of rotors with journal bearings has been presented and applied to a specific naval turbocharger. Trying to combine only the advantages of some very popular computational tools such as the theoretical simulations, measurements and signal processing together with the Neural Networks, the proposed method promises better physics- based and more detailed diagnosis, with quantification of the corresponding faults. Several difficulties to set up the Neural Networks and the development of theoretical models have been encountered due to complexity of the problem but the results' are very promising. However, it is essential that the full non-linear hydrodynamic equations should be also included in order to attain a more realistic model of the transient journal bearing behaviour.

ACKNOWLEDGEMENTS

This research was funded by the Greek Secretary of Research and Technology and the Maritime Company of Lesvos under PAVE-97-BE-121.

REFERENCES

1. Childs, D. (1993). Turbomachinery rotordynamics. Phenomena, Modelling and Analysis, J. Wiley and Sons, Inc, NY, 1993.
2. Pantelelis, N., A. Kanarachos and N. Gotzias (1999). Neural Networks and simple models for the fault diagnosis of naval turbochargers, Mathematics and Computers in Simulation (in print).
3. Lund, J. (1966). Self-Excited Stationary Whirl Orbits of a Journal in a Sleeve Bearing, Ph.D. Thesis, Rensselaer Polytechnic Institute, Troy, NY.

Condition monitoring of earth moving vehicles – a novel approach

S RAADNUI
Department of Production Engineering, King Mongkut's Institute of Technology North Bangkok, Thailand

ABSTRACT

Most earth moving machinery operate with a circulating lubricating fluid *i.e.* engine oil and hydraulic oil etc. This fluid is very similar to human blood that circulates around our body. By checking our blood quality at regular interval, condition monitoring of our health can be established. Likewise, lubricants normally carry vital information of machinery that can be used to assess machinery condition. Not only the quality of the used oil that can be used in setting up condition monitoring program for machinery but also the wear debris and/or contaminant analysis can furnish us with additional important information in assessing the machinery condition. In this particular work, a case study of condition monitoring of earth moving machinery through used oil and wear particle analysis will be presented and discussed. The benefits of used oil analysis are also detailed with practical examples. They also show how regular trend plotting of the results will give warning of the loss in effectiveness of the lubricant. In addition, wear debris and contaminant analysis provides useful information in identifying wear mode and/or mechanism involved. Lastly, but not least, the concept of "total maintenance system" is implemented utilizing small group activity. Hence, corrective actions may be established. This leads to the achievement of optimum system performance, cost and reliability.

1. INTRODUCTION

Engine oil condition monitoring is normally carried out not only to monitor, either physically or chemically, the condition of the oil but also level of wear particles and contaminants either suspended in used oils or trapped by oil filters (1 to 3). In addition, the analysis of failure history data, statistically in particular, provides useful information in the process of Failure Mode and Effect Criticality Analysis (FMECA) (4). Careful examination of this information thoroughly can give specific information about the condition of engines. Normally, used oil test results tend to a constant rate of change. If this rate change alters, then some change in operating condition may be expected and possibly incipient failure may be detected. However, only used oil analysis result give little information about engine condition. Further, by constantly monitor the wear debris composition, quantity, size and morphology can give specific information about the condition about the condition of the moving surfaces of engine components from which they were produced, the wear mode/mechanism may be assessed. In addition, Filter Debris Analysis (FDA) of the oil filter provides further information of large size wear debris and contaminants normally over 100 micron in particular. Lastly, but not least, the FMECA utilizing failure statistics leads to a better selection of maintenance strategies in preventing similar failure occurring in the future.

2. MULTIVARIATE MONITORING CHART FOR USED OIL ANALYSIS RESULTS ASSESSMENT

Regular examination of the oil from an engine crankcase can give vital information on two quite different front. Firstly, simple routine test for physical and chemical properties of the oil indicate whether or not it is fit to perform its function. Secondly, since mechanical wear is inherent in all engines with the generation of metallic particulate matter, examination of these suspended wear particles spectrographically or morphologically give means of monitoring engine condition. As most used oil analysis data normally obey the normal distribution law, statistical parameters and control limits can be extracted which if exceeded suggest that the engine oil should be changed (5). This work in particular, failure history and lubricants' condition of a group of forty-five earth moving vehicles, namely rear dump truck, is gathered within three year period. The Utilizing of Upper and Lower Control Limit (UCL: mean value plus three standard deviation, LCL: mean value minus three standard deviation), multivariate monitoring chart can be established. Figs. 1 and 2 illustrate typical examples of these charts. As simple as in the implementation of statistical process control chart, once the value of parameter(s) reaches UCL/LCL (in some specific parameters *i.e.* oil viscosity) or goes beyond its limits, this can be interpreted as "abnormal" condition of the engine.

3. FILTER DEBRIS ANALYSIS (FDA)

As it has long been recognized that wear debris and contaminant trapped by oil filter are invaluable to condition monitoring analysts. Filters of lube oil of earth moving vehicles were collected and consequently contaminants and wear debris were extracted and assessed visually under conventional microscope and, in some cases, scanning electron microscope (SEM). Two sets of oil filters from different engine operating modes, namely, run-in and overhaul period, were assessed to demonstrate the distinction between debris characteristics. The filters to be analyzed are cut by a special tool to prevent additional metallic particles from its housing (6). Debris is separated from the oil filter media by submerging the oil filter in proprietary solvent and applying ultrasonic cleaning for 15 minutes. The debris is then captured on a 0.4 micron (μm) absolute, polycarbonate Millipore filter using a vacuum to expedite the filtration process. Figs. 3 and 4 show typical debris characteristic found on the Millipore filters. Elongated wear particles are generally found in the filter collected from run-in period. On the other hand, chunky and/or spherical type wear particles are normally presented in those collected from overhaul period.

4. FAILURE ROOT CAUSE ANALYSIS

Further analysis can then be applied *i.e.* the implementation of Pareto diagram, Cause and Effect diagram etc. that can be employed to identify the root cause of the failure problems. This technique is normally called Failure Mode and Effect Criticality Analysis (FMECA). Generally, Pareto and cause and effect diagram is part of quality control's seven tools that are always used in solving quality control problems in production environments in particular. Fig. 5 shows Pareto diagram of the first most frequent failures of the group of rear dump trucks under investigation. Typical cause and effect diagram for the relevant is shown in Fig. 6.

5. CONCLUSIONS

By analyzing/knowing the root cause of the symptom, appropriate maintenance policy for earth moving vehicle can then be selected.

1. for predictable failure : Predictive Maintenance
2. for unpredictable failure : Preventive Maintenance and/or Corrective Maintenance
3. for misuse/mishandling : training for proper use and proper maintenance tasks
4. for improved/better autonomous maintenance and
5. for the basis of reliability calculation.

In summarize, it has been conveyed the importance of the total maintenance system in a novel way. Used oil condition and critical wear production rates can be recognized enabling corrective action to be taken, prior to a catastrophic equipment failure and subsequent vehicle operation stoppage.

6. REFERENCES

1. Bowden, R., Scott, D., Seifert, W.W. and Westcott, V.C., "Ferrography", Tribology International, 9, No.3, 1976, pp.90-115
2. Hunter, R.C., "Engine failure prediction techniques", Aircraft Engineering, 47, No.3, 1975, pp. 4-14
3. Roberge, P.R., Selkirk, C.G. and Fisher, G.F., "Developing an expert system assistant for filter debris analysis", Lubrication Engineering, September, 1994, pp. 678-683
4. Raadnui, S., "Statistical Failure Analysis: its significant in reliability improvement of earth moving machinery", Proceedings of the 53rd meeting of the Society of Machinery Failure Prevention Technology (MFPT), Virginia Beach, Virginia, USA, April, 1999, pp. 209-216
5. Raadnui, S. and Subunpawong, P., "Multivariate monitoring chart for used oil analysis results' assessment", presented at the 52nd Annual Meeting of the Society of Tribologists and Lubrication Engineers (STLE), Kansas City, Missouri, USA, May, 1997
6. Raadnui, S., "The analysis of debris in used oil filters", presented at the 54th Annual Meeting of the Society of Tribologists and Lubrication Engineers (STLE), Las Vegas, Nevada, USA, May, 1999

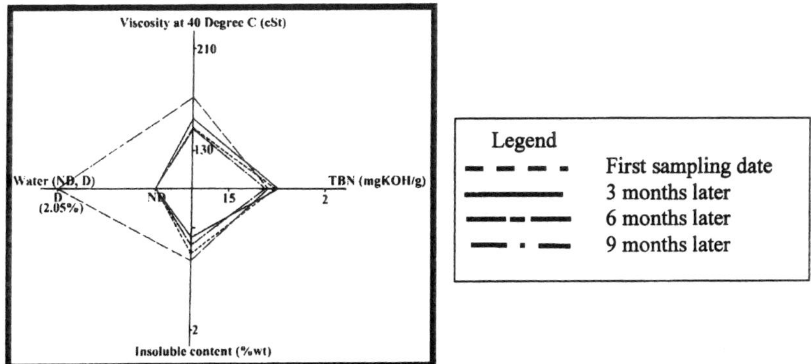

Fig.1 Multivariate monitoring chart for physical/chemical used oil properties.

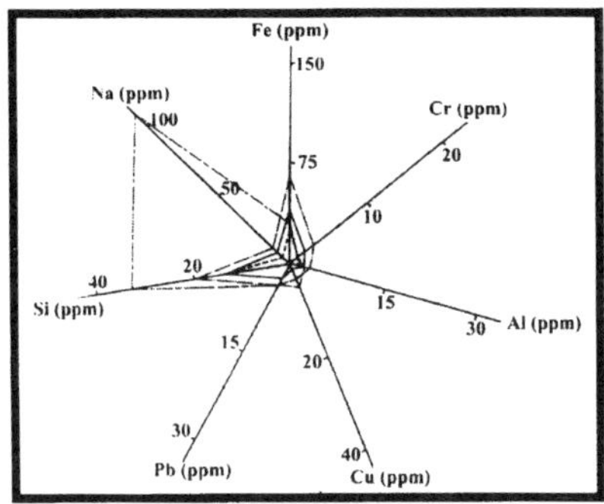

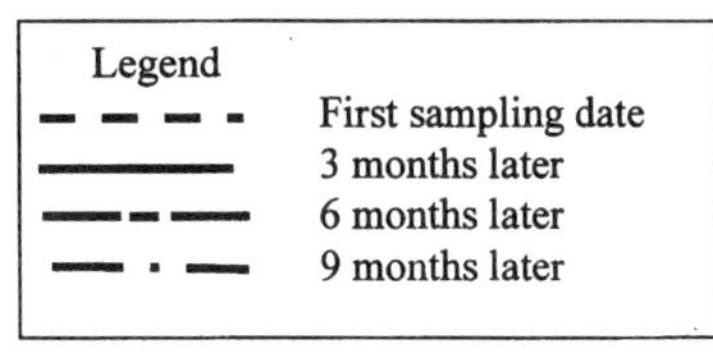

Fig.2 Multivariate monitoring chart for elemental analysis of used oil.

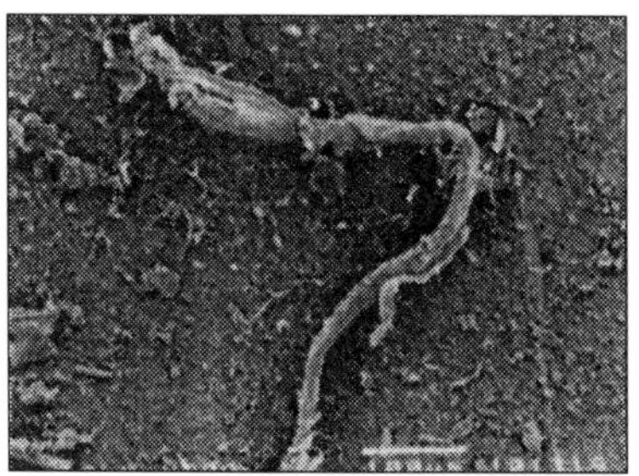

Fig.3 Elongated shape wear particle.

Fig.4 Spherical wear particles.

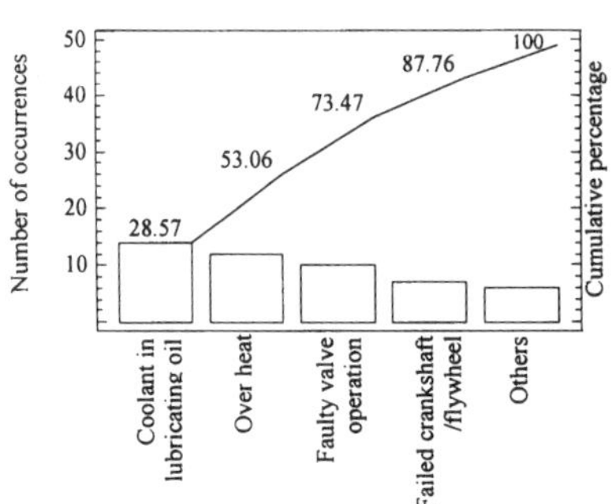

Fig.5 Pareto diagram of failed engines.

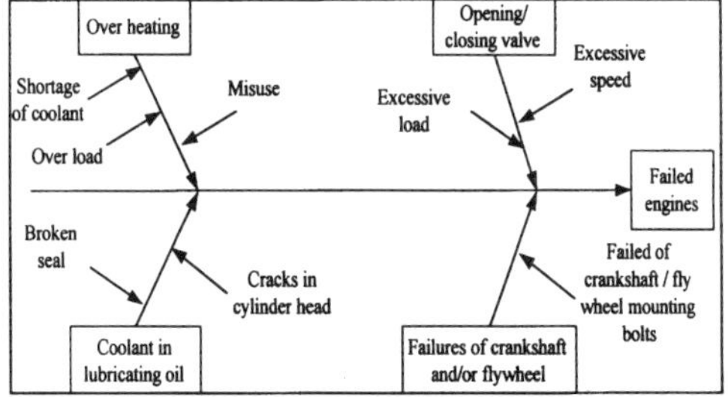

Fig.6 Cause and effect diagram of failed engines.

Technical standards activities in the field of COMADEM

R B K N RAO
School of Computing, Engineering, and Technology, University of Sunderland, UK

ABSTRACT

Condition Monitoring And Diagnostic Engineering Management (COMADEM) multidiscipline is now regarded globally as a very effective solution provider to many real-life industrial problems. As we move into the 21st century, we must be prepared to face both old and new challenges posed by the highly competitive international markets. We are all concerned with bringing prosperity, improved quality of life and better living environment and these are on top of everyone's agenda. There is now a paradigm shift in the way we should prepare ourselves and to meet these challenges in order to conserve and utilise our limited assets to maximum benefit. To this end, we need to develop sound and reliable industrial practices, which meet the customer's requirements. We urgently need good, workable and harmonised national/international standards in the fast growing field of COMADEM. This paper highlights the up-to-date progress being made in the development of appropriate technical standards in this field.
(**Keywords:** COMADEM, technical standards, machinery failure diagnosis)

INTRODUCTION

All industrial processes employ both physical and human assets to various degrees and proportions. It is the top management's responsibility to acquire, maintain and effectively utilise these assets to create wealth and to add value to their businesses. In order to fulfil their objectives businesses must ensure (a) wastage minimization, (b) minimum downtime, (c) customer satisfaction, (d) sound health and safety practices, (e) employing appropriate proactive measures, (f) judicious application of the available latest tools, techniques and strategies, (g) knowledgeable work force, and (h) continuous improvement and innovation. A number of facilities are now available on the market for industries to fully exploit and explore the potential benefits of employing them. Many enlightened organizations are taking full advantage of these facilities and are reaping maximum benefits. There are still many firms who for one reason or the other are lagging behind and missing the opportunities to be productive and prosperous. There are many valid reasons for this, chief amongst which are (a) a conflict of interest between various parties, (b) poor guidance and advice from vendors, consultants and contractors, (c) lack of budget, (d) poor procurement policies, (e) pressurised salesmanship, (f) lack of awareness and expertise in the latest technological/management developments, (g) poor after sales customer care service from vendors, (h) poor adherence to total quality management practices, and (i) lack of appropriate technical standards.

Need for technical standards in the field of COMADEM
Physical assets encompass different kinds of plants and machinery, ranging from very simple to very complex, expensive and sophisticated components, subsystems and systems distributed in different locations (sometimes in hazardous environment). Undoubtedly, each of these possesses a life-cycle of its own. To keep these assets in a fully productive and operational condition, various prescribed performance parameters need to be measured, monitored, inspected, calibrated, trended, compared, diagnosed, predicted, prognosed, controlled at regular intervals. While some world class organizations have established their reputation and benchmarks in maintaining and extending the life-cycles of their valuable assets by doing the right things, many companies are still practising old and outdated habits and are very reluctant to adapt to rapidly changing situations, even though they are fully aware of the dire consequences. A literature survey carried out by the author indicate (a) poor and dangerous maintenance practices, (b) lack of investment, (c) 'quick fix and forget' culture, (d) short sighted vision, (e) bad politics, (f) increased mobility, and (g) poor education/training policy implementation. Modern day machinery, processes and manufacturing operations are complex systems. The days of monitoring the 'health' of these systems with simple tools and human senses reliably and in unfriendly environments is now a thing of the past. With the current trend in the growth of high speed machinery, high degree of automation, high performance requirements, zero maintenance, increased awareness to health and safety regulations, satisfying global customers insatiable demands for quality and reliable goods and services, and globalization of markets there is a growing market for smart sensors, intelligent signal processors, intelligent software, intelligent data management systems, intelligent diagnosis and control systems, intelligent manufacturing systems, etc. By employing such user-friendly systems it is now possible to monitor the 'health' of plants and machinery and to take integrated proactive measures to minimise costly and risky breakdowns and major catastrophes which may ruin one's business abruptly. Various tools, techniques and strategies are now affordable at competitive prices. By judiciously selecting them it is possible to monitor (sometimes non-invasively) various performance parameters such as, speed variations, machine tool cutting conditions, temperature variations, pollution levels, vibration, noise, shock, corrosion, oil contamination, misalignment, out-of-balance forces, motor current harmonics, wear and tear of critical components, etc speedily, efficiently, cost-effectively and intelligently. Other successful strategies such as TPM, RCM, CBM and TQM are now implemented by a number of companies to maximum effect. By employing these many businesses have extended the life-cycles of their assets and have convincingly demonstrated that the "whole works better than the sum of its parts". But everything is not as rosy as it seems to be. The results of a recent pilot study conducted by the author revealed some interesting facts, viz, (a) many of the companies are not fully aware of the available technology, (b) some have doubts about the claims made by vendors about the cost-effective benefits of employing their monitoring system, (c) some are quite happy about their existing outdated monitoring facilities, (d) some end users had raw deals with vendors of soft and hardware, (e) some users complained about the reliability of the hard and software they bought, (f) some voiced that the monitoring equipment and software are expensive and not user friendly, (g) some admitted that they don't have the technically skilled personnel to implement the high-tech monitoring facilities, and (h) many felt very strongly that there are no reliable and authoritative advice and guidelines by way of appropriate technical standards published by British Standards Institution (BSI), European Standards Agency, or International Standardization Organizatioin (ISO) in the field of COMADEM. Many even admitted that they are unaware of any standards activities taking place at National, European and International levels.

A brief introduction to BSI and ISO activities
Many standards have been developed by the BSI and ISO in co-operation with industry bodies, trade associations, governmental and non-governmental bodies to allow everyone concerned to compete on equal terms. Consumers too value the reassurance of safety and quality that conformity to a standard brings. Standards can lead to cost-effective supply of products, services and processes. Many standards have saved money and improved the quality of life. It provides unambiguous forms of communication of value to industry and consumers at large. Without adhering to standards, markets will become increasingly difficult to penetrate or even sustain. Through it's various Technical Committees (TCs), Sub-Committees (SCs), and Working Groups (WGs) both BSI and ISO prepares national and international standards. All European Standards are automatically adopted as British Standards and all International Standards become British Standards as well. Today, there are in excess of 15,000 British Standards, with over 1500 new standards produced every year.

Technical standards activities in the field of COMADEM
The BSI has established the Technical Committee GME/21 that deals with Mechanical Vibration and Shock. The current Sub-Committee GME/21/7 is concerned with the development of technical standards in the field of condition monitoring activities. This Sub-Committee is chaired by Mr. S.R.W.Mills of UK. The author is a co-opted member of this Sub-Committee and also the UK Expert on the ISO/TC 108/SC 5/WG 7 which deals with the standards activities related to Training and Accreditation in the field of Condition monitoring and diagnostics of machines. Other Sub-Committees which are relevant to the field of condition monitoring are, GME/21/3 on Vibration and Shock affecting Structures, chaired by Dr.R.Skipp; GME/21/5 on Vibration of Machines chaired by M.McGuire, and GME/21/8 on Balancing of Machines chaired by Prof. Parkinson. The ISO has set up a Technical Committee No. 108 (ISO/TC 108) under the title "Mechanical Vibration and Shock". This TC comprises six SCs and each of these SCs are made up of a number of WGs .The scope of ISO/TC 108 defines the TC's areas of interest and overall responsibility in part as "Standardization in the field of . . . . Condition Monitoring and Diagnostics of Machines, including, ... methods of measurement, handling and processing of the data required to perform condition monitoring and diagnostics of machines". The SC 5 is now charged with the responsibility of issuing standards in the area of condition monitoring and diagnostics of machines. The scope of ISO/TC 108/SC 5 is "Standardization of the procedures, processes and equipment requirements uniquely related to the technical activity of condition monitoring and diagnostics of machines, in which selected physical parameters associated with an operating machine are periodically or continuously sensed, measured and recorded for the interim purpose of reducing, analyzing, comparing and displaying the data and information so obtained, and for the ultimate purpose of using this interim result to support decisions related to the operation and maintenance of the machine". This SC is currently composed of three Advisory Groups (AGs) A, B and D and ten WGs. The details of these AGs and WGs are discussed in Reference 1. Table 1 shows the current responsibilities for GME 21/7 with respect to the ISO/TC 108/SC 5 activities. Recently the ISO has issued the following New Work Items (NWIs) proposals, viz, (a) ISO/TC 108/SC 5 N139 on Thermal
Imaging with Convenor Mr.L.Hitchcock of Australia, (b) ISO/TC 108/SC 5 N 142 on
Training and Accreditation in Condition Monitoring & Diagnostics of Machines with Convenor Dr.Ronald Eshleman of USA, and (c) ISO/TC 108/SC 5 N 145 with Convenor Mr.S.R.W.Mills of UK.

Table 1

Key: WD (Working Draft) CD (Committee Draft)

COMMITTEE	SUBJECT	CHAIRMAN/ CONVENOR	SECRET- ARIAT	UK EXPERTS	REFERENCE	CURRENT STATUS
ISO/ TC 108/ SC 5	Condition Monitoring & Diagnostics of Machines	J Mathew	ANSI	S R W Mills		
ISO/ TC 108/ SC 5/ AG A	Procedures & Instrumentation	J Niemkiewicz	ANSI	S R W Mills		
ISO/ TC 108/ SC 5/ AG B	Condition Monitoring of Gas Turbines	A Hess	ANSI	J H Worsfold		
ISO/ TC 108/ SC 5/ AG D	Condition Monitoring of Power Transformers	A Marques - Cardoso	IPQ	W T Stokes		
ISO/ TC 108/ SC 5/ WG 1	Terminology	D Muster	ANSI		ISO/ AWI 13372	WD
ISO/ TC 108/ SC 5/ WG 2	Data Interpretation & Diagnostics	D Le Reverend	AFNOR	S R W Mills	ISO/ CD 13379	CD
ISO/ TC 108/ SC 5/ WG 3	Performance	S R W Mills	BSI	C P B Pearson	ISO/ CD 13380	CD
ISO/ TC 108/ SC 5/ WG 4	Tribology	M H Jones	BSI		ISO/ WD 14830	CD
ISO/ TC 108/ SC 5/ WG 5	Prognostics	B T Kuhnell	SAA		ISO/ AWI 13381	WD
ISO/ TC 108/ SC 5/ WG 6	Data Communication Formats	A Mukherji	ANSI		ISO/ AWI 13374 ISO/ AWI 13375 ISO/ AWI 13376	
ISO/ TC 108/ SC 5/ WG 7	Training & Accreditation	R Eshleman	ANSI	B K N Rao		
ISO/ TC 108/ SC 5/ WG 8	Condition Monitoring - Guidelines	S R W Mills	BSI	J H Worsfold	PWI	WD
ISO/ TC 108/ SC 5/ WG 9	Life Usage Monitoring		ANSI	J D Wilson	PWI	
ISO/ TC 108/ SC 5/ WG 10	Electric motors & Generators	A Marques - Cardoso	IPQ	W T Stokes	PWI	WD

CONCLUSION

This paper has highlighted the importance of COMADEM to industrial applications and the urgent need to develop suitable technical standards to help the industry to become world class leaders and to add wealth and prosperity to the nations economy.

REFERENCE

1. Rao, B.K.N. (1999). Development of Technical Standards in the field of COMADEM. Proceedings of COMADEM '99 published by Coxmoor Publications Ltd, pp. 469 – 479, ISBN 1 901892 13 1.

Express-method for monitoring the state of turbogenerator rotor end bells

G G ROGOZIN and **V A KOVJAZIN**
Donetsk State Technical University, Donetsk, Ukraine

ABSTRACT

Investigation is carried out with a view to improve the performance reliability of turbogenerators through the use of the induced eddy-currents in the solid constructional elements of the rotor for diagnostics of its mechanically conjugated end bells. Results being expected may be considered as an addition to the ultrasonic inspection of turbogenerator rotor end bells since the method being proposed requires a short time for putting into practice. The paper presents the results of mathematical simulation and experimental research of the turbogenerator rated value 241,3 MVA carried out at the turbogenerator factory.

1 INTRODUCTION

Among the most dangerous damages of a turbogenerator is the complete destruction of the rotor end bells caused by cracks on the supporting part of their internal surface. The modern well-known out-of-service monitoring destined for checking the rotor end bells (1) needs comparatively long time for its implementation (some six hours at dismantled upper end shields of a machine). Meanwhile there is a need to supplement the above mentioned inspection of the rotor end bells with the rapid shortcut technique of the same purpose which will allow to carry out the current check of the rotor end bells when the turbogenerator is disconnected from the electrical system in compliance with the generation schedule.

The new approach for solving the task set eliminating any labor-consuming preparatory works is based on inducing the eddy-current in the rotor constructional elements (RCE) including the mechanically conjugated end bells and the extreme slot wedges of the rotor. Weakening the fit of rotor end bells produced by corrosion fatigue is reflected in changing the contact pressure between the outlined elements and in redistributing the paths of the eddy current flows. Hence alterations in the course of decaying the induced eddy currents in the massive RCE reflect the state of the rotor end bells. The process of forming the input impulse signal needed for exciting the trailing transient in the massive RCE is based on using a procedure of field extinguishing by means of a capacitor (2). When the turbogenerator rotor is motionless or it revolves being out-of-service the impulse transient function of voltage measured at the rotor slip-rings during damping the eddy currents in RCE is used as a diagnostics time response of voltage $u_f(t)$.

2 SIMULATING THE TRANSIENT PROCESS IN THE ROTOR

Calculating the transient process in the massive rotor under the test conditions was carried out with the use of the d-axis equivalent circuits of the turbogenerator ТГВ – 200M (S_N=241 300 kVA, U_N=15,75 kV, I_N=9,06 kA) shown in Figure 1,a. The non-saturated quantities of the machine expressed in the per-unit system, are: X_{al}=0,177, R_{al}=0,001083, X_{fdl}=0,198, R_{fd}=0,000886, X_{kdl}=1,036, R_{kdl}=0,017, X_{kd2}=0,118, R_{kd2}=0,0505, X_{fkdl}=0,0659, X_{ad}=1,647.

The initial equations for the voltage drops in the loops of the equivalent circuits were written in the operation form. The results derived from the set of equations in the form of the operational functions of currents and voltage and their originals at the real time found by using the known expansion theorem are sufficient to allow the following conclusions:
- Influence of the power unit transformer at no-load conditions is negligibly small.
- A compromise between the rotor equivalent circuit quantities and characteristics of the eddy currents excited in the solid RCE by the action of field extinguishing by means of a capacitor is reached when its capacitance is in the range from 0,6 to 0,9 mF.

Unfortunately, the equivalent circuits shown in the Figure 1,a can't be put to use for mathematical simulating the changes in the given constructional elements of the rotor. This raises the problem of determining the rotor quantities adequately reflecting the physical elements of the rotor construction by using the so-called structural equivalent rotor circuit (see Figure 1,b). Determining the rotor parameters including the active and reactive impedance components of branches which represent the solid rotor ($Z_{Fe\,b}$), nonmagnetic rotor slot wedge system ($R\omega$), the edge portions of the core teeth and the end bells ($Z_{Fe\,e}$) as well as the leakage reactance of the RCE not associated with the skin effect (X_{adl}) was founded on geometrical dimensions of a turbogenerator.

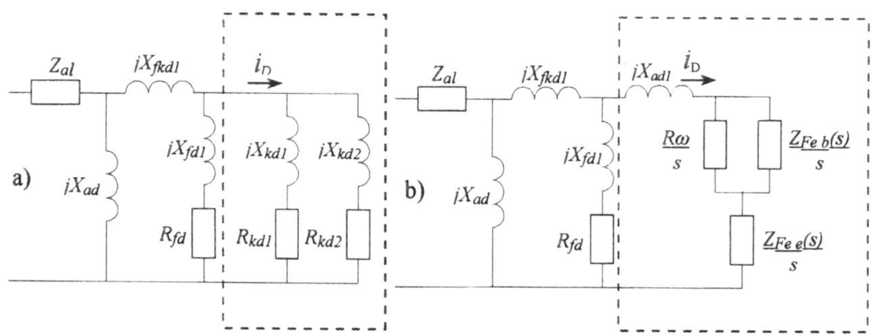

Figure 2. Equivalent circuits of a turbogenerator in the d-axis
(a) Universally adopted equivalent circuits. (b) Structural equivalent circuits for the massive constructional elements of the rotor.

The fault simulation of the RCE was carried out for the turbogenerator TBB – 200 – 2 (P_N=200 000 kW, U_N=15,75 kV, I_N=8,62 kA) by changing the RCE parameters in the structural equivalent circuits (X_{ad} = 1,675; X_{fkdl} = 0,05; X_{adl} = 0,03; R_ω = 0,0182; $R_{Fe\,b}$ = 1,05; $R_{Fe\,e}$ = 0,174). The alteration level of the RCE was set according to the expression

$$\Delta R_j = \pm R_j \lambda n, \quad n=1,2,\dots,m$$

where j = Fe b, Fe e, Rω; λ = increment measure of the parameter being varied.

progress of wear for one lifetime cycle of one bearing set, approx. 1 year.

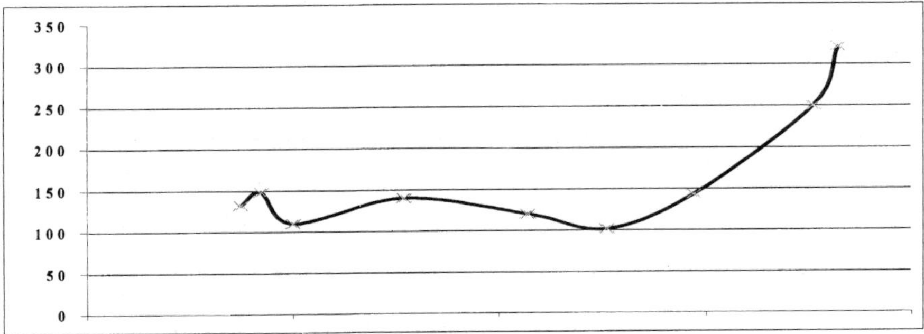

Figure 2

The following methods tend to analyse the bearing condition by identifying the kinematic frequencies of the roller bearing as a function of the oscillating speed. The results for one exemplary data set with the operating conditions described below are discussed to demonstrate the effectiveness of the methods.

The data set was recorded during rolling at a speed of 70 rpm. This corresponds to an oscillation frequency $f_{osc,n} = 0.144$ Hz, see Figure 1. The kinematic frequency of the inner race (the race with the highest failure rate) is $f_i = 3.466$ Hz. The data set corresponds to 5 revolutions of the universal joint shaft.

The first method is the <u>Short-Term-Fourier-Transform (STFT) of the envelope signal</u>. The results of this method are shown for three bearing conditions (Figure 3). The left graph shows the new bearing condition, the middle graph shows the bearing condition after six months operation and the right graph shows the bearing condition before changing the universal joint (approx. after 1 year of use). The amplitude of the envelope signal is plotted over time at the top of each graph. The STFT-spectrum is shown synchronously at the bottom.

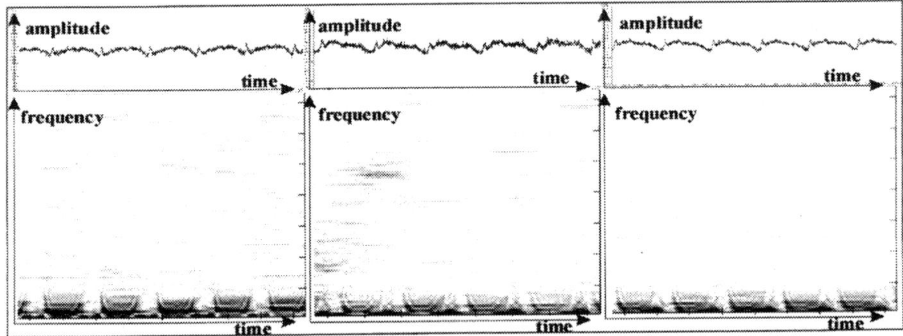

Figure 3

Due to the rapidly varying bearing speed, the STFT analysis does not show the expected kinematic frequency, characteristic for the wear of the inner race. The relatively high rotational speed of 70 rpm in particular leads to the problem of the Heisenberg Uncertainty

Principle when using STFT. Better results are to be expected at slower speeds.

The second method is the <u>Wavelet-Transform (WT) based on the envelope signal</u>. The results are shown for three bearing conditions (Figure 4). The layout of Figure 4 is similar to Figure 3, the only difference being that the WT-spectrum is shown at the bottom.

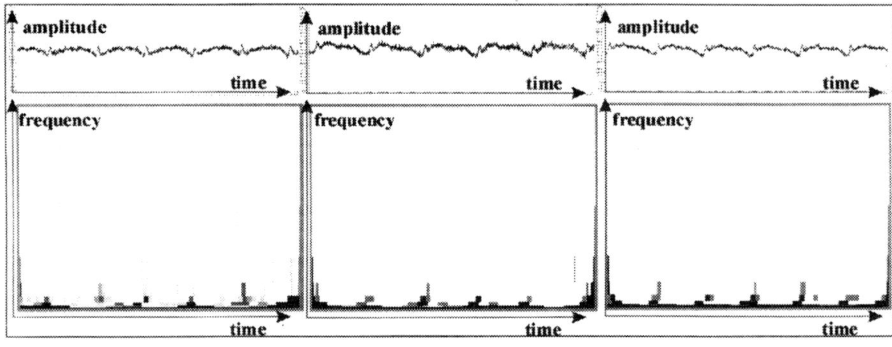

Figure 4

The increasing wear of the roller bearing races has a noticeable effect on the amplitudes of the wavelet coefficients, which rise consistently from the beginning to the end of operation (higher amplitudes of the wavelet coefficients are represented by darker pixels of the sonogram plots in Figure 4).

A direct comparison of the results from the STFT and WT analysis shows that the WT leads to a more distinct detection of the kinematic frequencies (2). Moreover, the WT-method provides a better read-out of the state of wear, resulting in higher amplitudes of the kinematic frequencies. Nevertheless, the results of both methods still do not allow a reproduction of quantitative condition monitoring. The complementary use of the analysis of trends for the RMS value is recommended. The incertitude of the time-frequency based methods may be the result of slippage between rollers and races (3).

The bearing set was changed after the life-cycle (approx. 1 year). During an inspection of the roller bearings, damage was detected to one inner race. The other bearings had no or only very minor defects.

REFERENCES

(1) Geropp, B: Schwingungsdiagnose an Wälzlagern mit Hilfe der Hüllkurvenanalyse, Dissertation RWTH Aachen, 1995

(2) Newland, D. E., Random vibrations, spectral and wavelet analysis, John Wiley & Sons, 1993

(3) Wolfgang Zientz, Untersuchung des Betriebsverhaltens oszillierend drehend bewegter Wälzlager in der Ghelenken von Mechanismen und Getrieben, Dissertation Siegen 1989

The application of artificial neural networks to identify on-line surface finish

T S SIHRA, J K MOHAMMED AL-HASHIMI and **D HOLIFIELD**
School of Product and Engineering Design, University of Wales Institute, UK and Bahrain Training Institute

ABSTRACT

In this paper, a brief description is provided of a comprehensive experiment conducted to investigate the affect of the vibration characteristics of a lathe as the tool tip is worn through repeated machining tests. The characteristics are measured as vibration spectra and visual pattern techniques are employed to detect trends in these spectra which indicate the variance in surface finish quality. Artificial Neural Networks (ANNs) are then employed to provide a more powerful tool for feature extraction and combined with surface finish measurement taken during the cutting operation, a correlation between these two parameters is undertaken. The results show that ANNs are capable of complementing existing tool wear monitoring techniques and potentially they could be applied to a wider spectrum of condition monitoring strategies developed to date.

1. INTRODUCTION

Traditional machining methods utilised the skills and expertise of the machine operator to assess product quality on-line during manufacture. As these are no longer feasible with modern machine tools, due to guarding and high cutting speeds, there is a clear need for the development of suitable monitoring systems to replace such skills in the full range of common machining processes. Attempts to date, focus on relating the signal; from the transducer with some aspect of tool wear, typically flank or crater wear. Many commercial systems available tend to be based upon force or power measurement where the assumption is that quality is dependant upon tool wear. This is not readily the case as there are occasions when machining parameters can be altered to achieve tolerable component quality without initiation of tool replacement. Experienced machine operators supply this vital input based upon their hearing and touch during the cutting process. It therefore makes sense to relate the signal from the transducer to the operator's expertise, rather than to direct tool wear. After discussions with several machine operators it became clear that their expertise was based on assessing surface finish on the workpiece during machining and consequently became a parameter of study in the programme.

those of surface finsih (see Fig1) bear a good correlation. A threshold vibration limit of 0.8mG may be used to indicate when the surface finish on the workpiece has exceeded 3.5 microns

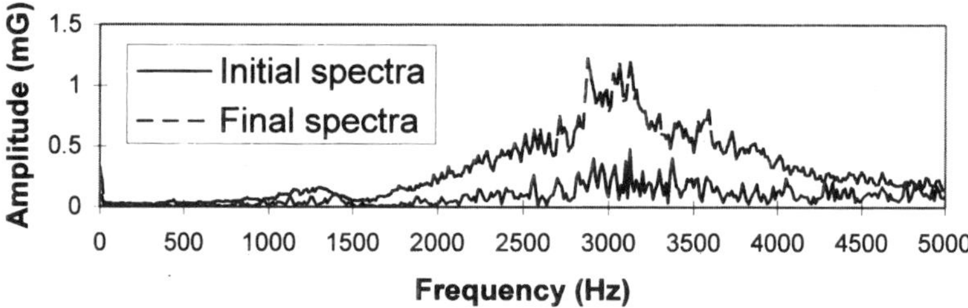

Figure 2 : Comparison chart of initial and final vibration spectra

To establish which region best illustrates how the vibration increases as the cutting time progresses, a more detailed analysis of the vibration spectra was completed. For this intent, trend plots were constructed showing the relationship between the amplitude changes in a particular region of the spectra and cutting time. While several regions in the spectra displayed changes that had the potential to indicate the trend in surface finish, the largest sensitivity occurred around 3kHz. This can be seen in Figure 3. A threshold vibration limit of 0.8mG may be used to indicate when the surface finish on the workpiece has exceeded 3.5 microns.

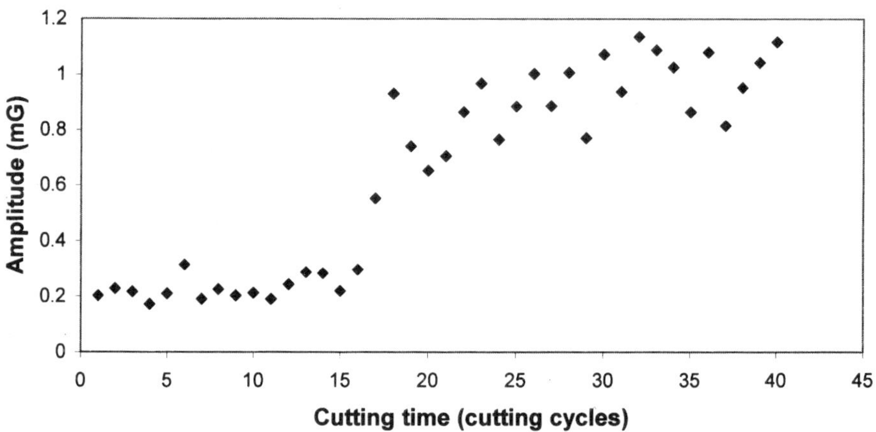

Figure 3 : Trend plot for range 2750 - 3250 Hz

2. EXPERIMENTAL WORK

A comprehensive series of machining tests were completed on a Colchester 2000 lathe. A vibration transducer was attached to the tool holder which incorporated a sintered carbide cutting tool. Mild steel material was machined using suitable machine speeds and feeds to obtain an acceptable level of surface finish. During the process of machining the workpiece, the vibration generated was measured using a signal analyser. At the end of each cut the resulting surface finish was measured using a portable surface analysis machine (Surtronic 3P). The roughness present was quantified using the centre line average parameter, a standard widely used in UK industry. When the level of surface finish had exceeded 5 microns the test was regarded to be complete at which point this process was repeated for the next material specimen.

3. RESULTS

The correlation between the vibrations generated during cutting and the resulting surface finish was examined using the vibration and surface finish measurements. The amplitude changes present in the spectra were analysed with centre line average measurements to establish whether any correlation with these two parameters were present.

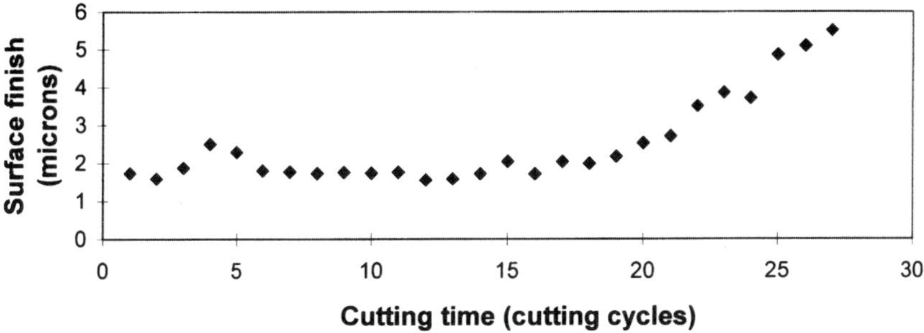

Figure 1 : Surface finish versus Cutting time

Figure 1 shows the trend that can be expected based on data from a typical test. One feature of importance evident in this relationship is that the roughness increases with progressive cutting time. It is also apparent that when a cutting time of 22 minutes has been reached the surface finish is of an unacceptable level. In relation to this the spectral measurements were next examined. In the first instance the nature of amplitude changes present were assessed with the view of establishing frequencies or frequency bands that visually showed correlation with changes in surface finish. Fig 2 shows the spectra obtained for surface finish measurements of 1.8 and 3.5 microns.Comparison of the two spectral measurements show that while several regions display changes in amplitude, the largest senitivity occurred around 3kHz. Succesive measurements in this region of the spectra were then examined more closely to establish the correlation between the amplitude changes in this region with measurements of surface finish. The results of the analysis are shown in Fig. 3. The changes present when compared with

The above results show that vibration measurement has excellent potential to be used as a reliable indicator of product quality. The method of visual inspection used to assess the relationship between spectral changes and resulting surface finish is, however, cumbersome for commercial applications. The time required to select the appropriate frequency region is too long and labour intensive. Furthermore, the analysis requires an "expert" to analyse the data before the developed system can be put to any practical use. One way around this problem was to employ intelligent software to remove the need for such an "expert". There is evidence from many literrature reports that Artificial Neural Networks (ANNs) [1] are ideal for just such a applications and upon this basis were used in the next phase of the study.

The vibration measurements collected in the experimental programme were therefore used in the development of an Artifical Neural Network. As the prime aim of the work was to assess when surface finish had exceeded the acceptable limit of 3 microns, it was considered that ANN need only classify each vibration spectrum into one of two categories; lower or greater than 3 microns. The experimental data was therefore split into two groups, the larger group was used to train the ANN and the smaller group was used as the test set. A suitable network size and architecture was chosen through empirical knowledge; namely a Backpropogation ANN with sixteen inputs, one hidden layer, 5 hidden layer nodes and a single scalar output. The ANN was trained sufficiently making note, not to over train the network and diminishing its ability to generalise. Once trained the test set was presented to the ANN and the resulting surface finish predictions were compared with the measured surface finish values taken during the machining operation. The test stage illustrated that the ANN was capable of classifying correctly 90% of the vibration spectrums held in the test set.

4. CONCLUSIONS AND FURTHER WORK

This investigation shows that visual inspection techniques prove that vibration spectra are useful in predicting the surface finish generated. However, to use this data to more effect requires tools more powerful than the time consuming and subjective process of visual inspection. Marrying vibration spectra with surface finish measurements and incorporating them into ANN design allows for much greater flexibility and impact of the data. Once trained ANNs are able to produce on-line predictions of surface finish quality and dispense with the requirement of expert supervision. More robust ANNs are possible with the integration of parameters such as tool tip force, power consumption or acoustic emissions. The benefit of this multi sensor approach is to provide a more universal system which could readily be used in a range of commercial processes to identify on-line tool wear.

REFERENCES

[1] CHESTER, M., "Neural Networks: a Tutorial"; *PTR Prentice Hall, Englewood Cliffs, New Jersey 07632, 1993.*

Strategies in data fusion for condition monitoring

A STARR, J ESTEBAN, P HANNAH, and **R WILLETTS**
Manchester School of Engineering, University of Manchester, UK
P BRYANSTON-CROSS
School of Engineering, University of Warwick, Coventry, UK

ABSTRACT

Applications of data fusion are found in areas where a required parameter cannot be measured directly. Those in condition monitoring and diagnosis include vibration analysis, non-destructive testing (NDT), and redundant sensor arrays and sensor validation. This paper reviews some of the frameworks, and examples of applications of data fusion, especially in the field of condition monitoring. It is apparent that there is a need for a generalised approach to data fusion. The paper characterises a strategy for creating a generic scheme for data fusion problems.

Keywords: Condition monitoring, data fusion.

1.　INTRODUCTION

Data fusion is a process of blending, or the process of combining data and knowledge from different sources with the aim of maximising the useful information content, for improved reliability or discriminant capability, whilst minimising the quantity of data ultimately retained.

The sensor and signal processing communities have been using fusion to synthesise the results of two or more sensors for some years. This simple step recognises the limitations of a single sensor but exploits the capability of another similar or dissimilar sensor to calibrate, add dimensionality or simply to increase statistical significance or robustness to cope with sensor uncertainty. In many such applications the fusion process is necessary to gain sufficient detail in the required domain. Traditionally the condition monitoring community has focused on a single-sensor, single parameter approach, but increasingly feature extraction is used on high density data, or multiple simple sensors of similar or disparate types are used.

2.　STRUCTURES IN DATA FUSION

The lay out of frameworks varies in relation to the field of application. In 1984 the US Department of Defence established the Sub-Panel for Data Fusion Joint Directors of Laboratories (JDL) in an effort to consolidate this analytical field among researchers. The framework developed by JDL characterises the process from the source signal level to a

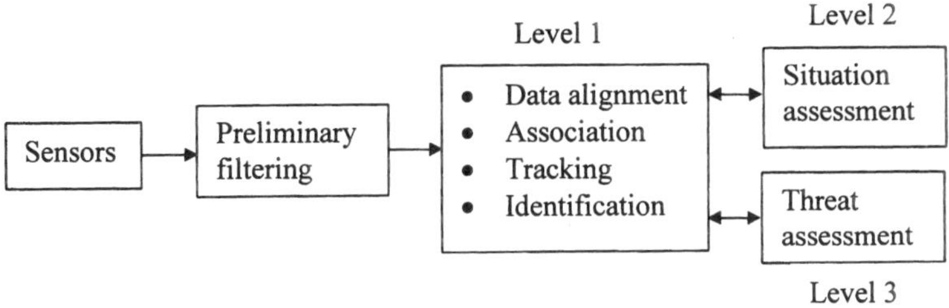

Figure 1. JDL data fusion framework.

refinement level (1). Fusion of information takes place in terms of data association, state estimation or object classification. Situation assessment may proceed, at a higher level of inference, to fuse the object representations provided by the refinement and draw a course of action (figure 1). This framework may be adapted to accommodate condition monitoring.

The strategy to implement data fusion varies from one application to the next, but three stages can commonly be identified. Depending on the problem, it is not always necessary to apply all the stages:

- ***Pre-processing***, i.e. reduction of the quantity of data whilst retaining useful information and improving its quality, with minimal loss of detail. The pre-processing may include feature extraction and sensor validation. Some of the techniques used include dimension reduction, gating for association, thresholding, Fourier transform, averaging, and image processing.
- ***Data alignment***, where the techniques must fuse the results of multiple independent sensors, or possibly features already extracted in pre-processing. These include association metrics, batch and sequential estimation processes, grouping techniques, and model-based methods.
- ***Post-processing***, combining the mathematical data with knowledge, and decision making. Techniques could be classified as knowledge-based, cognitive-based, heuristic, and statistical.

The fusion of data can take place at different levels of representation, i.e. ***raw data*** fusion at the signal/pixel level, where the data is robustly and redundantly merged or sensors are validated, ***feature*** level fusion, where a characteristic is extracted before fusion occurs, or ***decision*** fusion at the symbol level, where measured data with or without pre-processing is combined with processed data or *a priori* knowledge.

3. APPLICATIONS OF DATA FUSION

Practical applications of data fusion have necessarily been those areas in which the required output of an analysis may not be measured directly. This is important for non-destructive testing and condition monitoring, where the purpose is to detect faults and the degradation of machine health (2-7). Work at Manchester has applied data fusion methods to the detection, location, severity assessment and diagnosis of faults in fuel injectors, manufacturing systems, and gear faults (8-10). Techniques used include neural networks, system identification, optimisation, parameter estimation, fuzzy logic, and statistical methods (11-13). This range of applications has led to a deep understanding of particular techniques but moreover a comprehension of the differing frameworks of problem solution configurations and their unique characteristics.

4.   A CONDITION MONITORING PERSPECTIVE

Data fusion is used in condition monitoring because a large amount of data is needed to assess machine health. On-line or off-line data is processed sequentially or by batch. Data for consideration may contain vibration, temperature, pressure, oil analysis, and other measurements. An important aspect of condition monitoring is the quality of information. The data acquired must be consistent and low in noise. Sensor should be complementary, rather than redundant. These aspects should be considered at the source level to reduce pre-processing. Sampling must be frequent enough to catch machine faults and allow warning.

Source data passes to the pre-processing unit for digital conversion and manipulation. At this stage spectral analysis, correlation, image processing, time averaging, thresholding, and dimension reduction techniques are implemented. The processed data is passed to the fusion centre and routed according to the level of fusion sought, i.e. *raw data*, *feature*, or *decision* level fusion. Data will latterly reach the data alignment stage or the post-processing stage. No single fusion technique is proposed: selection must be made depending upon the application, based on the information available and the level of inference sought. Table 1 exemplifies some applications and some of the most commonly recommended fusion methods. The methods are discussed in more depth in (14).

For condition monitoring purposes, the output from the fusion centre should contain explicit information that can lead towards the health assessment of the machine. A fault indicator with measure of severity can aid in decision making. This information is derived from the best estimate, based on decision logic, in the form of a probability measure. A detailed example is shown in (14).

Table 1. Data Fusion techniques and applications.

Application	Fusion method
Combine signals to enhance information	Best-fit functions, Kalman filter (if a model exists)
Signal interpretation (data rich)	Neural network
Feature extraction	Extended Kalman, Gauss-Markov
Decision making	Classical inference, Bayesian theory
Decision making with belief intervals	Dempster-Shafer, Evidential reasoning
Handle vagueness	Fuzzy logic
Pattern recognition (knowledge rich)	Expert systems

5.   CHARTING THE WATERS

Data fusion has been used in many disparate fields, and must be regarded as a superset of data processing algorithms, parts of which are well classified and documented for particular fields and applications. The difficult part is to generalise the strategy. A necessary prerequisite for an engineering solution is a full statement of the problem: its definition and classification. This enables a specification to be drawn up and a solution devised. Indeed, it is often agreed that the problem definition is half the battle. In data fusion, individual problems have received thorough treatments. It is not easy, however, to approach a new data fusion problem. Even relatively experienced users of data processing methods have difficulty selecting the best approaches, and the expertise used in such selection is far from uniform. These issues have been characterised in the "application pull" from industrialists in the Faraday INTErSECT programme (15).

The limitations lie in the focus, to date, on particular advanced applications and specific techniques, combined with the understandable reluctance to report negative results. A generic framework is required which allows selection of best practice methods based on problem and required solution characteristics. Work at Manchester aims to characterise the range of known problem definitions with known good solutions to provide guidelines for problem definition and classification and an application route map. With known problem characteristics, it is possible to move towards a solution, which avoids the traps of trial-and-error and poor solutions which:

- do not reveal as much information as they might;
- are unreliable or not robust;
- retain excessive raw or processed data in an uninformative state.

6.   CONCLUSIONS AND FURTHER WORK

Data fusion is widely used by scientists of many disciplines. There are many individual examples of successful application. There is a need, however, for a clear overall strategy with which to define and classify the problem and hence select fusion techniques.

A picture is emerging of a flexible strategy, which incorporates a number of steps. This needs to be refined by encapsulating expertise, from a variety of sources, which defines the patterns and interconnection between solution steps, and matches the solution to the previously defined problem characteristics.

Details of the content of the fusion map are already established. It is their pattern and interconnection, which is yet to be charted. The guidelines can be tested by demonstration of the methodology in application to known case studies, application of the methodology to new problem data and application of the methodology to define *preferred* data characteristics, i.e. use data fusion as a part of the design process.

These will be achieved by collaborative work with the INTErSECT partners, and in particular will focus on spatial and temporal temperature measurements in harsh environments.

1.3.2   Advantages of thermography

The major advantage of thermography is the large number of possible industrial applications, **Table 1**. Also it does not contact with surface to be measured is largely non-hazardous to personnel and workplace, immune to electromagnetic noise, applicable to explosive environments and invariably real-time, limited only by the process time required and reliable, the components have a semi-infinite life time expectancy.

Machine Fault	Temperature	Pressure	Flow	Oil	Vibration
Electrical – cooling systems, earth faults, circulating currents, laminations, cracking insulation, commutators and brushes	X				X
Commutators, brushes and sliprings	X				X
Ancilliary equipment – fuses, loose connections, overload or unbalanced load, pitted relay contacts, switchgear, distribution boards, transformers	X				
Mechanical - Misalignment, bent shaft	X				X
Damaged rolling element bearings, gears	X			X	X
Inadequate or insufficient lubrication	X				X
Damaged journal bearings	X	X	X	X	X
Loose components	X			X	X
Energy systems - boilers, steam systems, flues, heat exchangers and regenerators Refractory insulation, buildings and roofing	X X	X	X	X	X
Electronic systems - discrete components, printed circuit boards and bonding	X				

Table 1   **Temperature condition indicators and faults**

2   INDUSTRIAL APPLICATIONS

There are many documented applications of IR monitoring in the Steel, Power Generation, Building Services, Automotive, Paper, Cement, Offshore, Glass and Electronic Industries (Thomas, 1999).

2.1   Electrical systems

When two conductors come together a contact is made and due to the roughness and deformation of the surfaces this contact is limited to a certain number of points known as 'elementary contacts'. This condition of contact is also affected by the construction resistance (this is the reduction in contact area causing increased resistance to current flow) and the film resistance (development of resistive oxide layers as a result of oxidation). The deterioration of contact through an increase in electrical contact resistance (the sum of the constriction and film resistances) will produce 'thermal hot spots'. However the rate of deterioration substantially increases with time. Ideally the trended temperature measurements should be performed in conjunction with current measurements where the load is variable. This will enable the temperature rises to be corrected to a reference current value for a valid comparison (British Standards Institution, 1989):

$$\Delta T_r = \Delta T_m \left(\frac{I_r}{I_m} \right)^2$$

where ΔT_r is the temperature rise at the reference current value (K), ΔT_m is the measured temperature rise (K), I_r is the reference current (Amperes), I_m is the measured current (Amperes).

2.1.1   Electric Machines

Rotating electrical machines are an essential item within many industries. They are however subject to a number of possible failures. These failures have been identified (Tavner and Penman, 1987) as: Rotor body defects, rotor winding faults, water coolant faults, stator winding faults, winding insulation defects, stator core defects. When applying condition monitoring to electrical machines there are three sources of possible problems:

- Mechanical sources; bearings, rotor unbalance, looseness, misalignment, end winding damage, brush / brush holder vibrations.
- Aerodynamic sources; turbulence, blade-passing frequency.
- Electromagnetic sources; static air gap eccentricity, dynamic air gap eccentricity, air gap permeance variations, open or shorted windings, unbalance current phase, broken rotor bars, torque pulses and magnetostriction.

The rating of electrical machines are generally set by the maximum permissable temperature with which the insulation can withstand and is therefore an excellent point to set temperature alarms.

2.2   Mechanical systems

Mechanical systems represent a large proportion of equipment in most industries. These systems include mechanical rotating and reciprocating machines that consist of a large number of mechanical components and have in common that heat is generated as a result of friction caused by defective components. Possible reasons for mechanical machine problems include:

- Increased loading on the bearings thereby reducing bearing life
- Increased stress on the machine components leading to fatigue problems.
- Increased forces applied to a machine, such as loose foundations.
- Inertia effect leading to imbalance in rotating parts.

2.2.1   Hydraulic components

Hydraulic component condition monitoring using thermography is generally not common, perhaps one of the reasons for this is the high flow speed, fast pressure changes and fast movements make diagnosis complex and difficult. However temperature in hydraulic systems can vary to some degree during the cycle of the system due to wear, malfunction and increases in leakage and have been documented (Pietola et al, 1995) as illustrated in **Table 2**.

Hydraulic component	Thermal problem
Cylinder	Leakage of piston or rod sealing.
Valves: Pressure valves	Leakage through failure, leakage, jammed spool.
Directional valves	Solenoid failure, leakage, jammed spool.
Check valve	Brake valve, etc. leakage.
Valve packages	Leakage, solenoid failure.
Valve blocks	Leakage near surface
Pumps and motors	Leakage (volumetric efficiency), failure, bearings.
Pipes, hoses, fittings	Leakage, clogging.
Complete system	Failure, wrong settings, too high warming of system, small leakage of oil outside of system.

Table 2   Thermal problems in hydraulic systems [Pietola et al, 1995]

3   ENERGY SYSTEMS

There is continued pressure for improved energy conservation particularly in light of 'global warming' and to protect the environment particularly electrical power generation. This can only be achieved through a further commitment to energy conservation and pollution control. For example the efficiency of space heating is reliant on adequate maintenance to ensure good insulation and to reduce the possibility of leaks or blockages to the system. Faulty insulation and fluid leaks are readily visible as local increases in temperature whereas blocked pipes are detected as differential temperatures across a pipe.

4   CONCLUSION

The technological development associated with IR cameras has resulted essentially in two types of systems. Firstly those that are extremely accurate and fast with large amounts of analytical capability probably for research purposes and those where extreme accuracy is not necessary but are driven by a simple to use software system for predictive maintenance purposes.

REFERENCES

British Standard Institution (1989) PD 6524: Guide to specifying permissible temperature and temperature rise for parts of electrical equipment, in particular for terminals.

Davidson, J (1998) The concept of mechanical reliability; *The Reliability of Mechanical Systems.* Edited by John Davidson; IMechE Guide for the Process Industries, p 10.

Epperly, R.A., Herberlein, G.E. & Eads, L.G. (1997) A tool for reliability and safety: predict and prevent equipment failure with thermography, *IEEE App. Soc. Procs.*, pp 59-68.

Pietola, M., Makinen, R., Vayrynen, P,. Kesanto, J,. S. & Varrio, J. (1995) 'Using a high resolution thermograph in predictive maintenance and fault diagnosis of fluid power components and systems', *The Fourth Scandinavian International Conference Procs. on Fluid Power*, pages 719-727, September 26-29, Tampere, Finland.

Ross, W.H. (1995) Condition monitoring of electrical machines in Scottish Power. *IEE Colq. On Cond. Mon. of Elect. Machs.*, Digest No. 95/019.

Tavner, P.J. & Penman, J. (1987) Failures on real machines, *Condition Monitoring of electrical machines*, pages 38-54, 1987.

Thomas, R.A. (1999) *Thermography*. Oxford: Coxmoor Publishing Company.

Root cause analysis through a proactive lubricant management programme

M K WILLIAMSON
Entek IRD International, Chester, UK

ABSTRACT:

This paper highlights the potential financial benefits of an on-site oil analysis strategy, and describes an economical approach to ensuring a focussed and effective lubrication strategy.

Keywords: oil analysis, condition monitoring, proactive maintenance, particle counting.

1. PROACTIVE MAINTENANCE – ROOT CAUSE ANALYSIS

What is Proactive Maintenance? By addressing the root cause, the conditions promoting ultimate failure are either eliminated or reduced to minimise the onset of failure. The result is that the system enjoys an extended life and the maintenance costs are thus reduced.

Why oil analysis as a tool? Without lubricant, it could be said that industry would grind to a halt. Many companies are aware of the benefits of predictive wear debris analysis as a means of determining a potential failure. Similarly, the proactive monitoring of the lubricant has been seen as an ideal method of ensuring the lubricant is fit for continued service.

There is, however, a third dimension, which is where the major benefits exist. It is a known fact that contamination of the lubricant is a root cause of both lubricant and system failure and therefore contamination control is critical to the financial success. Contamination can be defined as any component of the lubricant that the manufacturer did not intend to be included. This is typically solid contaminant, air, moisture and excessive temperature. Research has shown that the life extension of lubricants and systems through reduced contamination is often more than three times the historical life. The cost benefits become apparent when the equipment replacement cost could be reduced by at least 50%.

2. JUSTIFYING OIL ANALYSIS

It is not easy to calculate the benefits incurred by oil analysis. There is the investment in education, improvement of current lubricant and maintenance practices and the analysis equipment required. More complex are the savings, such as the reduced labour costs, the productivity improvements, the reduced handling and inventory costs and overall improved quality of product. It is not a simple matter to identify a saving when the consequence of not taking action is only predictable on past experience.

With the restrictions of environmental policies, the handling of lubricants is becoming increasingly more complex, thus it makes financial sense to ensure that the lubricant remains in service for longer periods. In order to qualify for the latest ISO 14001, many companies will have to tackle their lubrication strategy, and this will involve ensuring appropriate storage, minimising leakage, and correct disposal.

For many companies, oil analysis has been extremely successful. The main reason for this is that they have focussed on the three critical fundamentals with an emphasis on the proactive aspects. Firstly, the proactive role of monitoring the contamination. Secondly, the proactive role of monitoring the lubricant quality. Lastly, the predictive role of monitoring the wear debris.

However, for other companies, oil analysis has also failed because of the failure to;
1. Identify the focus of the programme as being proactive rather than predictive,
2. Get quality data from either correct sampling location or extraction practices,
3. Identify the most appropriate tests relevant to the systems,
4. Set realistic and achievable targets, and to constantly fine-tune these,
5. Sample at the required frequency or to perform exception testing,
6. Demand or pay for quality service from commercial oil analysis laboratories,
7. Understand the system design, and to play a proactive role in the design process,
8. Use basic human sensory skills with the analysis data for additional confirmation,
9. Educate to ensure action, or compliance from others,
10. Monitor the progress or to advertise the financial success to the management.

3. GAINING FROM AN ON-SITE OIL ANALYSIS PROGRAMME

The most effective way of conducting an oil analysis programme is to bring the operation and control on-site. Whereas many companies will not sub-contract their vibration analysis for the obvious benefits that control is maintained in-house, the same should be said for oil analysis.
On-site analysis offers the following benefits:
- Enables proactive control of root-cause conditions that lead to lubricant and machine failure.
- Identifies components at the onset of the severe wear stage.
- Allows immediate re-testing to confirm abnormal readings.
- Aids trouble-shooting and problem localisation on the system.
- Allows frequent scheduling during the predictive phase at minimal costs.
- Ideal as a complementary technique to vibration or infra-red thermography.
- Minimises the cost of laboratory analyses – sample on condition only, or less frequently.

There are four basic parameters that should be measured on-site; these are the levels of contaminant, the concentration of wear metals, the viscosity and the moisture content. These will provide the critical 'first-line-of-defence' to the maintenance team and guide them in their remedial activity. This needs to be backed up by an effective and regular laboratory analysis programme to assist with more in-depth analysis of the lubricant's remaining life, troubleshooting and problem identification, and wear debris analysis.

Cleanliness of oils is critical to extended lubricant and system life. The National Research Council of Canada found that across a range of industries, as much as 82% of system failures were the result of particle induced failure. Gearboxes and bearings, for example, will benefit from life extensions factors of at least 10 times where proactive cleanliness is practised.

Viscosity is an excellent indication of the ageing process as indicated by a steadily increasing trend over time. Monitoring the viscosity will also identify where the incorrect lubricant has been used, either by a sudden shift up or down in the trend.

Monitoring the moisture level will also ensure that free or emulsified water is not damaging machine surfaces or causing additive scrubbing. SKF state that "It is well-known that free water in lubricating oil decreases the life of rolling element bearings by ten to more than a hundred times." Free and emulsified water will cause valves and orifices to silt more rapidly, icicle formation causing valve seizure, and shortens filter life. Therefore the moisture level should be maintained well below the saturation point.

Monitoring the machine wear by the simple expedient of analysing the ferrous content is a good indication of the wear process occurring. A rise in the wear taking place is a direct result of the contamination in the system.

6. SELECTING THE RIGHT TESTS FOR ON-SITE

A number of devices or instruments are available to the oil analyst. Each has their merits and some are application specific. A word of caution is required; oils are toxic and some tests involve visual examination, therefore protective equipment should be worn at all times.

6.1 The simple tests
These checks should be used as a screening tool to call for higher level laboratory testing:
1   Visual (inspection of tanks, site glasses, levels, aeration, moisture (cloudiness) etc.)
2   Moisture crackle test (hot plate crackle indicating presence of free water.)
3   Smell/odour (indicating oxidation and trace leakage.)
4   Blotter spot test (indicates failure of dispersancy additive and oxidation.)
5   Filter usage (sudden decreased element life, debris inspection etc.)
6   Magnetic testing for ferrous wear (wear debris on plugs etc.)

6.2 Instrumentation
Generally, portable oil analysis systems are cheaper to purchase than vibration analysis tools. The following points cover the typical system requirements of most programmes:
- User-friendly, portable and easy to operate, preferably free of the need for solvents.
- Capable of analysing most fluid types irrespective of conditions.
- Inexpensive to maintain, service and calibrate or validate
- Provide rapid throughput of samples with results in industry standard classifications.
- Allow for analysis of particle sizes, numbers and composition, as well as ferrous content.
- Allow for viscosity measurement over a wide range of viscosity grades.
- Allow for detection of moisture below the saturation point.

- Provide flexibility for either on-line or bottle sampling.
- Proven accuracy, repeatablility and reproducibility.

Generally, a compromise of the above is required to meet the needs of the user.

With contaminant monitoring, light blockage particle counters are more common but are restricted by fluid conditions, hence pore blockage technology is better suited to a broad range of fluids and conditions for trending purposes. Viscosity should be trended in absolute values to avoid problems with shifting specific gravity. Ferrous analysers should allow for the entrapment of the debris to analyse under the microscope. Moisture sensors should be capable of measuring below the saturation point in order to trend the relative humidity so that action can be taken prior to the onset of free water in the oil.

Online monitoring is an emerging technology, currently capable of particle counting, viscosity, and moisture. However, the latest trend is to low cost permanent mount sensors that will have a major benefit for oil analysts.

7. WORKING WITH THE SUPPLIERS

One of the most effective approaches is to work with the commodity suppliers and explain the aim of the programme. It can also be economical as often the lubricant and filter companies will offer these value-added services such as lubrication and cleanliness audits as well as a laboratory analysis service.

8 CONCLUSION

The topic of oil analysis is a very simple concept obscured by the variety of equipment available and the differing approaches and philosophies. The focus has to be on maintaining beneficial conditions and avoiding the root causes of lubricant and system failure. If the proactive approach is adhered to, then the greater benefits of oil analysis are achieved. The predictive approach will provide some success in avoiding unplanned stoppages, but does not address the potential gains of improved system and lubricant life. Control of the programme has to be in-house, and is the responsibility of every employee to ensure that these beneficial conditions are maintained. With the right focus on the analysis, ie., some effective and low-cost tools, immediate remedial action can be taken when the root cause conditions exceed the target levels. Education of all personnel is key to the success, maintenance personnel must be proactive in their activities.

9 REFERENCES

Jones, M H (1998), A Pragmatic Approach to Lubricant and Wear Debris Analysis. Maintenance & Asset Management, October, 1998.

Troyer, D (1995), Three Dimensions of Equipment Condition Monitoring with Oil Analysis. P/PM Technology, April 1995.

Williamson, M K (1999), A Low Cost On-Site Oil Analysis Strategy. Comadem 99, Sunderland, UK.

4.0 Results and Discussions

Figure 1 represents the monthly energy heat and power which is required to drive absorption and vapour compression chillers having the same output capacity when either of them is used to cool a building in Sheffield. The input energy in the graph shows how they are proportional to the COP's of 4.0 and 0.7 for vapour compression and absorption respectively [7]. The graph also shows that if the heat used was free or very cheap the calculation will be in favour of the absorption chiller.

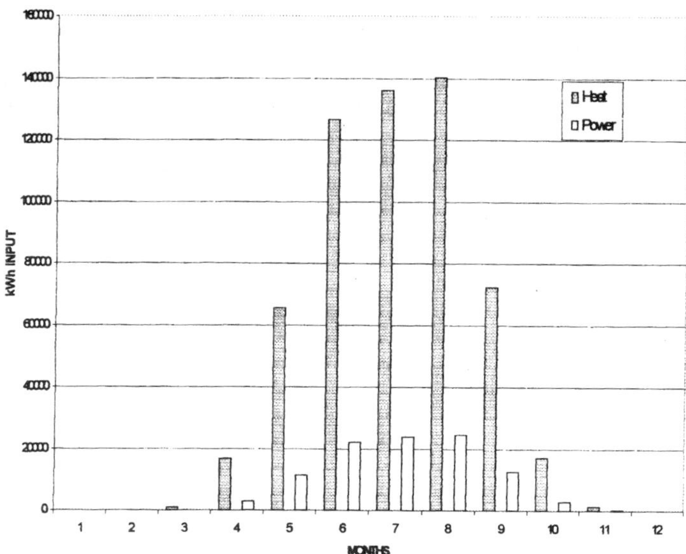

Figure 1. ENERGY (HEAT & POWER) REQUIRED TO DRIVE CHILLERS (ABSORPTION & VAPOUR COMPRESSION) TO COOL A BUILDING IN SHEFFIELD

In this calculation an average price of electricity charged to a large commercial company is approximately 5p/kWh [8] and the average price of heat charged to large commercial building is 1p/kWh [9], also 3p/kWh [9] has been used in certain applications. In this case it was found that for the absorption machine to be competitive in running costs the heat must be priced at less than 0.8749 p/kWh, in the ratios of the COP's and the units prices. The capital cost of absorption machines is considerably larger than that of vapour compression machines. To be an attractive alternative the absorption machine running costs should be even lower in order to offset the higher capital costs and ancillaries including cooling tower fans, water treatment, condenser pump, and electrical power for the small solution and refrigerant pumps.

Table 1 represents the most recent Carbon dioxide performance index for fuels [10]. The main advantage of absorption chillers using waste heat is that they can displace the CO_2 produced by the power stations used to provide energy for vapour compression machines. Both types of machines will be able to meet the cooling loads satisfactorily but the absorption machine will be more environmentally friendly. In the present case it is estimated that savings in CO_2 emissions could be 44,361 kg/year.

Types of fuels	kgCO$_2$/kWh
Electricity (1998 - 2000)	0.44
Natural Gas	0.19
Gas/Diesel Oil	0.25
Renewables	0

Table 1 Carbon dioxide performance index for fuels

The proposed energy tax (climate change levy) on fuel, which aims at encouraging energy efficiency, will provide an incentive for the adoption of environmentally friendly systems, especially when exempting from the levy electricity generated from new forms of renewable energy like solar, wind power and good quality combined heat and power plants [11]. And that is going to increase the financial advantages in the favour of the absorption systems.

5.0 Conclusions

The object of this work was to demonstrate how energy could be saved in cooling buildings using waste heat, as well as assisting the designer to select equipment and balance energy parameters relating to environmental needs. These aims have been realised and can be summarised as:

- It has been shown that from national tariff considerations (8) has reduced this by 80% of cost. The present research has shown how this can be further reduced by 12% .
- The designer has been offered a more environmentally friendly option. And
- Improved accuracy using part loads has been achieved using the program.

6- References

1. D A Reay, (1997), "Heat powered cycle research – A European perspective".
2. CIBSE Guide, "Energy Efficiency in Buildings", (1998), The DETR
3. DETR, "Absorption cooling", (1999), The Department of the Environment, Transport and The Regions.
4. Abdul Kadir, F. S., July 1998, MPhil thesis," Determination of building cooling loads using real weather data", **Sheffield Hallam University**.
5. Firas S. Abdul Kadir and Dr Ian W. Eames, (1999), (Unpublished research report), Sheffield Hallam University and University of Nottingham.
6. This discussion is based on an unpublished paper "Energy Analysis Using Part Load Performance Curves", by Firas S. Abdul Kadir and Ian W. Eames.
7. Edward G. Pita, (1984), "Refrigeration Principles & Systems".
8. Charles Moor, Personal Communication, (phone call discussion with F. Abdul Kadir), (8/11/99), Energy Manager, **Sheffield Hallam University**.
9. Nigel Garrod, Personal Communication, (phone call discussion with F. Abdul Kadir), (5/11/1999),Operation Director, Sheffield Heat & Power.
10. Anne Tope, Personal Communication, (phone call discussion with F. Abdul Kadir, FAX, 23/Nov & LETTER,25/Nov/1999), ETSU.
11. HM Treasury 7 (Report), 9 November 1999.

An update on the Nigerian environment

P B U ACHI
Department of Mechanical Engineering, Federal University of Technology, Owerri, Nigeria

ABSTRACT

About 100 industries in Nigeria are randomly selected and their environmental protection awareness and waste handling techniques are analysed. While the foods/Breweries, civil/marine, service/maintenance industries operation are more environmentally friendly, the petroleum/chemical industries still present serious hazards to the Nigerian environment. The study follows a similar one five years ago and the findings indicate considerably increased environmental protection awareness represented by increased environmental watch/posts but there is no significant corresponding improvement in the previous poor state of the Nigerian environment!

1. INTRODUCTION

In October 1997 a Russian tanker dumped 33,000,000 litres of toxic and foul smelling petrol in Lagos. Many people took ill with throat and stomach ailments on inhaling the fuel which actually killed a commercial driver in Ibadan after he sucked the fuel in order to funnel it into his tank! Like the Koko dumping of toxic wastes of an Italian Company (1) in 1988, sporadic external attacks on the Nigerian environment continues. Earlier studies on Nigerian environment (2) listed some major constraints on Nigeria's Regulatory Institutions as the following:
- Limited Funding
- Inadequate Staffing
- Weak monitoring and enforcement capacity
- Lack of environmentally sustainable development emphasis
- Corruption!

The studies did not pay adequate attention to the problem of gas flaring in the Niger Delta environment. The Qua Ibon river area of Uqua Ibeno of Akwa Ibom State of Nigeria is polluted with every rainfall when toxic effluents from gas flaring are washed into the river and the whole farm environment of the flaring.

In September 1997, 200 families were rendered homeless in Ogbogu Community of Ogba local government area of Rivers State in Nigeria when the deposits of continuously flared gas totally corroded the zinc roofing of their homes.

Since the 1991 study (1), many more industries in Nigeria have appointed industry based environmental safety officers in effort to maintain environmental quality but there is no corresponding improvement on the Nigerian environment.

2. *CONTINUED TOXIC AND OTHER WASTE DUMPING*

In 1988, the Federal Government released Decree 42 on the environment: "The Harmful Waste (Special Provisions) Decree" which prohibits the purchase, sale, importation, deposit and storage of harmful wastes in Nigeria. Also by Decree 58 of 1988, the government established the Federal Environmental Protection Agency (FEPA) charged with the regulatory and "watchdog" duties in the Nigerian Environment. Since it was established, FEPA has succeeded to make many Nigerian industries just aware of their responsibilities to protect the Nigerian environment but is unable to enforce the environmental rules . What is more serious is that dumping of toxic wastes in Nigeria goes on.

On October 6 1996, there was one death and one serious injury from toxic wastes dumped at Izombe in Imo State by two major multi-national oil companies operating in Nigeria. . One of the oil companies paid the medical bills of the victims of its secret dumps but pays little attention to the maintenance of a clean environment.

In November 1996, as President Clinton of USA was raising a fresh alarm over global warming arising from ozone layer depletion from automobile emissions, Nigeria's Oil Minister was also warning oil companies against pollution of the Nigerian environment from gas flaring. That day was the world environmental day in 1996 during which oil spillage threatened the livelihood of one community after another in Nigeria's farming Communities!

3. *ANALYSIS OF THE STATE OF NIGERIA'S ENVIRONMENT*

Environmental pollution by Nigerian industries include(1):
(i) dumping indecomposable solid and liquid wastes on soil surface (DS),
(ii) channeling inert liquid waste into open collection points (CO),
(iii)channeling inert and decomposable liquid waste into rivers or streams (CR)
(iv)open air burning (uncontrolled) of solid and gaseous waste (OB)
(v) dumping of decomposable solid/liquid waste on soil surface (PDS),
(vi)emission of inert fumes (coloured smoke) into open air (ISA),
(vii)emission of hazardous (toxic) gas into open air (TGA),
(viii)channeling hazardous liquid waste into rivers/streams (TCR),
(ix)channeling hazardous liquid waste into open gutters (TCO),
(x) pollution of coastlines and farmlands due to oil spillage (OC)
(xi)burying hazardous wastes in pits (TBP),
(xii)burying inert solid/liquid wastes in pits (IBP).
Table 1 illustrates the updated situation of pollution of the Nigerian environment by the eleven categories of Nigerian industries.

4. *THE PRESENT STATE OF THE NIGERIAN ENVIRONMENT*

When the percentages of industries polluting the Nigerian environment (Table 1) are compared with those in the previous report (1) five years ago it is evident that considerable improvements have been recorded in the civil/marine, service/maintenance and food/breweries industries. However the state of the Nigerian environment has not witnessed equal considerable improvement because the Petro-chemical industries which dominate the industrial

Table 1: Pollution By Nigerian Industries

S/NO.	Type of Industry	% Of Plants Polluting Environment	Pollution Type
1	Petro/chemical	85	All except (v)
2	Civil/Marine	60	DS, CO, CR, IGA, TCR
3	Non-metals	45	CR, OB, IGA
4	Metals	85	DS,CO,CR,OB,IGA,TGA
5	Textile/rubber/leather/ foam	70	OC, DS, OB, IGA, TCR
6	Cement/asbestos	100	DS, CO, TCO
7	Food/breweries	70	DS,CO,CR,IGA
8	Electrical goods/power	75	DS,CO,IGA,TGA,TCR
9	Agricultural Production	-	
10	Service/Maintenance	80	DS,CO,CR,OB
11	Vegetable Oil	-	

Table:2 MD's, Environmental and Quality Control Officers and Reprocessing In Nigerian Industries

S/NO.	Type of Industry	%With Environmental Officers	% With Product Quality Control Officers	% MD's With Arts Degrees	% MD's With Science Degrees	% Partial Or Complete Reprocessing
1	Petro/chemical	70	70	2.5	over 70	45
2	Civil/Marine	40	40	-	60	20
3	Non-Metals	30	70	45	30	85
4	Metals	30	70	-	80	85
5	Textile/Rubber/ Leather/Foam	50	50			50
6	Cement/Asbestos					
7	Foods/Breweries	65	80		50	65
8	Electrical Goods/ Power	25	-		90	-
9	Agricultural Production				100	
10	Service/ Maintenance	25-	25	20	40	25
11	Vegetable Oil				-	-

environment have not made commensurate effort at environmental maintenance. Although most of the industries have considerably increased reprocessing of their wastes, they do so mainly to recover certain economic materials in the wastes. Many of the reprocessing activities create secondary environmental pollution and so the effect of reprocessing of wastes is yet to have an impact on environmental protection in Nigeria. The industries lack the effective technologies to make reprocessing achieve its usual objectives (1) fully. The Petro-chemical Industries in Nigeria should have invested more on research in reprocessing to maintain environmental quality more effectively. The continious monitoring and reassesment of effluents and safety provisions will also reduce the level of atmospheric, terrestial and aquatic pollution from petro-chemical operations.

In the area of environmental watch posts, most of the industries have appointed environmental officers to enable them comply with the directives on environmental impact assessment from FEPA. The major achievement of FEPA lies in the area of awareness of environmental protection nationwide. In Nigeria, last Saturdays of every month used to be observed as nationwide environmental clean up days. The end of the month clean-up of homesteads and premises created an additional problem. The refuse from premises dumped on roadsides would remain there for weeks owing to the severe shortage of vehicles to clear them to appropriate places.

The 24-hour gas flaring by most of the oil companies in the Niger/Delta area continues unabated. A lot of the reported clean-up of the Nigerian environment especially in the Niger/ Delta area by oil companies are spurious! In fact a particular oil company which buried tons of toxic spent paint products in 1987 was recently ordered by FEPA to exhume and incinerate the pollutants .

It is true that certain oil companies in Nigeria have commenced to execute plans to protect the environment but the process is too slow to make the required impact on protection of the Nigerian environment. Some companies, for example, have set the year 2010 as target date for ending the flaring of precious natural gas. What happens before the year 2010 is the burning of billions of dollar worth of Nigeria's natural resource, continuing attacks on the already depleted ozone layer over us, increased adverse environmental pressures on the villages and communities most inhabitants of which are poor farmers.

5. CONCLUSION:

Five years after the first study, the Nigerian environment is still in bad shape inspite of the environmental awareness which is now nation wide. The awareness is certainly stemming the tide of earlier identified pollution activities but there is resurgence of other new ones and so much cleaning up to be able to restore the health of the Nigerian environment. The lack of the necessary technologies and the poor state of the economy are militating against the protection of the Nigerian environment. The limited experience of and commitment of the environmental "watch dogs" most of whom were appointed only recently constitute another problem. Meanwhile a World Bank report stated recently that Nigeria loses $5.1bn annually through environmental degradation by deforestation, bush-burning and indiscriminate construction work. When the losses from gas flaring and human capital (through bad health) are added, Nigeria's losses will exceed her present GDP considerably! With the presently increased environmental watch posts, let us hope that the next five years will witness significant improvements in the maintenance of the quality of the Nigerian environment.

6. REFERENCES

1. Achi, P. B. U: :Management of Industrial Wastes in Nigeria" Environmental Protection Engineering No. 3-4. Vol. 17 1991.
2. World Bank: West-Central Africa Dept.,"Defining an Environmental Development Strategy for the Niger Delta" Vol. II.May 25th 1995.

Gas turbine fuel system robust design

D BONGINI and **P CITTI**
Dipartimento di Meccanicae Tecnologie Industriali, Università degli Studi di Firenze, Italy
V MEZZEDIMI and **L TOGNARELLI**
Technology Department, Nuovo Pignone SpA, Firenze, Italy

1. ABSTRACT

Design For Six Sigma (DFSS) is a quality initiative used at Nuovo Pignone to develop new products, in order to meet Customer expectations over time.
Application of DFSS at the design of the gas fuel system for a Dry Low NOx (DLN) combustor used on a Nuovo Pignone 5.5 MW gas turbine is discussed.

2. INTRODUCTION

Design For Six Sigma (DFSS) is a systematic methodology to predict and improve quality before building prototypes; it utilises statistical methods and tools to understand the effects of variation on system performance, focusing on defects prevention and producibility (1).
In this paper the system process capability assessment and design optimisation of the gas fuel system will be described for a Dry Low NOx (DLN) annular combustor used on a 5.5 MW gas turbine. Identification of the Customer requirements (CTQs) and the formulation of the conceptual design are reported in (1).

3. GAS FUEL SYSTEM DESIGN OPTIMIZATION

The DLN gas fuel system regulates the distribution of fuel delivered to a multi-nozzle single annular combustor. Functionality of the gas fuel system relies on the high-speed gas shut-off valve and the gas control valves (to regulate the desired gas fuel flow delivered). A Pilot flame is used to obtain flame stability while a Premix of air and fuel provides minimum NOx and CO emissions. Correct split between Pilot and Premix is achieved by measuring the gas flow rate through temperature and pressure transducers.
There are some basic modes of distributing gas fuel to the combustor. At fire condition the Premix valve is completely closed and remains in this position until Full Speed No Load (FSNL) condition is reached. Then the percentage of Premix valve fuel flow is increased from 0% to 50% of total gas flow. At the initial loading of the turbine the Pilot flame accounts for 50% of total fuel, a percentage that is progressively reduced to 10% at full load when 90% of the total turbine fuel is used in the Premix mode.
During the gas fuel system design, control valves are sized versus the following requirements:

- Valve effective area calculations, in order to provide the right gas fuel flow to the turbine combustion system, and valve size selection.
- Valve travel determination for full control range.

A conventional design approach is based on worst case flow conditions and each valve is sized independently. Not all sources of uncertainties are systematically taken into account and trade-offs between requirements are not thoroughly analysed. As a result this type of approach might not allow to have a robust valve configuration at system level. Using DFSS, simultaneous selection of the three gas fuel control valves has been assessed, taking into account the various operating and environmental performance conditions of the machine.

By simulating the gas turbine behaviour during Fire, Full Speed No Load (FSNL), 50% Load, Full Speed Full Load (FSFL) conditions in different temperature environments (ISO, -30°C, +30°C), Pilot and Premix downstream pressures have been calculated in order to find the worst case flow requirement for the three valves (Main, Pilot, Premix). This also set the valve effective area requirements.

Regression equations (transfer functions) representing valve flow metering behaviour are used. The position of the valves (%Travel) is expressed as a function of effective area at different pressure ratios using a log transformation to increase the regression model fit (2,3,4). Transfer functions have been assessed for various valve sizes (port areas). As an example, the travel equation for 0.1 inches port area follows:

$$\%Travel = 10^{\left(1.9689 + 0.0925\,\log_{10} PR - 0.3144\,\log_{10} EA + 0.0706\,(\log_{10} PR)^2 - 0.2705\,(\log_{10} EA)^2 - 0.0221\,\log_{10} PR\,\log_{10} EA\right)}$$

where $PR = p_2/p_1$ (pressure ratio), EA = effective area.

Then a statistical design approach for the selection of the valve size has been considered.

To measure the ability of a process to meet its requirements notwithstanding variability in operating conditions and uncertainties, process capability index can be used, comparing the distribution of the process in relation to the specification limits (2,5):

$$C_p = \frac{USL - LSL}{6\sigma} \qquad C_{pk} = \min\left(\frac{\mu - LSL}{3\sigma}, \frac{USL - \mu}{3\sigma}\right)$$

where USL = Upper Specification Limit, LSL = Lower Specification Limit, μ = process average, σ = process standard deviation.

While C_p relates the spread of the process relative to the specification width, it does not look at how well the process average is centred to the target value. The index C_{pk} instead measures not only the process variation with respect to the allowable specification, but it takes also into account the location of the process average.

A true "six sigma" process, by definition, is obtained when $\mu \pm 6\sigma$ equals the product specification tolerance interval and when the process mean is centred between the upper and the lower product specification limits (2). With reference to its capability, a process is considered to be at six sigma quality level if $C_{pk} = 1.5$, that means 3.4×10^{-6} probability of defects. A 1.5 shift is used as a compensatory offset in the mean to generally account for dynamic non-random variations in process centring. This assertion is justified from Motorola's Six Sigma initiatives and it represents the average amount of change in a typical process over many cycles of that process (6).

To determine whether the valves are capable of meeting their first requirement (allow flow capability), process capability of Main, Pilot and Premix valves have been assessed.

Thus valve size configurations capable to reach at least a six sigma quality level ($C_{pk} \geq 1.5$) for all the operating and environmental conditions have been selected.

Mean and standard deviation of the %Travel for Main, Pilot and Premix as well as C_{pk} index have been calculated using the transfer functions ($y = f(x_i) + \varepsilon$) assessed with response surface regression. Considering, as an example, FSFL ISO performance condition, table 1 shows the process capability levels of the three valves.

Table 1. Valves process capabilities at FSFL ISO condition

Description	Mean	Std Dev	C_{pk}
Travel 0.3 Main	63.67	1.13	10.74
Travel 0.3 Pilot	19.44	0.38	17.18
Travel 0.3 Premix	65.06	1.15	10.15

Variance of %Travel transfer functions is estimated by Taylor series expansion (see table 2 as an example) (7,8).

Table 2. %Travel calculation: standard deviation assessment

Main Gas Fuel Control Valve (Sizing Calculations) 0.3 inch FSFL.							
Variable	Name	Mean	Std Dev	LSL	USL	Type of distribution
Valve Inlet Pressure	P1 [psia]	580	1.933	574.20	585.80	Normal
Main Valve Discharge Pressure	P2 [psia]	443	1.477	438.57	447.43	Normal
Main Valve Metered Flow	Gf Main [pph]	3180	10.600	3148.20	3211.80	Normal
Gas Temperature	T [Deg R]	581.7	1.939	575.88	587.52	Normal
Gas Specific Heats Ratio	K	1.31	0.033	1.21	1.41	Normal
Gas Compressibility Factor	Z	0.95	0.033	0.85	1.05	Normal
Gas Specific Gravity	Sg	0.61	0.033	0.51	0.71	Normal
Pressure Ratio	P2/P1	0.763888				
R(K)		0.543927				
Effective Area	EA	0.100322				
Transfer Function Uncertainties	ε	0	0.00596			
Response	%Travel	63.66844	1.127975	0	100	

The following equation provides estimators of the standard deviation for functions of independent random variables:

$$\sigma_y \approx \sqrt{\sum_{i=1}^{n} \left(\frac{\partial y}{\partial x_i} \right)^2 \cdot \sigma_{x_i}^2} \qquad (3.1)$$

Transfer function regression uncertainties are taken into account by introducing a statistical error (ε), that is assumed to have a normal distribution with mean zero (3). The estimate error standard deviation has been considered into the previous equation with a sensitivity coefficient $\frac{\partial y}{\partial \varepsilon} = 1$.

Since a statistical description of the independent variables is very difficult to obtain, with reference to the tolerance of $\pm \Delta x_i$, for each variable x_i, the following assumption on the standard deviation σ_{x_i} has been made:

$$\sigma_{x_i} = \frac{\Delta x_i}{3}$$

To minimise cost, the following valve size configuration has been selected among all the configurations that were capable to reach a six sigma quality level ($C_{pk} \geq 1.5$) with reference to the travel calculation:

- Main Valve: 0.2 inches
- Pilot Valve: 0.1 inches
- Premix Valve: 0.2 inches.

It is also recommended, as a further requirement, that no less than 60% valve travel range has to be used for the full control range of each valve in worst case flow conditions (see table 3).

Table 3. %Travel range calculation at ISO condition

Description	Mean	Std Dev	C_{pk}
%Travel 0.2 Main Max (FSFL)	86.925	2.9054	1.5
%Travel 0.2 Main Min (Fire)	13.662	0.4957	9.2
%Travel 0.1 Pilot Max (50% Load – 50% Pilot / 50% Premix)	88.874	2.4724	1.5
%Travel 0.1 Pilot Min (50% Load – 10% Pilot / 90% Premix)	16.993	0.9267	6.1
%Travel 0.2 Premix Max (FSFL)	77.497	2.0758	.3.6
%Travel 0.2 Premix Min (Fire)	0	N/A	N/A
Travel Range 0.2 Main	73.263	2.9474	1.5
Travel Range 0.1 Pilot	71.882	2.6404	1.5
Travel Range 0.2 Premix	77.497	2.0758	2.8

Tests on a gas turbine prototype were conducted to validate the design. Results are verifying the goodness of the predictions. Endurance tests and field results are then necessary to assess product quality over time and to refine the models used during the design phases.

4. CONCLUSIONS

Product quality prediction and improvement are important design considerations that have to be taken into account before building prototypes. Design For Six Sigma enables us to identify problems in the early design phases where the cost to change is low, evolving from a situation where quality is determined by how well we react to problems to a situation where product quality is by design.

5. REFERENCES

1. D. Bongini, P. Citti, V. Mezzedimi, L. Tognarelli, "New Product Introduction and Design For Six Sigma Processes Integration in Gas Turbine Design", QRM 3rd International Conference, Oxford, 2000
2. W.J. Kolarik, "Creating Quality: Concepts, Systems, Strategies and Tools", McGraw-Hill, 1995
3. R.H. Myers, D.C. Montgomery, "Response Surface Methodology: Process and Product Optimization Using Designed Experiments", John Wiley & Sons, 1995
4. N.R. Draper, H. Smith, "Applied Regression Analysis", John Wiley & Sons, 1998
5. H.M. Wadsworth, "Handbook of Statistical Methods for Engineers and Scientists", McGraw-Hill, 1998
6. M.J. Harris, "The Vision of Six Sigma", Sigma Publishing Company, 1994
7. E.B. Haugen, "Probabilistic Mechanical Design", John Wiley & Sons, 1980
8. E.B. Haugen, "Probabilistic Approaches to Design", John Wiley & Sons, 1968

New product introduction and design for six sigma processes integration in gas turbine design

D BONGINI and **P CITTI**
Dipartimento di Meccanicae Tecnologie Industriali, Università degli Studi di Firenze, Italy
V MEZZEDIMI and **L TOGNARELLI**
Technology Department, Nuovo Pignone SpA, Firenze, Italy

1. ABSTRACT

A Design for Quality methodology (Design for Six Sigma – DFSS) based on the utilisation of statistical methods and tools is embedded in a New Product Introduction (NPI) process to develop new gas turbines able to meet Customer performance expectations.
To show how this process is worked out at Nuovo Pignone, reference will be made to its application at the preliminary design of the gas fuel system for a Dry Low NOx (DLN) combustor used on a 5.5 MW gas turbine.

2. INTRODUCTION

Customers are demanding without compromise higher levels of product quality, at lower cost. Customer satisfaction is closely tied to the extent of certainty a Customer has that quality, cost and delivery standards will be complied with: it is the degree of confidence a Customer has that his product and service related expectations will be met by the producer (1). This message has profoundly affected Nuovo Pignone strategic goals and objectives, as well as daily business practices.
Two integrated quality initiatives are used at Nuovo Pignone with the goal of maintaining a competitive position in the global market: Design For Six Sigma (DFSS) and New Product Introduction (NPI).
The object of New Product Introduction process is to establish the critical milestones (fig. 1) or Tollgates that must be passed during the introduction of a new product, in order to manage risk, to early detect problems and to ensure the introduction of competitive, winning products into the market. Each Tollgate, that has a scope, deliverables and responsible personnel, is summarised by a checklist that schedules all the activities to be completed prior to the final Tollgate review. Tollgate reviews are held with business leaders that make the decision whether to proceed on toward the next Tollgate. Risk abatement plans are then developed and translated in detailed action plans to be completed.
Design For Six Sigma is a key initiative, embedded into the NPI process, that has become part of Nuovo Pignone business practices. It is a rigorous approach to designing products to ensure they meet Customer expectations and can be defined as the utilisation of statistical methods and tools to predict and improve quality before building prototypes. Testing is then used to

verify and validate the predicted quality. The goal is to drive and predict quality early into the development phases where cost to eliminate defects is low.

Figure 1. The NPI Tollgate process

3. THE DFSS APPROACH

The outline of the Design For Six Sigma methodology can be represented by the following steps:

- <u>Identify</u>. It consists in understanding the Customer requirements and in analysing and deriving those technical requirements that are critical for the Customer (Critical To Quality – CTQ).
- <u>Design</u>. It is the conceptual design phase, where potential risks are identified. The influence of the critical design parameters on CTQs is determined through the development of transfer functions: the outputs (product requirements) are mathematically expressed as a function of the inputs (design variables).
- <u>Optimize</u>. Design is iterated until the desired quality results are obtained. For a given transfer function, analyses on how inputs variations impact on the outputs are performed. Simulation techniques are used to implement Robust Design. Reliability tasks are also considered.
- <u>Validate</u>. Predictions are verified by tests, understanding how to make better models and putting in place the process controls to verify that product quality is sustained over time.

To show how Design for Quality and DFSS are worked out at Nuovo Pignone, we refer, as an application, to the preliminary design of the gas fuel system for a Dry Low NOx (DLN) annular combustor used on a 5.5 MW generator drive gas turbine.

The DLN gas fuel system regulates the distribution of fuel delivered to a multi-nozzle single annular combustor. The major components of the gas fuel system are the high-speed gas shut-off valve and the gas control valves (to regulate the desired gas fuel flow delivered). This system utilises a Pilot flame (to obtain flame stability) and a Premix of air and fuel (to reduce NOx and CO emissions). To establish the correct split between Pilot and Premix, temperature and pressure transducers are used to measure the gas flow rate during engine operation.

With reference to the gas fuel system design, the following DFSS process steps will be described:

- Customer CTQs and product requirements identification (Quality Function Deployment).
- Conceptual design formulation, design parameters and transfer functions identification (Quality Function Deployment, Pugh Concept, Engineering Analysis).

4. CUSTOMER AND PRODUCT REQUIREMENTS IDENTIFICATION

Quality Function Deployment (QFD) is the product process specific tool that has been used to identify Customer requirements and connect them with product requirements and design parameters (2).

A cross-functional team has populated the two houses of quality (fig. 2) that link Customer CTQs (Critical To Quality) to design parameters.

Figure 2A develops Customer demands in the left vertical column and then identifies product requirements in the first horizontal row. The degree of importance to Customer requirements is given by a weight column. A score from 1 to 9 in the body of the matrix quantitatively assess the relationships between the Customer demands and the product requirements (e.g. "accuracy/repeatability" have a big impact on "emissions" in order to control them). For each column, we can multiply the cell score for the relative weight and sum down, calculating a product requirement priority score. As we can see from the final ranking, all the product requirements have approximately the same importance on the CTQs; so they all have been considered during design.

The design parameters are defined using the second house of quality (fig. 2B). The input of this house is the output from the previous one. Customer demands are thus translated into design parameters and then prioritised.

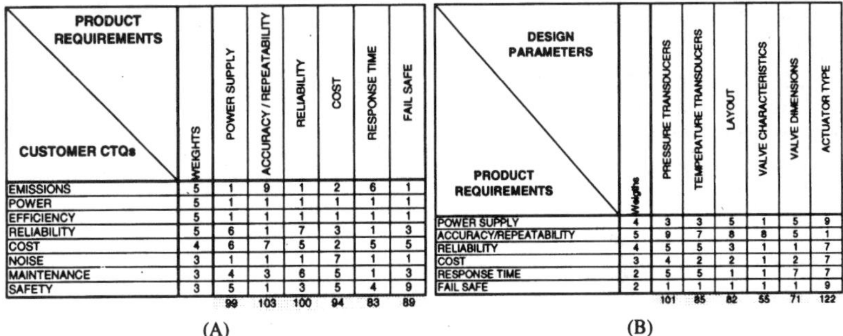

(A)

CUSTOMER CTQs	WEIGHTS	POWER SUPPLY	ACCURACY / REPEATABILITY	RELIABILITY	COST	RESPONSE TIME	FAIL SAFE
EMISSIONS	5	1	9	1	2	6	1
POWER	5	1	1	1	1	1	1
EFFICIENCY	5	1	1	1	1	1	1
RELIABILITY	5	6	1	7	3	1	3
COST	4	6	7	5	2	5	5
NOISE	3	1	1	1	7	1	1
MAINTENANCE	3	4	3	6	5	1	3
SAFETY	3	5	1	3	5	4	9
		99	103	100	94	83	89

(B)

PRODUCT REQUIREMENTS	Weights	PRESSURE TRANSDUCERS	TEMPERATURE TRANSDUCERS	LAYOUT	VALVE CHARACTERISTICS	VALVE DIMENSIONS	ACTUATOR TYPE
POWER SUPPLY	4	3	3	5	1	5	9
ACCURACY/REPEATABILITY	5	9	7	8	8	5	1
RELIABILITY	4	5	5	3	1	1	7
COST	3	4	2	2	1	2	7
RESPONSE TIME	2	5	5	1	1	7	7
FAIL SAFE	2	1	1	1	1	1	9
		101	85	82	55	71	122

Figure 2. Houses of Quality

5. PRELIMINARY DESIGN

At the beginning of the design phase the choice of the actuator type has been considered, as it results the design parameter having the biggest impact on Customer CTQs. The "Pugh Concept" selection process is a simple method of selecting the best alternative among several design options and it has been utilised to choose the actuation type of the fuel valves (fig. 3).

	Hydraulic (High Pressure Oil)	Electric	Note
Technical Requirements			(1) Elimination or reduction (IGV) of the hydraulic power supply allows a reduction of maintenance associated with fluid quality monitoring, control and filtration system servicing.
Maintenance (1)	0	+1	
Simple Design (2)	-1	+1	(2) An electric actuator allows a system more simple than with hydraulic components.
Dimensions (3)	-1	+1	
Oil Leakage (Pollution Risk) (4)	-1	+1	(3) Reduction of on engine hydraulic piping connections with the associated space, weight and cost penalties.
Fire (Safety) (4)	-1	+1	(4) Elimination of concerns related to hydraulic fluid leakage, both on engine and in storage, including fire and pollution hazards.
Fail Safe Capability (5)	0	+1	
Proven Solution (6)	+1	0	(5) Reliable fuel shutoff has to be provided.
Other Issues			(6) Validation of the system due to previous designs.
Preservation Procedures Issue (7)	0	+1	(7) Achievement of unlimited unit storage life without preservation procedures.
Reduced Variety of Stocked Parts (8)	-1	+1	
Cost Issues			(8) Reduction of the variety of system spares to be tracked and stocked. Readily available replacement spares.
Expected Overall Cost	-1	+1	
Overall Performance	**-5**	**9**	

Figure 3. The "Pugh Concept" selection process (Valve Power Supply)

Each alternative is compared in the following manner: "+1" means clearly better, "-1" means clearly worse, "0" indicates about the same.

Brainstorming and past experience suggested several possible design choices. A subset of these design choices has been downselected by experienced engineers using again the "Pugh Concept" (fig. 4).

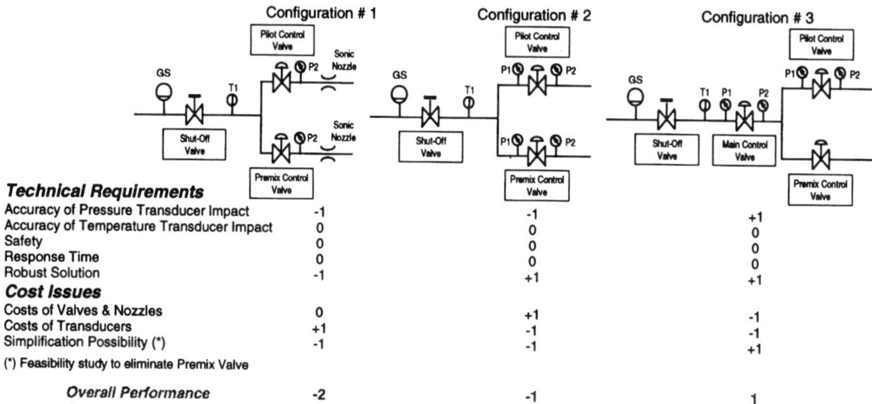

Technical Requirements	Configuration # 1	Configuration # 2	Configuration # 3
Accuracy of Pressure Transducer Impact	-1	-1	+1
Accuracy of Temperature Transducer Impact	0	0	0
Safety	0	0	0
Response Time	0	0	0
Robust Solution	-1	+1	+1
Cost Issues			
Costs of Valves & Nozzles	0	+1	-1
Costs of Transducers	+1	-1	-1
Simplification Possibility (*)	-1	-1	+1
(*) Feasibility study to eliminate Premix Valve			
Overall Performance	-2	-1	1

Figure 4. Gas fuel system layout selection

Then valves are sized, using appropriate thermodynamic formulations, by calculating their effective area, in order to provide the right gas fuel flow to the turbine combustion system. To choose the proper valve size, the effective area required to meet worst case flow conditions must be determined.

6. CONCLUSIONS

The discussed approach aims to enhance product quality during design. Design For Six Sigma and New Product Introduction methodologies are systematic and efficient ways to meet this challenge. Understanding the Customer needs and identifying the most critical parameters to be considered are important design considerations that have to be taken into account. Design for Quality tools provide a real advantage, improving the overall product quality and shortening the global time to market.

A further work of the authors (3) illustrates how DFSS can be used to optimise design, in order to meet Customer performance expectations (CTQs).

7. REFERENCES

1. M.J. Harry, J.R. Lawson, "Six Sigma Producibility Analysis and Process Characterization", Addison-Wesley, 1994
2. W.J. Kolarik, "Creating Quality: Concepts, Systems, Strategies and Tools", McGraw-Hill, 1995
3. D. Bongini, P. Citti, V. Mezzedimi, L. Tognarelli, "Gas Turbine Fuel System Robust Design", QRM 3rd International Conference, Oxford, 2000

Modelling of parametric relationships in jet fans

L D KOUNIS, J M O'SULLIVAN, and **I R McANDREW**
Department of Manufacturing Systems Engineering, University of Hertfordshire, Hatfield, UK

ABSTRACT

This paper outlines the difficulties of using both Signal-to-Noise Ratios (SNR) and Capability Indices(Cp_k) as quality improvement techniques. Having proved (1) in various areas the contradictory indication of aforementioned quality technique tools, a 2D geometrical model is described and applied on Jet Fans in a tunnel. The present work builds on preliminary attempts of a 2D geometrical model to indicate clearly the contradictions between the indicators, and to examine the model on jet fans in a tunnel. In particular, the following parameters were regarded as being of importance to establish the optimum setting for a jet in a niche: Niche depth, Niche Length, Recess Angle, and Area Ratio, and are used to calculate the Cp_k and SNR values, respectively.

1. INTRODUCTION

The pace at which changes occur in modern industry is increasing, particularly in the quality sector. This is attributed to the development of the BS EN ISO 9000 standards in 1987, and an ever increasing customer expectations rate which in particular forced car manufacturers to raise their quality standards. The start occurred when GM, Chrysler, and Ford, developed their own Quality Manual, the QS 9000 series.

The contemporary interest in Quality Audits and the concomitant development of fundamental quality systems especially in the wake of ISO 9000 has accelerated over this last decade. Now with the advent of QS 9000 the motor industry sector have developed the ISO 9000 to include continuous improvement techniques.

Many industries and specialist standard institutions adopt accepted indices such as SPC, FMEA, and SNR. It has been shown (1), that Signal-to-Noise Ratios and Capability Indices can lead to contradicting actions. Such contradictory actions may result in increased variability whilst maintaining a central target.

Although the car industry sector concentrates on Cp_k and SNR there is no guarantee that these two techniques will recommend the same conclusions. The results published in this paper investigate such contradictions and show that a balance can be achieved between Robustness and Capability Indices, when the results are viewed to an optimum condition; ideally meeting the target value, denoted as T, and achieving zero variation.

The number of companies subscribing and satisfying QS 9000 requirements is indeed increasing, with companies like FESTO AG & Co. in Germany, and AVON Vibration Management Systems Ltd. in Great Britain. It is therefore becoming apparent that the application of a variety of quality technique tools, as directed in QS 9000 3rd. Edition will force companies to implement them, as robustness of designs and processes are also a requirement. Continuous improvement will result in a company trying to find the best settings for robustness. The most used approach is to measure the SNR value and subsequently select the parameters to this value. SPC, FMEA, QFD, Taguchi and Capability Indices are generally accepted as a key indicator to predict reject levels (2). Motorola's 6σ approach has focused the aims of reaching Cp_k indices of at least 2.0, which is a difficult task, and might not be possible, if tolerances are set at unachievable levels. For a number of manufacturing processes, Cp_k indices are stated in the contract stage, and subsequent evaluation may require proof of this achievement.

The aim of this research is to:
- Identify where such contradictions occur and why,
- Find a common solution which will satisfy the requirements of capability indices and robustness, by minimising the detrimental effects on each other.

2. COMPARISON OF TECHNIQUES

Although different types of SNR ratios exist industry nowadays generally makes use of three basic ones: min.-the-best, max.-the-best, and nominal-the-best. Each of these SNR ratios requires individual treatment. The use of SNR were normally shown in industry to show a form of robustness but seldom as a long-term commitment to achieving optimum settings. It was observed in a variety of cases (1), that SNR ratios did meet their target, showing a considerable variation.

Another quality technique tool used by companies is the Capability Index (Cp_k), which is a method of describing the likelihood of rejects. This technique also focuses on the target value and variations that exist; however, the ends of the variation are related to specification limits. The Capability Index is calculated according to the following formula:

$$Cp_u = USL - \mu / 3\sigma, \text{ and } Cp_l = \mu - LSL / 3\sigma$$

Thus: $$Cp_k = \min.[\, Cp_u\, ;\, Cp_l\,].$$

Using 3σ for each side of the bell-shaped curve, 99.73% of the population variation can be explained with respect to the upper and lower specification limits, respectively. The following four Cp_k values are nowadays used: [1.0; 1.33; 1.67; 2.0], with the latter one becoming the basis at which companies are being judged. However, from industrial data gathered, it has been observed that Capability Indices of $Cp_k \geq 5.0$ are being achieved, as shown in an example in the "Analysis" section underneath.

3. ANALYSIS

One parameter influencing the positioning of jet fans in a tunnel is the Niche-parameter, which is described by an $L_9 (3^4)$ orthogonal array. Four factors at three levels are considered to be of importance:

- Niche depth: fan diameter (Factor A); Levels: I (0.5), II (0.75), III (1).
- Niche length: in mm (Factor B) ; Levels: I (F), II (F+1D), III (F+2D).
- Recess angle: in degrees (Factor C); Levels: I (10), II (20), III (30).
- Area ratio: Area of Fan divided by Area of Tunnel, namely: A_{FAN} / A_{TUN} (Factor D); Levels: I (0.015), II (0.0314), III (0.0559).

The SNR for max.-the-best is calculated as follows: $n = -10\log(\Sigma y^{-2} / n)$. The results are shown in the table underneath:

Table 1 – Experimental results from jet fans in tunnel

Exp	EFFICIENCY in %			y(=Average)	σ_{n-1}	SNR	Cp_k
1	76	75	77.2	76.067	1.10	37.62	0.67
2	83	82.2	83.5	82.9	0.66	38.37	4.32
3	91.2	93.2	92.2	92.2	1	39.29	0.33
4	85	86	85.6	85.533	0.5	38.64	5.03
5	87	86.4	87.7	87.033	0.65	38.73	2.82
6	83	82	83.5	82.833	0.76	38.36	3.71
7	79	80.1	78.2	79.1	0.95	37.96	1.12
8	83	82	81.2	82.067	0.9	38.28	2.29
9	78	79	79.1	78.7	0.61	37.92	2.25

From table 1, the contradictory results between SNR and Cp_k values may be observed; in particular, the highest SNR value of 39.29dB as per experimental run three, does not meet the highest Cp_k value of 5.03, observed in experimental run four. Additionally, this experiment shows those Cpk values of greater than 2.0 appear in 6 out of 9 experimental runs, thus contributing to the minimisation in variation. In order to compensate for the small loss in robustness of experimental run four, as opposed to the robustness of experimental run three, a 2D geometrical model can be applied in this case. It comprises of an x-y-axis and shows the following data:

- the [-x; x] axis, shows the different values of x, in ascending order,
- the y axis shows the values of the standard deviation σ_{n-1},
- the top negative value on the –x –axis, denotes the Lower Specification Limit, or LSL; in this case it is LSL = 75.00,
- the top positive value on the x-axis, denotes the Upper Specification Limit, or ULS; in this case, USL = 93.20,
- the five diagonally drawn straight lines, which meet the σ_{n-1} axis, and form a triangle, represent the different Cp_k values, starting from a $Cp_k = 1.0$ $Cp_k = 1.33$, $Cp_k = 1.67$, $Cp_k = 2.0$, down to a $Cp_k = 5.03$.
- the (0;0) stands for the difference of (84.1-84.1), with 84.1 being the average between the LSL, and USL, respectively.

Any given point on the 2D model can be described by trigonometric formulae, in terms of angular position and distance from the origin. It further suggests that the best setting is achieved by either a short and close distance from the origin, and/ or a high angle of the distance from the origin. This is mathematically justified, because, the higher that angle is, the closer it gets to the y-axis, i.e. the σ_{n-1} axis, minimising as such, variation, while the closer a point Σ is to the origin, the closer it comes to the target value T.

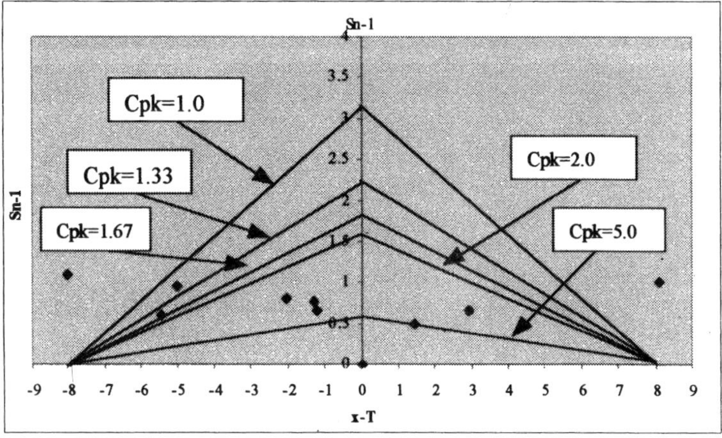

Figure 1: The 2D geometrical model applied for jet fans.

4. CONCLUSIONS

The 2D geometrical model shows the contradictory outputs of the SNR, and Capability Indices, on jet fans; particular, the highest SNR value of 39.29dB as per experimental run three, does not meet the highest Cp_k value of 5.03, observed in experimental run four. By determining the position of each point to the datum enables a setting to be selected which is a balance of capability and robustness. Additionally, the positions above the different Cp_k lines indicate the necessary improvement. Finally, it allows informed judgement in selecting the most appropriate settings.

5. REFERENCES

1). L.D.Kounis, J.M.O'Sullivan, I.R.McAndrew: 1999, *Preliminary Experiments on a model to relate Cp$_k$ and SNR indices.* 15th National Conference on Manufacturing Research (NCMR), Advances in Manufacturing Technology-XIII, University of Bath, pp 409-414, Ed: A.N.Bramley, A.R.Mileham, L.B.Newnes, G.W.Owen.

2). Besterfield, D.: 1991, *Quality Control,* 2nd Edition. Prentice Hall.

Design and development of maintenance measurement system for the offshore industry

U KUMAR and **J P LIYANAGE**
School of Science and Technology, Stavanger University College, Norway

ABSTRACT   In recent times there has been major paradigm shifts in maintenance philosophy for many companies in oil & gas industry. Maintenance is no longer regarded as a necessary evil, which costs what it costs, but an important part of the business process. If one could formulate a concise statement on the role of maintenance, it is primarily to reduce business risks on a continuous basis in a cost-effective manner. Therefore, it is essential to develop a system of measurements to assess the contribution of maintenance towards the ultimate business goals. This will also help business managers to allocate optimal resources to maintenance process. Any maintenance measurement system should be equipped with appropriate indicators by taking into account different inputs and resultant outputs from the process. This paper presents some of the critical issues in design and development of maintenance performance indicators for the Norwegian oil and gas companies.

1. INTRODUCTION

Maintenance process is critical for the economic viability of many of the Norwegian offshore and other engineering companies (1). Increased competition and rapidly fluctuating oil prices are resulting in a demand from the part of the management to cut the operations & maintenance (O&M) costs, and to downsize the staff. The effect of such actions would normally be short-term savings, but has the potential threat to induce significant losses in long-term. There are certain driving forces, which justify the need for the oil & gas industry to maintain a balanced perspective in every aspect of the business. The principle focal points in this regard can be classified into four major areas, namely financial, management, technical viability and statutory & regulatory regime (2).

In general, without proper performance indices or indicators for maintenance, it is almost impossible to assess the performance of maintenance organisation, its contribution towards the core business process, and its strengths and weaknesses. Most importantly, the modern maintenance manager, in oil and gas in particular, should have the true image of the strategically important relationships between the outputs of the maintenance process in terms of total contribution to business goals and the inputs to different maintenance sub-processes. The substantial maintenance budget and its subsequent impact for the overall business process, considerable amounts of maintenance resources, and emerging legislative requirements are too great for the companies to ignore the maintenance process. As a result, maintenance has become an integral and a critical part of the business decision process. If the maintenance process is vulnerable, it will eventually endanger the integrity of the internal processes and consequently the overall business. Performance indicators or indices are therefor, essential tools for the management hierarchy. Some of the strategically significant measures in assessing the value of the maintenance process contribution are illustrated in Figure 1. Performance indicators (PI) are necessary measures, which grow in value as their use is extended over the years. They are capable of promoting rationality and objectivity in

""

information that are dealt with, in addressing the challenging decisions through quantification mechanisms, and thus tend to remove the associated subjectivity to a certain degree that interfere with the effort to avoid business related risks.

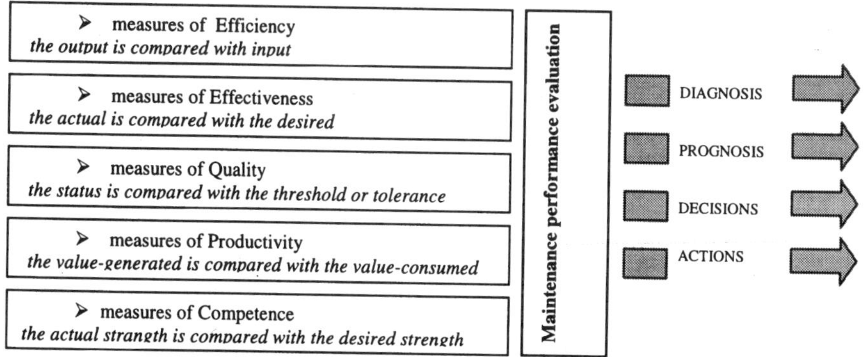

Figure 1. Some strategically significant measures for maintenance process

The PI s can be divided into two groups, namely *lag* and *lead* indicators depending on whether they are measures of outcomes (achieved cost per barrel) or performance drivers (for example, future targets for cost per barrel). It is practically difficult to obtain measures of maintenance inputs and outputs, outputs in particular, which are suitably quantifiable for the purpose of constructing meaningful indicators. The difficulties in this regard are attributed to:

- maintenance is a support process to the main stream and is intangible by nature
- shortcomings and merits are not immediately traceable
- outcomes are masked by other business related activities
- often a time-lag exist between the activities and the potential outcomes
- some of the outcomes are not apparent but exist within the work culture.

The indicators designed and developed for measuring maintenance and the responsibilities linked to them must be unambiguous, clear and be linked to the objectives, processes, visions and strategies stated. To deal with continuously volatile business circumstances, oil & gas companies are often recommended to adopt comprehensive systems of performance measurement through a series of performance indicators. The challenge for maintenance managers, within the present business context, is the development of a comprehensive performance measurement system, which is capable of projecting the intrinsic value and the identity of the process in the long run towards business excellence.

2. MAINTENANCE & PERFORMANCE INDICATORS

Any performance measurement system in general is a collection of performance indicators. Many companies have been using their own maintenance performance indicators focussing at different aspects of the business. The type and the effectiveness of these indicators are greatly influenced by the need of the company, the internal knowledge and the infrastructure available to design, develop, implement, analyse and manage them successfully within the organisation. In a generic framework, maintenance performance indicators are essentially a blend of economic, organisational and technical indicators (see Figure 2).

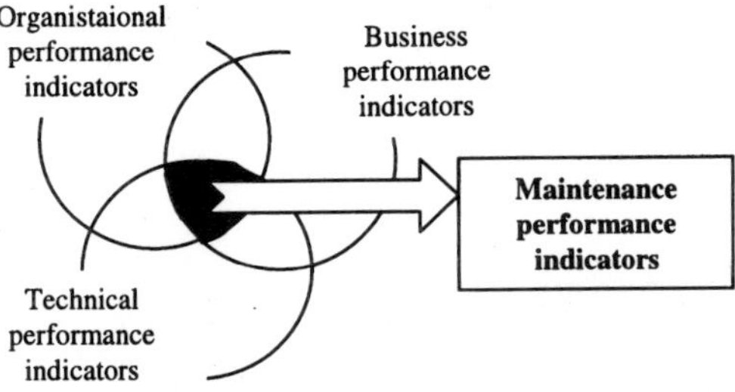

Figure 2. The core of maintenance performance indicators and a classification

One of the major drawbacks of the type of maintenance performance indicators found in practice is that, at best they only indicate that some action is necessary without being specific about the type of action required. Further, it is a question as to whether the present indicators take into account the process-specific constraints that are encountered in revealing the performance level.

3. DEVELOPMENT OF MEASUREMENT SYSTEMS FOR MAINTENANCE PROCESS

Many companies have developed some or other form of maintenance indicators. But seemingly, these indicators do not clearly relate the total inputs (e.g. resources) to the so-called outputs from the process. If designed properly maintenance performance indicators can:
- provide a basis to justify and deploy resources
- identify problem areas
- provide individuals and teams, some means to measure own performance
- provide individuals and teams to measure own contribution to business objectives
- facilitate benchmarking
- reflect performance trends
- indicate contribution of maintenance to overall business objectives.

Further, many complex models and indicators are available for measuring reliability, maintainability and availability. But models showing the value-added effect of O&M process is yet to be developed. A more recent attempt in the oil & gas industry is to allocate scorecards for the maintenance process (2) from the top-strategic to the bottom-operational level.

Any measurement system in general is most effective when it is used to drive a change within the business-process. In fact, an effective measurement system should induce a change within the process concerned. To transform any internal process, one needs a clear understanding of different perspectives, which are critical to the success of the core business process. For instance, the balanced scorecard introduced by Kaplan and Norton provides such a balanced perspective (3) & (4). Tsang (5) has discussed about implementation of the Balanced scorecard to measure maintenance process of an electric utility. Scorecards can be developed for maintenance process in oil and gas industry, as illustrated in Figure 3 where all the critical perspectives such as financial, internal processes, health and safety, human resources and technical perspectives are considered.

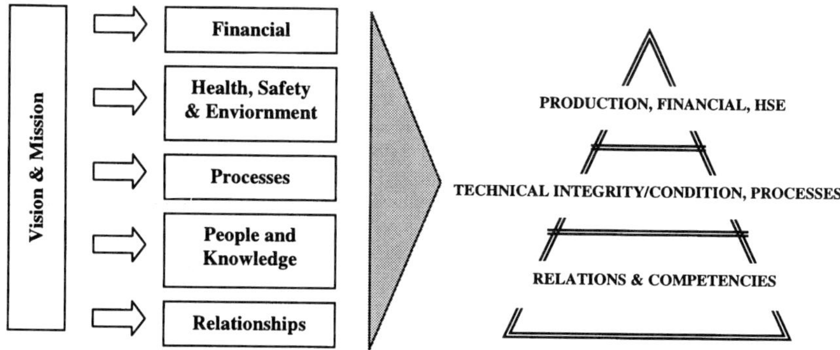

Figure 3. Some of the critical perspectives for a maintenance scorecard for the oil & gas industry

4. CONCLUDING REMARKS

This paper briefly examined some of the critical issues related to development of measurement systems for the maintenance process and its link to the overall business. *As the managerial activities get more complex and cognitively demanding, the managers need to get strengthened through organisational transparency, which enable them to view performance in several perspectives in its true nature"* (2). Maintenance performance indicators (MPI), provide such a transparency to run the business. Many companies have developed different types of indicators mostly related to financial measures, which lack the required transparency. A system of scorecards with a balanced view could be a better approach to overcome the existing drawbacks in maintenance performance measurement.

REFERENCES

1. Kumar, U. and Ellingsen, H. P. (2000). Development and implementation of maintenance performance indicators for the Norwegian oil & gas industry, Euro-maintenance 2000, Sweden.
2. Liyanage, J.P. and Kumar, U. (2000). Measuring maintenance process performance using the balanced scorecard, Euro-maintenance 2000, Sweden.
3. Kaplan, R. S. and Norton, D. P. (1992) The Balanced scorecard - measures that drive performance, Harvard Business Review, Jan-Feb 1992, pp 71-79
4. Kaplan, R.S. and Norton, D. P., (1996). Translating strategy into action: Balanced scorecard, Harvard Business School Press.
5. Tsang, A. H. C. (1998). A strategic approach to managing maintenance performance, Journal of quality in Maintenance Engineering, Vol. 4(2), pp. 87-94.

Tia cement-lined pipes for waterworks

Y L LEE and **A AZIZ ABU BAKER**
ITTHO, Parit Raja, Batu Pahat, Johor, Malaysia
M WARID HUSSIN
Civil Engineering Faculty, Universiti Teknologi Malaysia, Johor, Malaysia
K O KHOO
Petro-Pipe~Daido Concrete (KL), Negeri Sembilan, Malaysia

ABSTRACT

The abundance of Timber Industrial Ash (TIA) and Rice Husk Ash (RHA) and their escalating proliferation has resulted in much expense and increasing disposal problems. This paper aims at promoting the innovative utilisation of TIA and RHA in the production of cement-lined pipes and fittings for waterworks. The methods of materials preparation and test procedures are described. Results of materials characterisation are presented. A novel method of determining the water permeability of TIA composites is highlighted.

Keywords: TIA, RHA, cement-lined pipes, water permeability test

1. INTRODUCTION

A suitable blend of timber industrial ash has been used for the experimental production of blended cement bricks and high strength concrete [1, 2]. Rice husk ash (RHA) is another material with high silica content suitable for the production of calcium silicate composites. RHA containing reactive silica can be produced, depending on the oxygen content, time and temperature of incineration. An integrated process of cogeneration and the production of a quality assured indigenous microsilica is being developed. The blended TIA is processed and formulated to be an alternative to silica fume [3]. It has an average fineness of 400 m^2/kg with a silica content of around 70 %. The reactivity of the silica phase is suitable for the development of a blended cement for durable and enhanced performance concrete structures [4].

A holistic approach has been adopted to achieve enhanced performance of concrete structures while developing an indigenous ash utilisation technology. The microsilica enhanced the strength development and water impermeability of the cement lining due to the formation of additional calcium silicate hydrate gel. This paper describes the materials characterisation, processing and formulation for the experimental production of TIA cement-lined pipes and fitting for waterworks.

2 . MATERIALS PROCESSING AND FORMULATION

Due to the inherent properties of TIA and RHA there was a resultant high cost of handling and transportation. However, the result of the processes adopted in this research complied with international standards. The micronised and densified materials with an innovative integrated process reduced the cost of transportation. Lee, Y.L. [5] and Yeoh, A.K. [6] had examined the chemical composition of TIA and RHA and recommended methods for effective utilisation. The TIA and RHA were processed in accordance with the requirements of MS 1226: Part 1: 1991. Ordinary Portland Cement (OPC) conforming to MS522: Part 1: 1989, specially graded washed silica sand complying with grading zone 4 in Table 5 of BS 882: 1983 were also used.

3. EXPERIMENTAL PROGRAMME

Extensive water permeability testing was carried out on the samples in which NDT techniques [7] were used in the process. Water permeability tests system was developed in accordance with ISO/DIS 7031. The test cubes were conditioned at approximately 30°C and 75% RH for the water permeability test at the age of 28 days. For high performance durable concrete, it was specified for the Jubilee line project in London that the concrete samples shall be conditioned at 75% RH and 20°C and tested for water permeability at 35 days [8].

4. RESULTS & DISCUSSION

The materials formulations were designed to produce the cement-lined pipes and fittings with enhanced performance. The water: cement: sand ratio was 0.5: 1: 2. Preliminary tests on strength development of test cubes indicated that the specified minimum compressive strength of 17 N/mm^2 at 7 days and 31 N/mm^2 at 28 days had been achieved for both the TIA and the OPC control specimens. The target values of water permeability were around 1×10^{-11} m/s. The test specimens containing 10 % TIA were subjected to two different curing conditions, namely air-curing and moist-curing. The study focused on the combined effect of TIA and curing on strength development and water permeability of the cement mortar cubes which were taken to represent the performance of the cement mortar lining on the steel pipes.

5. CONCLUSIONS

1. This study was aimed at promoting the innovative utilisation of TIA and RHA to achieve enhanced early strength and durability of cement-lined pipes and fittings. An improved materials formulation has been experimented. A novel dual-test approach onto the standard 150 mm test cube with a nondestructive testing method for water permeability before the determination of compressive strength has been specified for quality control during production of the cement-lined pipes and fittings.

2. The use of a reliable, convenient and cost effective standardised test method as a tool for quality control during the mix design and production process will provide useful information on the durability of the product. The development of the ITTHO water permeability test system highlights the importance of design for durability and the quality control process throughout the life span of the product. The water permeability test system can be used in the laboratory and as a NDT in-situ. Its use can be extended to concrete test cubes for the determination of water permeability prior to the determination of compressive strength. Low value of water permeability is suitable for cement-lined pipes and concrete products or structures subjected to water pressure. Concerted efforts to exploit the utilisation of agricultural and industrial wastes in cement and concrete composites for infrastructure development is a positive move towards improving the quality of life and conservation of the environment.

ACKNOWLEDGEMENT

The authors wish to express their appreciation to the Malaysian government for the IRPA grant (72215), *Testmate (KL) Sdn. Bhd.* for jointly developing the water permeability test system, *Evergreen Fibreboard Sdn. Bhd.* and *Ban Heng Bee Rice Mill (1952) Sdn. Bhd.* for the contribution of TIA and RHA, *Blacktop Industries Sdn. Bhd.* and *Petro-Pipe~Daido Concrete (KL) Sdn. Bhd.* for the technical support.

REFERENCES

1.   Lee, Y.L., M. Warid Hussin and Siow, Ken (1999), "Experimental Production of Timber Industrial Ash Cement Bricks – Malaysian Experience" Proceedings of International Conference on Infrastructure Regeneration and Rehabilitation, University of Sheffield, pp. 287-296.

2. Lee Y.L., Khoo, K.O., Chong, S. and Hussin, M.W. (1999) "Strength development and water permeability of high strength TIA concrete" Proceedings International Congress on Creating with Concrete, University of Dundee, 6-10 September 1999

3. Lee, Y.L., Siow, K. & Hussin, M.W. (1997) "Agricultural Fly Ash (AFA) - A Potential Alternative to Silica Fume for High Performance Concrete", presented at the ITTHO-PERKOM-IEM Seminar on Concrete Technology (R&D-97) themed *Research & Development for Competitive Advantage* (2 October, 1997, Allson Klana Resort, Seremban).

4. Lee, Y.L. (1998) "Towards Achieving Quality Assured Maintenance - Malaysian Experience" Proceedings 2nd International Conference on Planned Maintenance, Reliability and Quality, University of Oxford, England, pp.143-148.

5. Lee Y.L., M. Warid Hussin and Khoo, K.O.(1999) "Timber Industrial Ash Concrete" Proceedings Sixth International Conference on Concrete Engineering and Technology (CONCET '99), The Institution of Engineers, Malaysia, MARA Institute of Technology, University of Malaya, (presented on 30 June 1999).

6. Yeoh, A.K. (1997) "Mechanisms To Commercialise Viable Research – SIRIM's Experience", presented at the ITTHO-PERKOM-IEM Seminar on Concrete Technology (R&D-97) themed *Research & Development for Competitive Advantage* (2 October, 1997, Allson Klana Resort, Seremban).

7.   Lee, Y.L. (1997) "Non-Destructive Testing of Concrete Structures - Some Practical Aspects" Proceedings 5th International Conference on Concrete Engineering & Technology. (CONCET '97), pp. 165-177.

8.   Cabrera, J.G. (1999) "Design and Production of High Performance Durable Concrete" Proceedings of International Conference on Infrastructure Regeneration and Rehabilitation, University of Sheffield, pp. 1-14.

Safe and economical operation of technical systems on ships in the wake of electrical power quality

J MINDYKOWSKI and **T TARASIUK**
Department of Ship Electrical Power Engineering, Gdynia Maritime Academy, Poland

Abstract: The presented paper refers to widely understood electrical energy quality in ship electrical power system. The basic information about ship electrical power system and electrical power quality have been laid. Their importance in ship exploitation processes has been underlined including the help of real voltage waveform. Concluding remarks are focused on preventive measures improving the electrical power quality in the considered systems.

1. INTRODUCTION

The presented paper focuses on mutual interaction between ship electrical power system and others ship technical systems in the wake of electrical energy quality. Electrical power system influences all important ship technical system, especially sensitive to supply disturbances, like navigation, steering or propulsion. So, there are undeniable connections between parameters of electrical energy and capability of carrying out ship tasks. The problem involves reliability of electrical power supply and suitable parameters of this energy. Every disturbance in the electric power system brings not only economic loss but increases the risk of ship disaster. Of course, every kind of ship disaster imperils human beings and environment and its extent is hardly to be foreseen.

2. SHIP ELECTRICAL POWER SYSTEM

The main task of ship electrical power system is to deliver and process electrical energy. The system contains autonomous free standing generating sets (typically synchronous generator coupled to distinct internal combustion engine), cable lines, switchboards and different kinds of loads. Sometimes a shaft generator or/and turbogenerator are present for the sake of energy savings. The characteristic attribute of ship electrical power system is an enormous ratio of nominal power of singular electrical energy receiver to one generating set. So, it causes flexibility of ship electrical network. It creates a great sensivity to load changes, which evoke notorious voltage and its frequency deviations from their nominal values. However, over the recent years, common development in electrical energy utilization processes brings new problems. Amid these broadly known distortion of waveform of electrical signals should be underlined.

3. ELECTRICAL ENERGY QUALITY

3.1. What does it mean: „Electrical energy quality in ship electrical power system"?

Common understanding of the term „electrical energy quality" is proper rms value and frequency of voltage, symmetry of voltage, and its sinusoidal waveform. Thereby, it equals to voltage quality. In order to achieve the goal of safe and economical exploitation of the ship electrical power system and the whole ship, it should be complemented with demand of proportionality of active and reactive power distribution between generating sets working in parallel. In ship electrical power plants there is often a necessity for free standing generating sets to work in parallel. For safety reasons it is necessary to maintain power surplus; as a rule, it is about 20% power of single generator at sea or more when manoeuvring (1).

Resuming, the electrical energy quality in ship electrical power system can be described by parameters of three-phase supply voltage on the bus bars of the main switchboard (rms value, frequency, asymmetry, waveform distortion) and proportionality of active and reactive power distribution (2).

3.2. Influence of the electrical energy quality in the system under consideration on correct running of the remaining ship technical systems

The phenomena occurring in real ship electrical power system can be divided into two basic groups: steady state and non-steady-state phenomena. For steady state phenomena, the following attributes can be used (3): amplitude, frequency, spectrum, modulation, source impedance, notch depth, notch area;
and respectively, for non-steady-state (3): rate of rise, amplitude, duration, spectrum, frequency, rate of occurrence, energy potential, source impedance.
The influence of the electrical energy quality in the electrical power system on correct running of the remaining ship technical system depends on many factors. At first, the character of disturbance should be taken into account. In shorthand, it can be stated that transistory, substantial worsening of electrical energy quality (non-steady-stated phenomena) has the greatest influence on safe operation of ship technical systems; especially, automation and navigation systems are in jeopardy. It may cause the false alarms, incorrect operation and even switch off some devices. At the same time, the sustained lowering of electrical energy quality (steady-state phenomena) in the first place brings economical effects. Additional waste heat occurs in cables, electric motors and generators as well as condensers. But it should be stressed here that these phenomena can be instrumental in decreasing the safety level of running ship technical systems by reducing their durability and at the same time also reliability of important system elements. In order to eliminate this influence it is necessary to increase the work expenditure to operate ship systems which obviously adds to the increase of crew and spare parts costs of ship service (1).
Taking into account aforementioned phenomena, it is required very high standards for electrical devices installed in ship electrical power systems. For example Polish Register of Shipping requires that all electrical devices installed on ships work correctly with voltage and frequency deviations from nominal values given in the Table 1 (4).

Table 1. Limiting deviations of voltage and frequency in the ship electrical power system

Parameter	Deviation from nominal values		
	long duration	short duration	
	value [%]	value [%]	time [s]
voltage	+ 6.0 - 10.0	± 20.0	1.5
frequency	± 5.0	± 10.0	5.0

Other problems are distortion and asymmetry. The distortion of voltage is typically described by means of total distortion coefficient THD (Eq. 1):

$$THD = \frac{\sqrt{\sum_{h=2}^{\infty} U_h^2}}{U_{rms}} \cdot 100 \quad [\%] \tag{1}$$

where: U_h – the rms value of the h-harmonic, U_{rms} – the rms value of whole waveform, taking into account also higher harmonics

Distortion of the supply voltage waveform has a negative influence on the work of almost all loads. It means increase of risk of a ship's disaster as well as additional energy loss. Equally important are distortions of currents in the system under consideration. Distortions of voltage and current curves have influence also on the work of measuring systems and voltage regulators of ship's generators.

Asymmetry of supply voltage influence on induction motors in chief. One of the indicators describing this voltage asymmetry is the voltage asymmetry coefficient C_{va} defined as the biggest voltage deviation of any phase Δu ratio from the voltage average value U_a to the voltage average value (Eq. 2):

$$C_{va} = \frac{\Delta u}{U_a} \cdot 100 \quad [\%] \tag{2}$$

According to the International Electrotechnics Commission settlements (IEC Report 892/1987) if the voltage asymmetry exceeds 5% the work of the electrical motor should be analysed with regard to negative-sequence voltage. At smaller voltage asymmetry load limitation to a degree dependent on this asymmetry is recommended. It is obviously impossible during the ship exploitation.

4. EXEMPLARY TIME-DEPENDENT CURVES ANALYSIS

Exemplary three phase voltage waveforms taken on the bus bars of the main switchboard of ship are depicted in figure 1 (5).

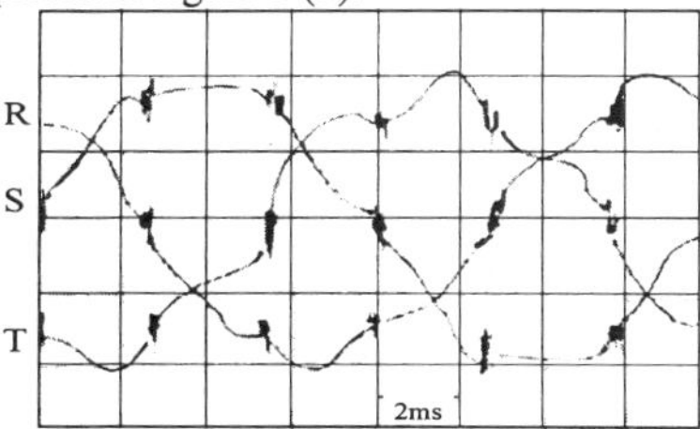

Fig. 1. Exemplary three phase voltage waveforms

The main problem of these waveforms is perceptible at the first blush. Of course, there is a wide range of distortions (harmonic and notching), including the low- and high frequency effect. However significant asymmetry is also included albeit less discernible. The greatest THD of above depicted curves exceeds 17% and C_{va} 3.7%.

Most classification societies require the THD less than 10%. However, the distortions exceeded by 20% often may be noted (6). Moreover, the level of distortions is practically not recorded neither on- nor off-line. An even more disturbing fact is the lack of any classification societies rules tied to voltage asymmetry, which is also often observed.

5. CONCLUDING REMARKS

Because the disturbances of electrical energy quality are impossible to avoid completely, the importance and complexity of electrical energy quality problem in ship electrical power system need suitable approach. At first, it should be taken into account during the projection stage. Appropriate electrical equipment choice and a suitable configuration of ship electrical power system is necessary. The safety sub-system should be set off. It can be supplied by a separate generator or by means of an electromagnetic power converter. The active harmonic filter can be used as well. That depends on economical and technical reasons. All-important and sensitive devices of relative small nominal power should be supplied by means of its own safety sub-system. It is due to navigation and automatic control system requirements. The remaining devices can be supplied directly from the ship's power plant. It contains proof and disturbing receivers. They need special carrying out of the design and use. It means relatively small magnetic flux density and surplus of power related to power demand.

The exploitation stage is equally important. The characteristics of important parts of the electrical power system vary during the period of ship exploitation. Some receivers are changed or added as well. Thus, the continuous control and estimation of electrical energy quality is required. So, the preventive measurements are advisable. This distinct problem involves appropriate choice of measured factors as well as measurement instrumentation. Steady-state and non-steady-state phenomena should be taken into account. Recent progress in microprocessor technology (hardware and software) enables to solve that problem (7). This new measurement instrumentation should be compatible with devices used for improving electrical energy quality as well as protection devices (8). This compatibility means full-adjustment of the measured parameters and the used algorithms of automatic control.

REFERENCES

(1) J.Mindykowski, T. Tarasiuk:, "The influence of electrical energy quality on economical exploitation of ship technical system" *4th International Conference „Electrical Power Quality and Utilisation"* Cracow, Poland, 23-25 September, 1997, pp. 329-334.

(2) J.Mindykowski, T. Tarasiuk, "Quality factors of electrical power and their measurements under ship's conditions", XIV *Imeko World Congress*, Tampere, Finland, 1-6 June, 1997, vol. IVb, pp. 19-24.

(3) R. Dugan, M. McGranaghan, H. Beaty: "Electrical Power Systems Quality", *McGraw-Hill Companies*, 1996.

(4) Przepisy Klasyfikacji i budowy statków morskich. Polish Register of Shipping, Gdańsk 1995.

(5) B. Dudojć, J. Mindykowski: "Diagnostic measurements of harmonic filters as an instrument of energy quality improvement in ship's networks", *Zeszyty Naukowe WSM w Gdyni* 1997, pp. 120-131 (in Polish).

(6) H. Reinecke, W. Schild: "Harmonics in main electric supply systems with semiconductor rectifiers and subsequent methods of compensation", IMECE'91 *China*, pp. 1-10.

(7) G. Bucci, C. Landi: "On-line digital measurement for the quality analysis of power system under non-sinusoidal conditions", IEEE *Instrumentation and Measurement Technology Conference, Brussels, Belgium,* June 1996, pp. 934-938.

(8) J. Mindykowski, T. Tarasiuk: "Specialized analyser for electrical energy quality estimation in ship electrical power system", IEEE *Instrumentation and Measurement Technology Conference, St. Paul, Minnesota, USA,* May 18-20, 1998, pp. 791-796.

Quality maintenance for building services engineering

L MITCHELL
Manufacturing Systems Engineering, University of Hertfordshire, Hatfield, UK

ABSTRACT

Our understanding of *"what maintenance needs to be done?"* in building services engineering is changing dramatically. As plant and equipment seems to be the least priority in most buildings, the consequences of equipment failure are becoming even more severe, as many building operations are outsourcing the maintenance function altogether. Some of the outsourcing strategies have found to be only effective in certain areas, or with a particular maintenance practice. However, *quality maintenance for building services engineering* uses, and applies maintenance performance measures to provide operational guide-lines for maintenance personnel, ensuring that the plant and equipment is maintained correctly, using the correct tasks at the correct frequency. It is therefore aimed at the functional operation of a plant room system. A selected methodology of quality maintenance for building services engineering is discussed, with examples to help bring a turbulent building services maintenance problem under control. Maintenance improvements have been established through the selected choice of maintenance performance measures. The selected measures are used to measure any critical maintenance weakness. The weakness is then minimised, and converted into strength. The chosen generic maintenance measures transfer easily into the functional maintenance approach. This is due to the low-abstraction level, limited complexity, and simplicity of use and application. The final outcome is also concerned with the allocation of resources i.e. that maintenance costs do not rise significantly!

1. INTRODUCTION

The purpose of building services maintenance engineering is to reduce the adverse effects of breakdown and to maximise the building availability at minimum cost. The need for quality maintenance planning is obvious. However, maintenance in most buildings is often a secondary process, where administration is the core business. The result is that quality maintenance does not receive enough management attention, and it is often the case that maintenance is outsourced to sub-contractors. Management therefore, often looks at maintenance as a necessary evil, not as a means to reduce costs. The work for this research project has included the choice and implementation of maintenance performance measurements in selected buildings in the London area. Information was gathered using personal interviews with management and engineers, and the use of both internal and external documentation, with a "hands on" approach. With the quality maintenance issues the author proposes a variety of selected maintenance performance measures to assess the reliability of a complex plant room system.

2. METHODOLOGY

Quality plant and equipment maintenance is a key process in building services engineering. It represents a significant component of the operating cost in these buildings. Much of the expenses are consumed by non-value adding activities such as energy, management control and related maintenance activities. Through the application of common sense maintenance performance measures, quality maintenance can be streamlined to eliminate waste, and any non-required functions valued by management. In an exercise of this nature it was found that measures of building services engineering performance are best selected on a basis of individual convenience. Overall, despite what is written about building services engineering, in practice they remain largely vague and ill defined, with day to day management decisions having a high subjective content. To improve this situation, it was necessary to think about quality building services engineering in simple, common sense terms, rather than to think of more sophisticated mathematical algorithms, models and frameworks. The research and its problems has been approached, not by attempting to use direct philosophies, e.g. Total Preventive Maintenance (TPM), and not by attempting to force fit the competing conceptual approaches. The work has been approached rather by searching out the fundamental strengths and weaknesses of the individual plant room system, then maximising the strengths and minimising the weakness. The emphasis was therefore placed on selected maintenance performance measures and simply asking the fundamental question of: *"what works in any plant room system?"* The answer is simply the common sense basics of: reducing waste, reducing confusion, increasing flexibility, improving quality maintenance, and anything else which improves criticality maintenance performance in the view of the engineer.

3. PERFORMANCE MEASURES

The objective was to identify a series of procedures which improved the plant and equipment's overall condition and reduced or eliminated the six big losses of: equipment failures and set-up adjustments known as *downtime losses,* idling, minor stoppages and speed reductions known as *speed losses,* defects in the plant system and yield reductions known as *quality losses.* The case study is a five-floor building in London. The building is the corporate headquarters of a national organisation. On the top floor of the building is the plant room, which is the nerve centre of the building. The plant room consists of: a main control panel, air handling unit, extract fans, humidifiers, gas boilers, pumps, calorifier, pressurisation units, chiller unit, condenser unit, cold water tank, switchgear panel, miscellaneous equipment supplying 250 radiators and 250 (VAV) units. The selection of critical plant and equipment for maintenance would be the items where failure would have significant effects on health and safety. To establish which systems were significant; *fault tree analysis* was used to trace all the causes/failures that are likely to cause a system to fail. *Process flow diagrams* were used to derive the criticality assessment of the system, and simplified quality assessments were derived from *assessment forms* and included: losses, utilisation, quality, failures, downtime and repair times. (In a complex process plant situation subjected to criticality, defining which maintenance measurement to use is not a simple process; therefore *decision tree logic* is best used. It is a formalised method, used to decide the best measurement for specific components with a known failure mode).

At the beginning of the project a planned preventive maintenance programme of work was implemented, with various schedules of work for the plant and equipment. The maintenance programme was carried out through weekly and monthly visits of varying duration. During each visit selected plant was inspected and minor adjustments carried out. All plant and equipment would receive a minimum of one major service, one minor service and several checks within the programme of work. A *"certificate of conformity"* would be issued on the completion of each individual plant task. In order to achieve the "certificate of conformity" then the maintenance performance measures were introduced at the beginning of the preventive maintenance programme. The following *fans and air handling unit* example is used to reinforce the value of the performance measures. The example had moving parts, safety implications, and was critical to the performance of the plant room and building function. The planned preventive maintenance programme was as follows:

1. Check fans and motor bearings, lubricate if required
2. Inspect belts
3. Check pulleys for damage and alignment
4. Inspect taper locks and shaft coupling
5. Inspect for air leakage, corrosion and security of fittings
6. Check for correct operation of controls

Initially, each component would be assessed for the need to eliminate minor defects, plus a cleanliness procedure implemented before a *criticality assessment* takes place. The measures would also include the criteria of: safety (risk assessment), time to repair quality of maintenance, plant availability loss, maintainability and costs. The criticality tasks are made easier if the criteria are *ranked in matrix form,* giving a score of say 1 to 10; 10 having the maximum impact. From the matrix, the criteria scores are multiplied together for each component part of the function, thus establishing a priority or criticality for the component and its function. The *maintenance reliability* would be related to the functional operation of that particular system, (or sub-system), and not the unit function, and that the maintenance was considered, or related to the in-built reliability of that system. The collection storage, and analysis of any failure data in the example, forms the basis of the quality maintenance. It has to be understood that that the example has to perform a function to some given performance criteria, when that criterion is not met, the plant or item is said to have failed. Implementing the maintenance performance measures is about assessing failures of operational plant, or units of plant.

The measures to look out for in the example, were *age dependant failure, random failure, and early life failure.* The identification of these failures are measured using *selected statistical analysis*, where it is possible to calculate *mean time to failure (MTTF),* or *mean time between failure (MTBF),* and this information is used to actually assess the preventive maintenance activities for the plant item, as follows:

1. *Failure rate or bath-tub curves,* where the infant, adult and wear out phases are constructed and identified.
2. *Weibull analysis,* where it becomes possible to calculate MTBFs, and establish a failure characteristic. From this analysis a shape factor beta then gives an accurate indication of failure modes.

3. ***Failure modes and effects criticality analysis (FMECA),*** where the reliability data has been collected (over a period of time), and after analysis the results are used to make decisions relating to the most cost effective maintenance operation, and planning interval. From the performance measurements functional failures are derived. From each functional failure or system failure it is then necessary to establish the type of failure (and its mode) at the equipment and components level of the system. This is an excellent opportunity to rank the items and their modes of failure, and adopt the best maintenance procedure, (dependant on the importance of ranking) from the functional failures or system failures.

4. ***Condition monitoring*** is the introduction of cost effective technology that measures or checks the condition or the health of the plant and equipment, and has excellent benefits for the maintenance function. An example of this was establishing the predictive maintenance, and to trend the condition of ball bearing life, in order to minimise the risks and economic impact of unexpected failure. The condition monitoring measurement parameters gauged the ball bearing health, and this valuable information/data allows quality maintenance to be planned in a cost-effective manner. Plant and equipment is only then repaired or serviced relating to its specific condition.

5. ***Statistical process control (SPC)*** which is complementary to the maintenance function.

4. IMPLEMENTATION ISSUES

Organisations and building services engineering considering quality maintenance should not be led into the situation that the best returns are to apply any maintenance performance measures on a plant-wide basis. This will have the effect of being quickly swamped by huge quantities of systems data, FMECA studies, decision tree-logic information etc., and the expected quick returns are stalled. It is advised that quality maintenance studies and measurements should be made on a ***pilot plant (Go/NoGo)*** procedure, with implementation towards the total plant and equipment being made over a much longer-time period. This methodology maybe a little slower than required, but it is a learning curve approach, and allows quality maintenance to be refined, before application to the whole plant room system. Due to the amount of rotating machinery items in a plant room, there needs to be some method of monitoring and maintenance measurement, to ensure safety and integrity of the plant and equipment. The ***fans and air handling unit*** example used in this project had been assessed for its level of criticality, its safety and its breakdown costs. Depending at which level the item was found, a performance measure was assigned to it. The plant and equipment had data collected and fed back for analysis. This alone did not by any means indicate the condition of the item, so it was necessary to additionally monitor the equipment according to the proposed preventive maintenance programme of work. Each item in the plant room had its own set of maintenance measures, which were designed to prolong the life of the item. A typical maintenance function was to change the oil, or check pulley operation, or ball bearing condition etc. This was achieved over a routine number of weeks and months, and was activity dependent. Nearly all measures identified a problem that frequently occurred. The maintenance measures ensured that the day to day running of the ***fans and air handling unit*** for example, was reliable, but care had to be taken to prevent actual breakdown. The maintenance measures are powerful tools that bridge a gap in quality maintenance for building services engineering. The maintenance performance measures provide useful data for guiding management decisions, if they are appropriately selected to fit the specific operating environment, will impact greatly on the quality maintenance engineering function.

Uncertainty in soft design

A DONNARUMMA and **M PAPPALARDO**
Department of Mechanical Engineering, University of Salerno, Fisciano, Italy

ABSTRACT

In this paper is analysed the problem of soft dependence of parameters in design systems, when the features are analysed in probabilistic or non-probabilistic terms. Is discussed, and analysed, the soundness of the axiom on S*oft Design* coming out from Soft *Computing*. A procedure is proposed for to check the independence or for to measure *the dependence* between sets of parameters of design and for to evaluate the influence of uncertainty of data.

1. INTRODUCTION

The soft dependence of parameters in design system is in contrast to traditional hard computing. A new form of computing, called *Soft Computing*, is recently used in many emerging disciplines. Soft Computing "*... is tolerant of imprecision, uncertainty and partial truth.* (Zadeh 1994)" In large-scale systems the precise models, often, are impractical while the soft models are more easy to use. The Soft Computing uses many computing disciplines as main components. For example, the Fuzzy Logic, the Probabilistic Reasoning, Neural Networks, etc. gives the mechanism for to evaluate the new outcome of system and to update the previous outcome. The uncertainty plays a fundamental rule in analysing of complex systems. If Ω is the field of application of the system, and ω a remarked event, the measure of ω will be always incorrect. The knowledge of ω is not given by its co-ordinates in Ω, but it is possible to assert only that ω is limited in a subset A_i of Ω. In field of application, one event is represented by a subset $A \subseteq \Omega$ with $\cap_i A_i = \varnothing$ and $\cup_i A_i = \Omega$.

The fundamental problem is to ascribe some information $J(A) \overset{def}{=} \Phi[P(A)]$ to the realised event A. It is possible to define values of information for probabilistic and non-probabilistic events using the definition of entropy $J(A)$. In many problems of engineering it is possible to utilise empirical definitions. The Shannon's expression Entropy, *on the partition* $\pi : \Omega = \underset{i}{\cup} A_i$

is given by $H(p_i) = -k \sum_{i=1}^{n} p_i \log p_i$. Entropy is a generic value with very large meaning not connected to probabilities. *Wiener* introduces the entropy on a single element $A \in \Omega \equiv [0,1]$ on interval [a,b] inside [0,1]

$$H = -\log_2 \left(measure\ of\ [a,b] / measure\ of\ [0,1] \right)$$

$$J(A) = c\ \log(1/P([a,b]))\quad c = (1/\log 2)\quad P(A) = P([a,b])$$

From the definition of $J(\underset{i}{\cup} A_i)$ came out that the information on the set of events $\underset{i}{\cup} A_i$ is given from the knowledge of information $J(A_i)$ on *single events*. With the definition of

$J(\cup_i A_i)$ the information is measured for probabilistic and non-probabilistic events. In the correct classification of a plan D, characterised by the set $\{D_i\}$ of criteria, in each phase, needs to know and appraise the contribution of new elements introduced in the planning. For each new introduced criteria D_n it needs to determine the effect, since the condition $D_n \supset D_{n-1}$ is not generally valid. The new effect could be evaluated with the calculus of probabilities. In presence of complex problems it needs to adopt some algorithms that allow simplifying and keeping the process under control. If we are dealing in a space of probability distributions

$$D(\mathbf{p}:\mathbf{q}) = \sum_{i=1}^{n} \left[(p_i - q_i)^{2p}\right]^{\frac{1}{2p}} \qquad G(\mathbf{p}:\mathbf{q}) = \sum_{i=1}^{n} p_i \, ln \frac{p_i}{q_i}$$

D *and* G represent the *distance* of Minkowsky and Kullback between two probability distributions. If $\mathbf{q}$ *is a priori distribution* and we need to select the distribution $\mathbf{p}$ *closeness to* $\mathbf{q}$, we must minimise the distances $D(\mathbf{p}:\mathbf{q})$ *and* $G(\mathbf{p}:\mathbf{q})$.

2. AXIOM OF SOFT DESIGN.

In design the condition of dependence, on a finite and limited set of independent parameters, makes possible to analyse and to find solutions. Otherwise the needs of knowledge will be so large that it's impossible to solve with valid conclusions. We have efficient valid solutions taking advice of the fundamental axiom of Soft Design:

Axiom of Soft Design: *Valid design has minimum values of information and depends on a finite and limited number of independent, or soft dependent, parameters.*

The entropy is a measure of uncertainty and an ideal solution has minimum entropy. A system has infinite probability distributions consistent with given constraints, *but only when has maximum entropy or minimum cross entropy all probabilistic entities adjust their values as to give the optimum solution.* An ideal solution *has the minimum entropy* compatible with *all* information. These restrictions involve the stability and reproducibility of the same system.

If the set of functional parameters is $X = \{x_i\}$ and the set of output parameters is $Y \in \{y_j\}$ then the relationship between data can be represented as $X = R(Y)$. A complete control involves the definition of inverse relationship $Y = R^{-1}(X)$.

If $S = (P, A)$ is the information system on design's parameters with:

- $P \neq \varnothing, \; A \neq \varnothing$,
- P set of parameters (as example the geometric parameters)
- A set of features of independence (as example the relations among parameters)
- $B \subseteq A$ subset $\{a_i\}$ of features

then for every $a_k \in B$ is associated a binary relation of independence then $x_i \, R \, x_j$ divide the universe U in two classes $\{$ $I = independent, \; D = not \; independent$ $\}$

$$[x_i]_{\{I, \neg D\}} = \{x_j \in U : x_i \, R \, x_j\}$$

The input set X can be analysed as a composite set $X = X^*_I \cup X_{\neg D} = \{x_i\}$ in which $X^*_I \in \{x_i\}$ is *the independent and soft dependent subset* and $X_{\neg D} \in \{x_i\}$ is the not independent subset. If $X^*_I \cap X_{\neg D} = \varnothing$ then the relationship between data are $X^*_I = R(Y)$ and $Y = R^{-1}(X^*_I)$. The parameter of design came out only from acceptable conditions if independence and soft dependence A of inverse $A \in Y = R^{-1}(X^*_I)$.

3. MEASURE OF INDEPENDENCE.

Let S be an information system

$$S = (P, A) = ((X \cup Y), A) = (\{x_i\} \cup \{y_i\}, A)$$

with $P = X \cup Y$ a set of parameters and A a set of features. If $B \subseteq A$ is subset $\{a_i\}$ of features of independence then, for every $a_k \in B$, the binary relation of check of independence $x_i R y_j$ defines the independence of parameters of system S.

$$x_i R y_j = \{ (x_i, y_j) \in (X \times Y), \forall a_k \in B. \ x_i \textbf{ is independent from } y_j \}$$

If the features are in *probabilistic terms* for all elements of the set $X = \{x_i\}$ the relation $(p_i R x_i)$ assign the probability $\{p_i\}$ to $\{x_i\}$

$$x_i R p_i = \{ (x_i, p_i) \in (X \times \{p_i\}), \forall x_i \in \{x_i\} \ x_i \text{ has the probability } p_i \}$$

For independent events the value of information is additive.

$$X = (p(x_i), ..., p(x_n)), \ Y = (p(y_j), ..., p(y_m)); \ \Sigma_{i=1}^{i=n} p(x_i) = 1 \ . \ \Sigma_{j=1}^{j=m} p(y_j) = 1$$

$$H(X) = \Sigma_{i=1}^{i=n} log_2 p(x_i) \ , \ H(Y) = \Sigma_{j=1}^{j=m} log_2 p(y_j)$$

$$H(X \cup Y) = H(X) + H(Y) \qquad ,$$

If the events are not independent then $H(X \cup Y) < H(X) + H(Y)$. There is common $I(X,Y)$ information in the sets X and Y.

$$H(X \cup Y) = H(X) + H(Y) - I(X \cap Y)$$

If we introduce $\alpha \in [0,1]$ the parameter of *measure of dependence* of the Bayesan distributions X and Y we get the influence of uncertainty of data.

$$\alpha = \frac{2 \ I(X \cap Y)}{H(X) + H(Y)}$$

The analysis the probability of events given from the set $(X \times Y) = \{x_i \cap y_j\}$ give the valuation of independence. From Bayes' theorem, the probability of composite events is $p(x_i \cap y_j) = p(x_i) \ p(y_j|x_i)$ with probability of y_j conditioned by x_i. If the events are independent then is $p(y_j|x_i) = p(y_j)$ and $p(x_i \cap y_j) = p(x_i) p(y_j)$. For non-independent events will be $p(x_i \cap y_j) \ne p(x_i) p(y_j)$. The value $I(X \cap Y)$ is the distance, given by Kullback's definition of cross entropy of distribution $\{\mu^I_{i,j}\} = \{p(x_i) \ p(y_j)\}$ on independent set $(X \cup Y)$ from distribution $\{\mu^D_{i,j}\} = \{p(x_i) \ p(y_j|x_i)\}$ on set $((X \cup Y) - (X \cap Y))$

$$I(X \cap Y) = \sum_{ij} \mu^I_{ij} \ log_2 \ \mu^I_{ij} / \mu^D_{ij}$$

Then by definition of α

$$\alpha = \frac{2 \sum_{ij} \mu^I_{ij} \ log_2 \ \mu^I_{ij} / \mu^D_{ij}}{\Sigma_{i=1}^{i=n} log_2 p(x_i) + \Sigma_{j=1}^{j=m} log_2 p(y_j)}$$

For the values of $(1 - \alpha) \cong 1$ for which the dependence of parameters has low influence on design process than the *dependence of parameters is soft.*

- $\alpha = 0$ *independent distributions.*
- $(1 - \alpha) \cong 1$ *soft dependence.*

The information X conditioned by Y $H_Y(X) = H(X) - H(X \cap Y)$ and Y conditioned by X $H_X(Y) = H(Y) - H(X \cap Y)$ are given by Kullback's distance

$$H_Y(X) = H(X) - H(X \cap Y) = \sum_j p(y_j) \sum_i p(x_i|y_j) log_2 (1/p(x_i|y_j))$$

$$H_X(Y) = H(Y) - H(X \cap Y) = \sum_i p(x_i) \sum_j p(y_j|x_i) \log_2\left(1/p(y_j|x_i)\right)$$

The parameter of soft dependence will be

$$\alpha = \frac{(H(X)+H(Y))-(H_Y(X)+H_X(Y))}{(H(X)+H(Y))} = 1 - \frac{(H_Y(X)+H_X(Y))}{(H(X)+H(Y))}$$

The parameter is in term of conditioned distributions.

From $\beta \in [0,1]$, the parameter of the *measure of additivity* of entropy

$$H_\beta(X,Y) = H_\beta(X) + H_\beta(Y) + (\beta - 1)H_\beta(X)H_\beta(Y)$$

we get the expression of non additive measure of information.

$1 > \beta > 0$ it shows the *non additive measure of entropy.*

$\beta = 1$ the information is additive.

$\beta \cong 1$ the information is soft additive.

If $\beta = 0$ and $H_\beta(X)H_\beta(Y) = I(X \cap Y)$ then the information is on a Bayesan distribution.

For soft additivity of information it is possible to extend the *Axiom of Soft Design* on sets of parameter with *soft dependence.*

4. CONCLUSION

For designing or analysing systems we suggest to use the term *Soft Design* coming out from, the recently coined, term *Soft Computing* describing the simultaneously use of many new computing disciplines *tolerant of imprecision, uncertainty and partial truth.* Using the values of parameters α *and* β it is possible to have the *measures of dependence* of parameters in *Soft Design* in presence of uncertainty and imprecision. The exclusion of weak parameters can be executed analysing the distance between probabilistic or non probabilistic distributions using the very strong definition of *non probabilistic entropy* of *J. Kampè de Feriet* for non repeatable events.

REFERENCES

1. De Finetti, B. *La prevision: ses logique ,ses source subjtives.*Ist.H. Poincarè 1-68 1937.
2. Wiener N. *Cybernetics* Paris Herman (Act. Sc. 1053) 1948.
3. Kampe de Feriet J. *Note di Teoria dell'Informazione.*Anna Accademico 1971-72-Annali dell'Istituto di Matematica Applicata dell'Università di Roma. 1972-(In Italian).
4. J.N.Siddal: *Probabilist Engineer Design* M.Dekker JHG N.York 1983.
5. Nam P. Suh: *The Principles of Design.* Oxford University Press New York 1990.
6. N.Kapur K.Kesavan: *Entropy Optimisation Principles,* A. Press Inc. Boston 1992.
7. Zadeh L.A. *Fuzzy logic and Soft Computing : Issues, Contentions and Properties.* In HZUKA'94-3rd International Conference on Fuzzy logic and Soft Computing. Izuka Jp.
8. Pappalardo M.: *Information in Design Process*-2nd International Conference on Planned Maintenance, Reliability and Quality - Oxford , England 2nd -3rd April 1998.
9. Donnarumma A., Pappalardo M.: *Designing in Many-Valued Logic.* Proceedings of IPMM99 Vol.I Conference on Intelligent Processing .Honolulu Hawaii- July 10-15 1999.
10. Donnarumma A., Pappalardo M.: *Introduction to Soft Design.* DIMEC- Reports n.1 January 2000. *UNISAIT-Salerno Italy.*

A prototype of measurement system for Agile enterprise

J REN and **Y Y YUSUF**
Department of Mechanical and Manufacturing Engineering, Nottingham Trent University, UK
N D BURNS
Department of Manufacturing Engineering, Loughborough University, Leicester, UK

Abstract

The measurement of an Agile enterprise is a complex process involving different decision levels and various criteria. These decision levels and criteria may vary depending on the type of company being considered. This paper critically reviews the current measurement process in the design of Agile enterprise and discusses why there is a need for developing a new approach. It particularly focuses on a generic measurement methodology . The prototype of measurement system developed in this paper examines how to choose agile attributes, proposes the measurement diagram, and then sets up the architecture of the system. The developed system demonstrates that decision-making technologies can be successfully integrated into the measurement process.

Keywords: Measurement system, Prototype, Agile enterprise, AHP

1.Introduction

Agility as a concept was first introduced in 1991. It describes a company that is able to change and adapt quickly to changing circumstances. "Agility is dynamic, context specific, aggressively change embracing, and growth oriented. It is not about improving efficiency, cutting costs, or battening down the business hatches to ride out fearsome competitive storms. It is about succeeding and about winning profits, market share and customers in the very centre of competitive storms that many companies now fear" [1]. Increasingly Agile manufacturing is attracting more and more attention from both academic and industrial communities. Extensive programs are being conducted on relevant issues to propagate agile manufacturing concepts, to build agile enterprise prototypes and to realise an agile industry eventually. While much has been written on the subject of Agile Manufacturing, the models which measure how agile an enterprise is are somewhat preliminary.

This paper critically reviews the existing measurement processes in enterprises and discusses why there is a need for developing a new measurement system. The measurement process is complex, involving different decision levels and various attributes. These decision levels and attributes may vary depending on the type of company being considered. This paper particularly focuses on the choice of agile attributes in order to suit generic enterprises. It is obvious therefore that effective measurement system for agile enterprise must deal with a host of different type of factors that perhaps are in conflict with one another. In developing a prototype for such a measurement system, this paper examines how to establish agile dimensions, decision domains and attribute , proposes the block diagram, and then sets up the AHP model that can effectively deal with different factors in multiple attributes.

2.Prior Studies

In contrast with the abundant literature dealing with Agile enterprise, previous studies on how to measure the level of agility are virtually absent. A method developed by Sharifi and Zhang [2], consists of general factors such as: how responsive is the company against changes in its business environment; how able is the company in proactively capturing the market and customers desire and in taking the competitive advantage of unpredicted opportunities in the market. But Agility is an integrated concept which consists of different dimensions [3], each dimension includes several decision domains, and each decision domain is associated with some attributes[4]. Obviously, these dimensions, decision domains and attributes should be considered simultaneously in order to gain a better understanding of how company performs.

With regard to the measurement method, traditional methods include the categorical method, weighted-point method, matrix approach, and multiple objective programming (MOP) such as goal programming. To elaborate, a categorical method rates companies on a number of equally-weighted factors and then allows the decision maker to evaluate the company with the total score. Although this method is simple and intuitively appealing, it often introduces subjective error into the decision and oversimplifies the measurement scenario by equally weighting all factors. Some authors suggested a matrix approach which evaluates companies based on weighted scores obtained from a set of pre-determined benchmarks[5]. To improve this matrix approach further, some authors suggested a performance matrix approach that can measure the company's performance under various unforeseen scenarios by estimating the probable deviations from the original goal that the company needs to obtain[6].
Regardless of their strengths, none of these approaches can systematically measure both qualitative and quantitative factors and structure complex problems with a large number of criteria, attributes and alternatives.

As depicted above, all the traditional methodology and approaches are confined to rudimentary measurement which cannot be said to be integrated. In addition, most of the prior analytical studies considered only a limited number of attributes. The current study goes beyond the previous work not only by considering different type of factors relevant to the measurement of agile enterprise, but also by proposing a measurement system that is capable of guiding decision maker in the identification of the more critical agile performance need.

3.Measurement system

A diagram of the proposed measurement system is shown in the figure1. This diagram identifies the company which is to be measured, then analyse the environment both externally and internally, trying to find the main factor(s) affecting the company. After the goal of the company is established, the next step will be involved in choosing agile attributes and examining a measurement mechanism of the multi-attribute decision-making techniques like Analytic Hierarchy Process(AHP) technique. Based on the previous steps, the structure of AHP will be formulated. After computing the priority of each level, using weighted-score method, the final score of the company will be calculated. The following step is to analyse the reliability of the measurement. If it is failed, the AHP structure and/or the agile attributes must be reviewed. If passed, the measurement information can be stored. The user also can obtain a final report on the status of the company.

4.Agile dimensions, decision domains and attributes

In terms of the proposed measurement system, the most fundamental module in the diagram is how to select agile dimensions, decision domains and attributes. In 1995, Goldman *et al.* [1] identified four key dimensions of agile competition. The first dimension is enriching the customer. This entails a quick understanding of the unique requirements of each individual customer and rapidly providing it. The second dimension entails co-operation (intra-organisational, inter-organisational co-operation such as supplier partnerships and perhaps emerging virtual relationships with competing organisations) in order to enhance competitiveness. The third dimension utilises new organisational structure(s) to master change and uncertainty through techniques such as concurrent engineering and cross-functional teams. The fourth dimension leverages the impact of people, information and technology and recognises the importance of employees as a company asset, placing greater emphasis on education, training and empowerment. To achieve agility in terms of the four dimensions, some decision domains with related attributes must be established. Team building, for example, is a key concept in manufacturing strategies and a critical decision domain. The associated attributes involve: *Empowered individuals working in teams; Cross functional teams; Teams across company borders* and *Decentralised decision making.* Empowerment enables workforce to make decision and work autonomously. That speeds up business processes and creates an environment in which innovation may thrive; Cross functional teams solves problems using a team of people from different business disciplines enabling better, broader and more robust solutions to be devised; Encouragement of teaming with other customers, even competitors allows companies to react quickly to market changes at low cost; Decentralised decision making speeds up operations, allows for workforce innovation, maintains information flow at a workable level.

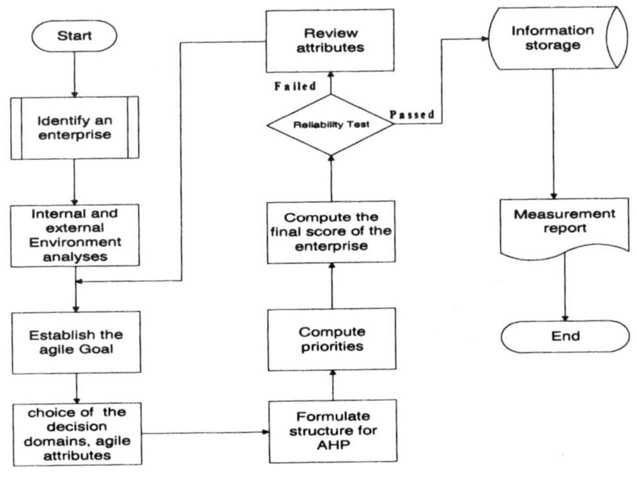

Figure 1. Diagram of the measurement system

Collectively the thirty-two agile attributes can be identified under the following ten decision domains:

(1).**Integration**: Concurrent execution of activities(using CEA to represent it, the same as the following) ,Enterprise integration(EI), Information accessible to employees(IAE);

(2).**Competence**: Multi-venturing capabilities(MVC), Developed business practice difficult to Copy(DBPD);

(3).**Team building** :Empowered individuals working in teams(EIW), Cross functional teams(CFT), Teams across company borders(TACB), Decentralised decision making(DDM);

(4).**Technology**: Technology awareness(TA), Leadership in the use of current technology(LD), Skill and knowledge enhancing technologies(S&KT), Flexible production technology(FPT);

(5).**Quality**: Quality over product life(QPL), Products with substantial value-addition(PVA), First-time right design(FRD), Short development cycle times(SDCT);

(6).**Change**: Continuous improvement(CONI), Culture of change(CLC);

(7).**Partnership**: Rapid partnership formation(RPF), Strategic relationship with customers(SRC), Close relationship with suppliers(CRS), Trust-based relationship with customers/suppliers(TBR);

(8).**Market:** New product introduction(NPI), Customer-driven innovations(CDI), Customer satisfaction(CS), Response to changing market requirements(RCMR);

(9).**Education**: Learning organisation(LO), Multi-skilled and flexible people(MS&FP), Workforce skill upgrade(WSU), Continuous training and development(CT&D);

(10).**Welfare**: Employee satisfaction(ES).

5. Analytic Hierarchy Process(AHP) Model

As the AHP method plays a key role in the whole diagram, it is necessary to discuss it with details. In fact, AHP is a simple decision-making tool to deal with complex, unstructured and multi-attribute problems. The strength of the AHP method lies in its ability to mimic the management judgement about the relative importance of different influencing factors and to structure a complex and multi-attribute system. The AHP consists of three basic steps:(1) design of the hierarchy;(2) the prioritisation procedure; and(3) calculation of results.

In order to measure the company in certain agile environments, a four level hierarchical model is devised (see Figure 2). The first level sets the main objective, here referred to as the agile company. The main objective (i.e. referred to as the main considerations) is divided into four main dimensions or sub-objectives, which are enriching the customer(EC) , co-operation(CO) , mastering change and uncertainty(MCU) and leveraging the impact of people, information and technology(LI). The third level of hierarchy includes ten decision domains. The fourth and the last level consists of thirty two attributes.

Once the AHP model is developed, each level of the hierarchy will be evaluated by means of the paired comparison rankings. The nine-point dominance-scaling approach is suggested by Satty[7]. In this approach, the experts are asked to rate the relative importance or preferences of each pair of items on a scale of 1 to 9. In this scenario, at the highest level, the expert must rate the four primary dimensions(i.e.,
enriching the customer , entails co-operation , master change and uncertainty and leverages the impact of people, information and technology) in terms of their importance in determining agile company measurement goal as viewed by the expert. The comparison data are converted to relative weights of the decision elements. This is accomplished by solving the eigenvalue problem of the matrix. At the rest of the levels, the same procedure is repeated. Finally, AHP procedure calculates overall synthesised weights of each attributes. Simply using weighted-score method, the final score of the agile company can be calculated.

5.Conclusion

An integrated system has been set up that can be used to measure an agile enterprise. The prototype system proposed is generic and guides decision maker in assessing a company step by step. An easily understandable AHP model has also been developed. In addition, the methodology not only helps in measuring desirability of agile enterprise but also in monitoring its health during various phases of its life cycle and thus ensures protection of investment.

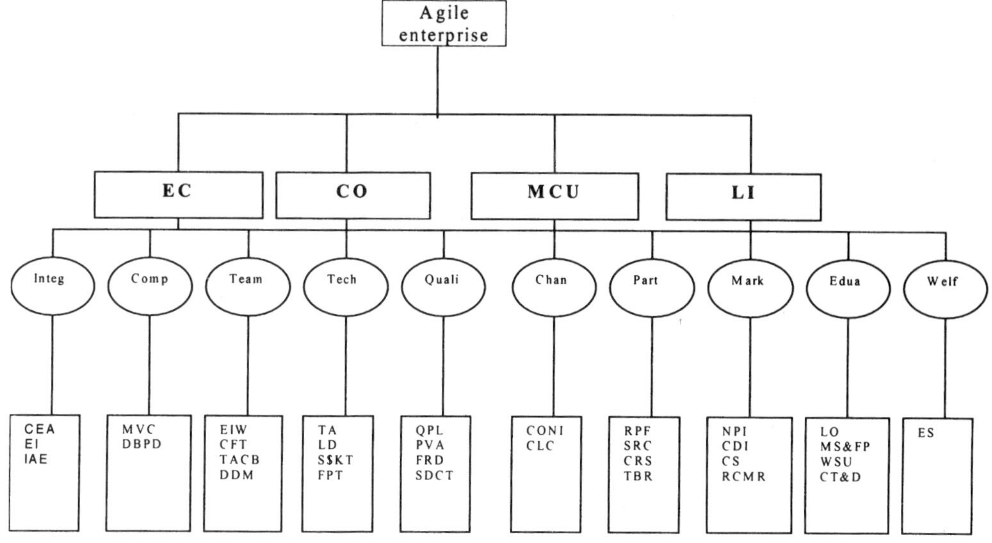

Figure 2 AHP Model of Measurement system

REFERENCES

[1] Goldman,S.L. and Nagel,R.N. and Preiss,K. (1995) Agile Competitors and Virtual Organisations: Strategies for Enriching the Customer, Van Nostrand Reinhold
[2] H.Sharifi and Z.Zhang (1999) A methodology for achieving agility in manufacturing organisations: An introduction, International of production Economics, 62(1999) 7-22
[3] Gunasekaran, A. Agile Manufacturing :A framework for research and development, International Journal of Production Economics, 62 (1999) 87-105
[4] Yusuf, Y.Y., Sarhadi, M. and Gunasekaran, A. Agile Manufacturing: Concepts, Drivers and Assessment, International Journal of Production Economics, 62 (1999)
[5] Soukup,W.R.,"Supplier Selection Strategies", Journal of Purchasing and Materials Management ,Vol.23 No.2, 1987,pp7-12
[6] Thompson,K.N.,"Supplier Profile Analysis", Journal of Purchasing and Materials Management ,Vol.26 No.1, 1990,pp11-18
[7] Satty,T.L.,The Analytic Hierarchy Process, McGraw-Hill, New York, NY,1980

Design for maintenance

J A SLATER
School of Engineering, Sheffield Hallam University, UK

ABSTRACT

This paper highlights the lack of consideration given to the maintenance function at the design stage. Three areas, which should be taken into consideration at the design stage, are identified and the opportunities for the designer to cater for these requirements are explored in power hydraulic systems.

INTRODUCTION

Design of a system can have far reaching effects on operation and maintenance. The operation and maintenance functions can be compromised by lack of provision at the design stage, arising from diverse factors; financial, design constraint, oversight, operation outside design parameters, error, and inadequate analysis of the process.

All good (successful) designs should appreciate and accommodate the maintenance function (1) (2) by acknowledging and preparing for the inevitable need to replace and/or repair no matter how reliable or carefully chosen the equipment may have been. The failure of the maintenance function to provide the operational function with operable equipment in acceptable time scales owes much to the design activity. Geraghty (3) illustrates that 35 % of failures in service are caused by deterioration in comparison to 20% failures which are designed in. Matthiassen(4) advises maintainability is becoming a more important factor in the success of businesses. The aim of the maintenance function is to maintain equipment to the appropriate level in an effective, economic and efficient manner (5). Designers should give maintenance more importance in arriving at the optimum design, three areas deserving particular attention are ; Access, Diagnostics and Isolation.

KEY AREAS

Access impacts on the mean time taken to maintain, repair and replace equipment. The provision of auxiliary equipment particularly for lifting can greatly facilitate access. The ease of access and the desirability of special tools for maintenance should be taken into consideration at the design stage when preparing the product design specification (1).

Provision of access to install temporary diagnostic equipment greatly assists the maintenance function in determining the cause of a fault efficiently and effectively, the use of diagnostic methods forming a central part of modern maintenance and reliability practise (5).

Isolation protects both personnel and equipment from inadvertent movement during operation and has become a crucial part of current safety methodology. Isolation offers the potential to operate equipment in a reduced capability whilst necessary maintenance operations are carried out safely.

Power Hydraulic Systems provide a number of examples where foresight on the part of the designer in addressing these key areas can vastly improve the performance of the maintenance function.

HYDRAULIC POWER UNITS

Hydraulic pumps are usually driven by an electric motor by directly coupling their shafts. A convenient arrangement uses the concentric spigots provided by the pump and motor manufacturers and a suitable housing (bell housing) having concentric bores accepting the mounting spigots. Usually the pump spigot has the smallest diameter often smaller than the outside diameter of the shaft coupling resulting in the need to remove the coupling from the pump shaft before the pump can be withdrawn from the bell housing. This may only be possible after removing the bell housing from the electric motor. However, if a bush is incorporated into the bell housing design, retained by a small shoulder, having an inside diameter to suit the pump coupling and an outside diameter larger than the coupling, the pump and associated coupling can be assembled directly to the bell housing (Figure 1).

Specifically indicated in Figure 3 are optimum locations for diagnostics (test points), locations of flexible hoses and modularization concepts. If this circuit is compared to those of power units in operation the advantages are readily seen.

Connection of the pump to the hydraulic system pipework via flexible hoses has many benefits including; Ease of assembly/disassembly of the pump to the system, Reduction in transmission of mechanical and fluid borne noise and vibration.

The test point is an important inclusion for the maintenance function enabling pressure measurement and the extraction of a dynamic representative fluid sample for analysis, which, can be vital to problem solving and preventative maintenance.

Incorporation of flexible connections eases assembly reduces transmission of noise and vibration and provides a ready means of access for temporary diagnostic equipment such as flow meters, data recorders etc.

Isolating valves located as shown provide facilities for test and set up of the pump controls and main system relief valve without subjecting the rest of the hydraulic system to unnecessary stress. The isolating valve between the main pump and the reservoir minimises the need to drain the hydraulic reservoir before replacing the pump and enables both activities to be carried out concurrently.

Hydraulic system fluid will need to be added to the fluid reservoir. The fluid contained in new sealed barrels is not clean enough to introduce into hydraulic systems without further conditioning. This situation is further compounded if the system designer has not provided a ready means of introducing the replenishment fluid. By providing a small independently driven pump and filter assembly arranged to circulate fluid around the reservoir and incorporating two three way ball valves these problems can be overcome (Figure 3).

HYDRAULIC SYSTEMS

Specifically indicated in Figure 2 are optimum locations for diagnostics (test points), locations of flexible hoses, location of isolating valves and modularization concepts. The advantages will be readily seen by comparison with equipment in current operation.
The hydraulic system also incorporates flexible connections, which offer advantages recognised earlier.

The facility to isolate the modules from the rest of the system and to isolate function by function improves; safety, availability of equipment and minimises loss of fluid from the system when replacing actuators.

The minimum permanently installed diagnostic equipment (test points) is indicated at convenient locations to enable each individual function to be monitored in addition to a global check on the control manifold.

DISCUSSION AND CONCLUSIONS

Relatively simple techniques and equipment have been outlined which if incorporated into an hydraulic power unit or system will dramatically improve the ability of the maintenance function to maintain the equipment to the appropriate level in an effective, economic and efficient manner. All three of the key areas identified early are incorporated in the measures discussed. The designers attention is specifically drawn to these small details for in these areas lie the successful design.

Much of the cause for the omission of the items described can be attributed to financial constraints and a lack of understanding on the part of designers of the maintenance activity. Ullman (2) suggests that designing diagnosability into a mechanical product is possible, recognises it takes extra effort and questions the value of incorporation into the design. Pugh (1) confirms that unfortunately recognising the importance of maintainability may result in an initial capital cost penalty that may be unacceptable at the project stage.

At the project phase the designers, specifiers and purchasers do not recognise or have to pay for the resources they are committing, so minimisation is rarely a consideration (3). However ultimately this additional expenditure at the capital stage will be recovered by reductions in the mean time to diagnose faults, the mean time to repair and the resulting improvement in the reliability of the equipment. In effect the expenditure will be recovered in the operational costs of the equipment.

REFERENCES

1. **Pugh, S.**, 1995 'Total Design Integrated Methods for Successful Product Engineering' Addison-Wesley Publishing Company, Wokingham, England. ISBN 0-201-41639-5

2. **Ullman, D. G.**, 1997 'The Mechanical Design Process' McGraw-Hill Companies, Inc., New York. ISBN 0-07-115576-7

3. **Geraghty, A.** 1998 'RCM Principles and Benefits' Reliability Centred Maintenance and RCM2 European Process Industries Competitiveness Centre Middlesborough

4. **Matthiassen, B.**, 'Robust Design of Mechanical Systems' 1st International Conference Planned Maintenance, Reliability, Quality. 6-7 April 1995. Clare College, Cambridge, England. ISBN 0 - 86339 - 698

5. **Carson-Mee, D.J., Wilkinson, W.J., Leonard, R.,** 'Maintenance Management Optimisation; an analysis of the National Grid Group'. 2nd International Conference Planned Maintenance, Reliability, Quality. 2-3 April 1998. St Edmund Hall, Oxford, England.
ISBN 0 - 86339 - 7867

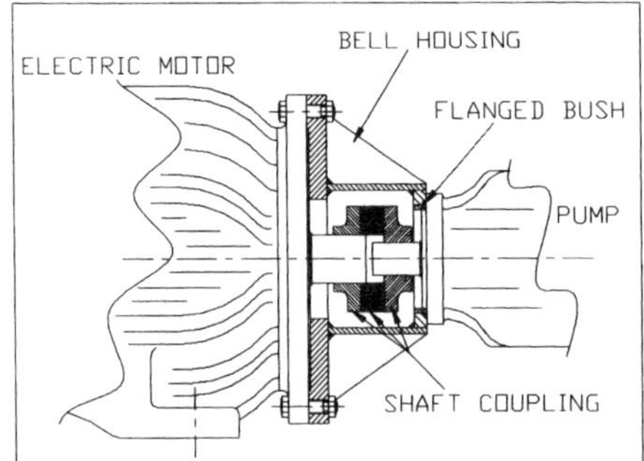

Figure 1

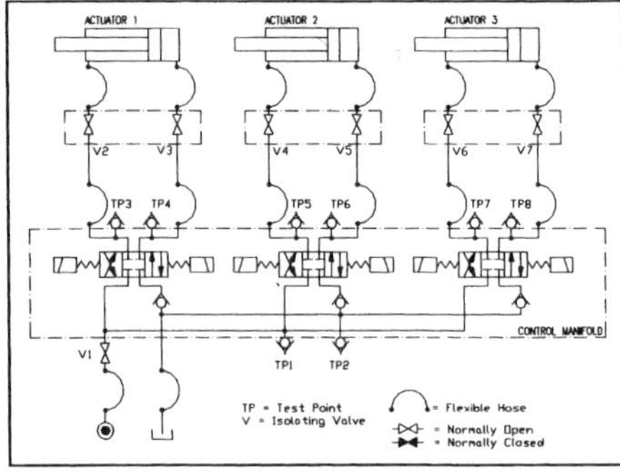

Figure 2

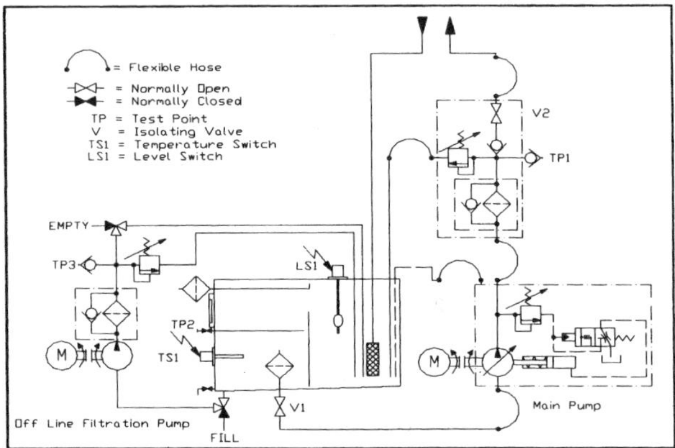

Figure 3

Machinery Directive in design process – a methodical approach

R VIGANO and G F BIGGIOGGERO
Dipartimento di Meccanica, Politecnico di Milano, Italy

ABSTRACT

Before a product may be sold in Europe, it must conform to certain technical, safety requirements and administrative procedures consistent with the European Union 'Directives'. This enables a certain product to be traded freely throughout the European Economic Area.
A procedure to reduce the amount of time for the application of the directives is presented and their impact cost on the product. Is presented in this paper. An approach is proposed, in agreement with methodical design philosophy, to apply the requirements of these directives into single steps of the design activity.

1. MACHINERY DIRECTIVE

For some years the concept of total market has been affirmed in the industry, intending with it that the old customs barriers and the limitations, imposed by national laws to the entry of deriving goods by other nations, have become increasingly transitory. On the one had the customer has increased choice, and on the other it imposes a endorsement of ability of the machine to perform its operating function correctly and moreover with guaranteed safety standards.

The European Union has approved some rules or *Directives* which will guarantee safety standards for the consumer. The design programme must adhere to the safety rules outlined in the directives.

The most important rules are contained in the Machinery directive(98/37/EC) (1). This directive is concerned with the machinery product. A machine being defined as an assembly of linked parts at least one of which moves. Some machines are excluded namely lifts.

There are components which have to satisfy rules dictated by other directives. However, such directives are formulated by a new strategy named *'New Approach'*. This allows the frequent changes of the directives as a consequence of the constant technological advances. For this motive the directives make reference to obligatory general requisite, while the technical dispositions of conformity of the product are formulated through the *'Harmonised Standards'* The Harmonised Standards' are elaborated by the European standards organisations (eg CEN,CENLEC). This provides the entire responsibility of the technical content of the standard. Products which comply with these directives are marked with the CE logo to indicate to enforcement authorities that they are conforming with the European standards.

All these standards require, aside both the manufacturer and of the distributor, a particular attention to the conformity of the product introduced on the market. Particularly, the manufacturer has to furnish, if required, a complete documentation on the product, through which all the choices effected for guaranteeing the conformity of the product and the specific procedures of use of the same are underlined. The creation of this documentation, together with the analysis of the correspondence of the product to the standards, asks for an addition of time to the manufacturing process, above all when it is necessary to define the design of new products. A correct implementation, inside the whole cycle of creation of the product, of the procedures of conformity is required and it results to be an effective solution to this time problem.

2 DESIGN FOR CONFORMITY

The practical application of the Machinery Directive, or of the other directives, add a large amount of time to the design process to control the product and to create the documentation required. In fact, all documents must be create before the product distribution. This documentation includes (Fig. 1):

- **Technical Construction File**: it contain design documentation, manufacturing, test reports and operation information to show conformity of the product.

- **CE Marking**: it must be shown on the device; it include the *CE* logo and the principal characteristics of the product.

- **Notified Body**: this document is required for specific device, as specified in *Annex IV*, and the proof of compliance with the directive must be certificate by indipendent certification bodies recognized in the EU.

- **Declaration of Conformity**: it is a declaration which the supplier has to sign to say that the machine meets the requirements of the directive.

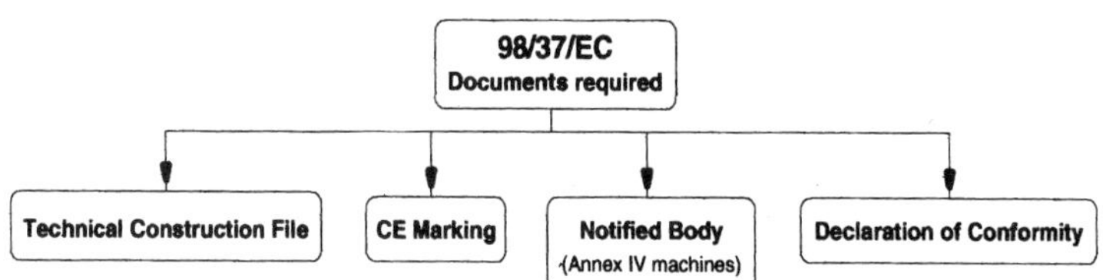

Figure 1: Documents required by Machinery Directive.

The cycle of creation of the machine includes all the phases that go from the definition of the requirements to the distribution on the market of the product (Fig. 2). Many companies have foreseen the phase of creation of the documentation required by the directive as following to the manufacturing phase (Fig. 3). This involves an increase of the *time to market* of the product with a consequent increase of manufacturing cost.

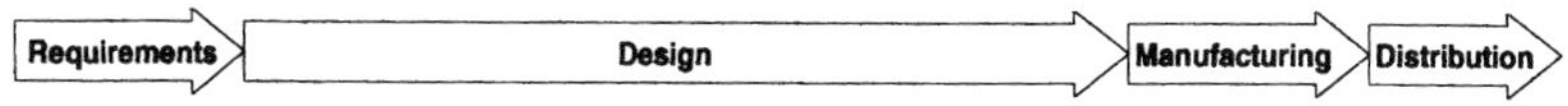

Figure 2: Cycle of creation of the product.

To reduce this time could be useful to analyze more in detail the documentation required by the directive. In fact many documents can be defined before the manufacturing phase. Then, a schematic subdivision of the directive and its requirements can be a useful aid to the designer work, so that he can control the conformity of the product during its development. Because the designer must be able to define the product conformity during the design steps, and not when the design activities are finished.

This subdivision mainly concerns the preparation of the *Technical Construction File*, because the information required is composed by many different documents. The same documentation is also necessary to define *CE Marking*, *Notified Body* and *Declaration of Conformity*. To better optimize the cycle of creation of the documentation required, a subdivision of the design phase can be useful too.

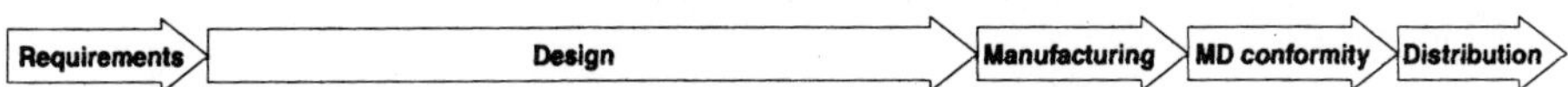

Figure 3: Machinery directive phase in the cycle of creation of the product.

3 METHODIC DESIGN

Many methods in engineering design science have been developed over the last years to describe the design process with a sequence of rational activities. Following this philosophy, the machine design can be structured in rational steps to create a systematic procedure allowing the attainment of the final goal. The *methodic design approach* is one of these [2] [3] [4]. This approach divides the design process phase in three principal steps (Fig. 4):

- **Schematic step**: allows the designer to analize the operating principle and scheme (frequently drafting scheme or layout) with which intends to achive his objective.

- **Qualitative step**: the designer define general shapes, families of materials and equipments pertaining to the configuration of the machine and corrisponding to a specific system.

- **Quantitative step**: corrisponding to the definition of all construction information (dimensions, admissible errors, detailed drawings) of the machine.

Every steps can subsequently be divided but, what is important it is the fact that the information created during these steps is substantially the same required by the Machinery Directive. Besides, the requirements of the project objective can immediately be compared with those related to the directive, allowing an immediate adjustment of the product and its equipments conformity.

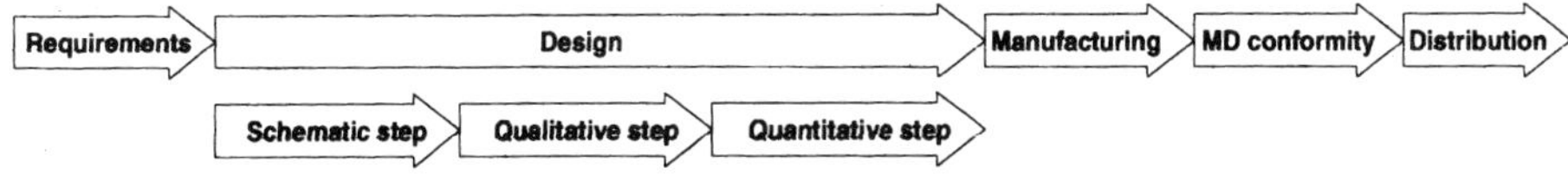

Figure 4: Design subdivision.

The union of the methodic approach with the subdivision of the requirements related to the directive allows to already appraise the compliance of the product from the initial phases of the cycle of creation of the product. This brings to the definition and creation of the documentation required by the directive during the whole cycle and not only when the product is ready to be distributed, allowing an analysis better both the single parts of the product and its equipments.

It is necessary here to underline that this approach requires the definition of a collection of design rules which, often, are not applicable easily for designers because the design processes are quite different from one another, and it is very difficult define an unique pattern for all products.

4 APPLICATION AND CONCLUSIONS

The authors have applied the method proposed in the analysis of the design process related to the definition of an industrial plant for the production of steel flat products. In this case the application of the directive was very difficult because the contractual relationship between supplier and user was complex. The application of the method has allowed to initially underline already the machines and the equipments that had to be subject to the directive. These are separately been analyzed during the cycle of definition of the plant also allowing the immediate creation of the documentation of conformity for that machines or equipments that were acquired by outside parties. This has allowed a reduction of the time to the creation of the documentation related to the Machinery Directive (Fig. 5).

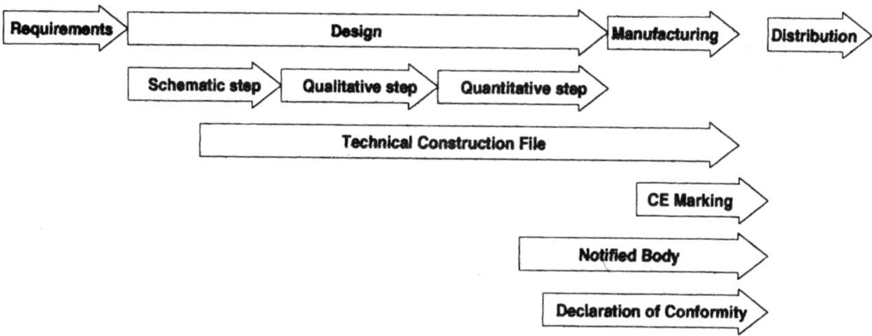

Figure 5: Final scheme of the product cycle proposed.

This reduction has probably been possible graces to the type of product developed; in fact the industrial plant is composed by machines that could singly be treated. However the proposed method is also retained applicable to other cases. The methodic approach allows an important assumption on the application of the Machinery Directive. It is than the documentation required by the directive can be partitioned into more specific part, and every single part of the information can be related to a particular step of the design process. This permits a more feasible approach to the preparation of the documents and a better control of the conformity of the product. With this method, the compliance of the product is not only decided by one company's supervisor but, all designers are allow to define it. This requires a better knowledge of the directive by all company's people that are dealt with the definition of the product.

REFERENCES

[1] European Parliament and Council. Directive 98/37/EC on the approximation of the laws of the member states relating to machinery (document 398L0037). *Official Journal of the European Communities*, 22 June 1998. Original Directive and its amendments: 89/392/EEC - 91/368/EEC - 93/44/EEC - 93/68/EEC.

[2] G. F. Biggioggero and E. Rovida. The formalisation of mechanical design process. In *Proceedings ICED 81 - Rome (Italy)*, 1981.

[3] G. F. Biggioggero and E. Rovida. Proposal for the methodic design in the mechanical field. In *Proceedings ISDS 84 - Tokio (Japan)*, 1984.

[4] G. F. Biggioggero and E. Rovida. Design for quality. In G. Q. Huang, editor, *DESIGN FOR X - Concurrent engineering imperatives*, chapter 16. CHAPMAN & HALL - London, 1996.

Education, Training, and Medical

ESIDE's bet was to develop, in collaboration with MIESA, real industrial systems. The prototypes already existing in the laboratory are: ion exchanger, stream boiler, heat exchanger and pH control of an effluent. Their key characteristics are the industrial instrumentation and the real industrial controllers used in their construction   "". It is also worthy of note that all of them include three optional control modes (industrial controllers, computer and PLC control) and that the supervision is implemented using the SCADA Factory Link. The laboratory lessons start from the modelization of the process, and are designed to familiarize the students with industrial control conception. The students learn to validate the model, to tune the controllers in the various control loops and to handle a real control system. The same equipment is also used for Industrial Instrumentation laboratory classes.

3.6 Committed participation of some of the most representative companies in Automation area.

At present the companies participating in the Automation area are Robotiker, Schneider, Miesa and Siemens. Robotiker is a technological centre specialized in Robotics and Communications. Besides to attend lectures on Robotics, Integrated Manufacturing Systems, and Industrial Perception Systems, Robotiker gives help in the development of common projects and favours the practical work of students in their installations. The actual collaboration of Schneider includes the donation of equipment with the updating engagement, lecturing and supporting with their experience in several projects. The collaboration of Miesa started with the development of the industrial systems aforementioned, and we agree on continuing to work on them and to develop further advanced industrial prototypes. Finally, Siemens supports the technological training of lecturers in the Automation area.

4. CONCLUSIONS.

In this paper, we explained our reflections in the process of designing the laboratory resources for the Industrial Automation area in ESIDE. We present the strategies adopted by ESIDE in order to offer our students a theoretical and practical education that will professionally qualify them. We described the practical curriculum corresponding to representative subjects and the equipment designed to reach the capacities required by industrial practice. Finally, we set out our experience in relation with the active participation of local companies in education. Although we have focused our attention on those companies collaborating in the Automation area, it has to be pointed out that several other companies are involved in education in ESIDE, such as Telefonica in the Telecommunication area and Iberdrola in the Industrial Mannagement area.

5.0 Bibliography

1. ESIDE. Laboratorio de Sistemas Electrónicos de Potencia. Memoria del proyecto de adquisición de equipamiento, Universidad de Deusto, febrero de 1997. Documentación interna.
2. ESIDE. Laboratorio de Sistemas de Medida, Regulación y Autómatas. Memoria del proyecto de adquisición de equipamiento, Universidad de Deusto, febrero de 1997. Documentación interna.
3. ESIDE. Laboratorio de Sistemas de Percepción y Robótica. Memoria del proyecto de adquisición de equipamiento, Robotiker, 1999. Documentación interna.

University–industry co-operation development factor

A BOUBAKEUR and **A OUABDESSELAM**
Ecole Nationale Polytechnique, Algiers, Algeria

Abstract

Through the experience of Ecole Nationale Polytechnique of Algiers, in its relations with Industry, we present in this paper, the importance of Industry-University cooperation. This cooperation has a great impact as well as on the engineering education quality than industrial products quality. We present different forms of cooperation, particularly in engineering education graduation and doctoral post-graduation levels.

1 INTRODUCTION

What and how we have to do, to realise industrial development.
We may admit, in a first analysis, that this realisation depends essentially on:
1. people: technicians, engineers, managers,
2. used equipment's and technologies,
3. manufactured products quality.
If we consider the first point, all carried out investigations show that men represent the main riches of a country. They are, as producers and consumers, the economic life basic elements, and development makers.
Therefore, one of the essential answer factors is education, as well as in quantity and quality at all levels. For that purpose, it is indispensable to establish a close co-operation between University and Industry. Beyond a doubt, that co-operation has been developed in industrialised countries, during the second world war. It has permitted the realisation of important scientific results, and after the war, its importance increased continually. In fact, the Industry became more and more performing and looking for many sophisticated innovations.
Through the experience of the Ecole Nationale Polytechnique of Algiers, in its relations with Industry, we present in this paper, the importance of Industry-University co-operation. This one has a great impact as well as on the engineering education quality than industrial products quality. We present different forms of co-operation, particularly in engineering education graduation and   doctoral post-graduation levels. We will present also some of this co-operation economic aspects.

2 FIRST LEVEL OF COOPERATION

This level corresponds to the period where the students pursue their graduate engineering education. In Algeria, the graduate studies are given in five years, two for basic sciences and three for specialisation. The students must accomplish during each year of specialisation, an obligatory training.   The first training called "worker training", impregnates the future engineers with their working environment. During the second and third training's, the students acquire a general formation and initiate them selves to production techniques in an industrial branch. At the end of each training, they must write reports to be evaluated. The evaluation is made by supervisors groups, from University teaching staff and Enterprise engineers or older technicians. The training evaluation score is taking into account in the final students classification.

An other Industry-University connection is realised during the engineering education programs conception, because it must consider the industrial reality and technology development.

Last of all, we cite the "end of studies projects", engineering projects made by students during the last specialisation year. When the topics of these projects are fixed conjointly by the industrial manufactories and university specialised departments, the students use frequently with profit the industrial installations (1).

These different forms of relations conduct sometime to good connections which permit engineers enrolment at the end of studies.

3 SECOND LEVEL OF COOPERATION

For the University, this level is situated in post-graduation and concerns first of all, doctorate theses preparation which topics could be defined in fundamental or applied research. In Algeria, post-graduation is organised in two steps. The first one is the "Magister" consisting of a two years research initiation with memory, and the second one the "Doctorate of State" of fore years duration with presentation of a theses at the end of investigations. As well as in Magister or Doctorate, the research projects could be made closely with industrial enterprises activities (1), (2). We have presented elsewhere examples of this kind of cooperation, in the field of High Voltage Engineering (3), (4), at Ecole Nationale Polytechnique (ENP) of Algiers, and many similar cases could be found in Mechanical Engineering, Civil Engineering, Industrial Engineering, Environmental Engineering, Metallurgy, Mining, Chemical Engineering, Hydraulics, Electronics and Controls. We are testing now at ENP, a new kind of relation regrouping Industrial Producer, Industrial Consumer and University. In a first time it could offer opportunity to pluri-disciplinary groups to work together at high research level. The investigations are carried out in industrial and University laboratories.

There is also specialised post-graduation which could be organised especially for engineers from Industry. It consists in a full year thorough studies including a half-year project. This kind of post-graduation represents a permanent engineering education organised for engineers, by the concerned institutions together (programmes, duration, costs...).

4 OTHER FORMS OF COOPERATION

Other forms of partnership could be situated at different levels with profits as well as to the Industry and the University. At ENP of Algiers, some confirmed engineers came from Industry to give lectures in specialised courses where examples of lived cases are great pedagogic contributions. The engineers works at the University as associate lecturers, and find with this manner an occasion to perform their knowledge and improve their efficiency in Industry. In an other way, lecturers from the University, could work partially as adviser engineers or consultants in Industry. And then, they will be able to illustrate their courses with real examples.

An other relation may be also established from research contracts authorising also the participation of best students as industrial research laboratory future engineers and/or future lecturers.

It must be also taken into account the possibility to associate University pedagogic staff in Enterprises Administrative Councils, and industrial engineers in Pedagogic and Scientific Councils in University Institutes. This form of connection leads certainly to a best working of the University and the Industry. Together, these institutions may also organise different scientific meetings: workshops, seminaries, conferences, congresses..., and create scientific and technical associations and societies, work groups and committees in standardisation organisations.

5 ECONOMIC ASPECT

The investigations carried out after the second war have demonstrate that a relation exists between the investment in education-training system and the economic progress. This system could be considered as an industry which activity may be at least, exerted in four directions.

1. It assures to the students, necessary acknowledges indispensable to their general or professional qualification, which they will reach later as engineers;
2. It increases the students cultural level;
3. It develops the scientific research;
4. It contributes to diffuse cultural, scientific and technical information on the whole population, by books and reviews publication, radio and television programmes and now using information electronic networks as INTERNET.

In many countries, University is the more dynamic sector of the education and training system. Its cooperation with Industry permits also to obtain important winnings in productivity and production at the national scale. It plays consequently a good evaluated part in economic growth.

6 CONCLUSION

Industry-University cooperation leads to a mutual enrichment: amelioration of education and manufactured products quality, doctorate theses and letters patent realisation, scientific and industrial innovations... In particular, the research make together in the field of computer science, control and communication, leads to a greater efficiency as well as in university education system and industry development research.

Evidently, this cooperation is fruitful only in the case of partners at sufficiently high development level, in such a way as the formed engineers could adapt themselves to technological evolution, and fully participate to the country development.

REFERENCES

(1) A.Boubakeur, A.Ouabdesselam, "Student Work Influence on the University-Industry Co-operation", 3rd UNESCO-UNISPAR World Network Seminar, UNISPAR'99, Polish Innovation Market, Nr 11, Lodz, Poland, September 1999, pp.29-28.
(2) A.Ouabdesselam, S.Benhadid, "Quelques aspects de la coopération E.N.P. d'Alger - Industrie, initiée par quatre projets UNESCO", World Congress of Engineering Educators and Industry Leaders, Congress Proceedings, Vol.I, Paris, France, July 1996, pp.449-451.
(3) A.Boubakeur, M.Bellag, "Impact de la relation Ecole-Industrie dans la formation d'Ingénieurs (Expérience E.N.P.-ENICAB)", World Congress of Engineering Educators and Industry Leaders, Congress Proceedings, Vol.II, Paris, France, July 1996, pp.55-58.
(4) A.Boubakeur, M.Bellag, "ENP-ENICAB Science Cooperation in the Field of High Voltage Engineering", World Congress of Industry Leaders and Educators, ENGIN'96, Lodz, Poland, October 1996, pp.6-9.

Activity through invention

L CANTEMIR, M I CARCEA, I CARCEA, and **A APARASCHIVEI**
Technical University of Iasi, Romania

The paper presents the results of the educational activity having as purpose the stimulation of students' technical creativity at the Electrical Engineering Faculty of Iasi, Romania.

The methodological context, briefly indicated, has been adapted during 9 years of experience according to the achieved results. The qualitative analysis of the results focuses on the themes and fields of activity approached in patents demanded by students from different sections and years of study. The main analysis criteria are: the technical field of the invention and its utility field. The result of the project is over 50% of the involved students turned into inventors.

1.0 INTRODUCTION

This hereby paper refers to the didactic activity stimulating the technical creativity of students at the Electrical Engineering Faculty of Iasi, Romania. It represents a side of the experience accumulated during the past 90 years from the apparition of Electrical Engineering Faculty of Iasi, Romania (1910) as well as the needs for training nowadays-future engineers. The authors believe that stimulating technical creativity is absolutely necessary besides traditional training. Accordingly, no human achievement was possible without a new idea at its foundation. Taking into account the fast development of our society, the necessity of training future specialists by developing their novelty creating skills becomes a requirement and an outstanding progress force.

2.0 TRAINING PROGRAM IN ORDER TO DEVELOP TECHNICAL CREATIVITY

The beginnings of a systematic activity for developing technical creativity took place in the academic year 1990-1991, when the curricula of the Electrical Engineering Faculty was enriched with a new course, "Inventions and Psychology of Creativity." It was introduced for the fourth and fifth years of study (full time and evening courses), major in Electrical Drives, Electrical Uses and Electrical Machines.

The fifth year students had the possibility to choose between "Inventions and Psychology of Creativity" and "Manipulation Devices and Industrial Robots." 53 students out of 86 chose the Inventions course. The course was compulsory for the fifth and sixth year students (attending evening courses). The subject had a two-hour weekly course and one hour of practical activity for one semester (full time courses) and a one-hour course for the fifth and sixth years. A professor with engineer training taught the course, as well as the practical activity.

The encouraging results obtained by the full time students, leading to a number of 20 inventors out of 53 course takers, represented an essential argument for introducing the subject as a compulsory one for all students. It is worthy to mention that among the 28045 graduates of the Technical University of Iasi, those of the Electrical Engineering Faculty i ranged among the best, due to their professional results. The mandatory character of the discipline changed the motivation support of individual and group activity. At the same time, we considered the necessity of stimulating through specific methods the psychical side of technical creativity during the practical activity. In this respect, next year, a psychologist with no technical training taught the practical activity. This new organisation created a sort of break between the course and the practical activity, accompanied by a lack of adjustment from some students triggering a bad reaction to the obligatory status of the course and to the psychological field that had no obvious relationship with the students' technical concerns.

Consequently, even if the number of course-takers rose to 93, the number of inventors lowered to three. Similar situations concerning requests for invention patent could be found during the next two years. Taking into account the insignificant number of students whose causes were supposed to be either a psychological inertia (shyness or lack of initiative) or the lack of knowledge about the object of a patent, we tried to change the training module. Thus, during the third year, during the course we organised a brainstorming session and at the practical activity classes we introduced complex self-knowledge sequences mediated by a psychologist and we diversified the content of the training sequence. Moreover, all students who took an active part in the brainstorming sessions received a bonus for the final grade. Taking into account the extremely dynamic participation of the students at the brainstorming session, this turned into a current practice for each academic year. At the same time, the modality of according marks to students was reconsidered. That is, the students that officially applied for a patent at the State Office for Patents and Trademarks received the highest mark (10) while the students that stood for the exam traditionally received nine as highest mark.
The results of teaching and practical activity development modifications started to become obvious since the academic year 1994-1995 as one can see from the table below.

	5th year, full time, Electrotechnics		4th year, full time Electromechanics		6th year, evening courses, Electrotechnics			
Academic year	Total	Inventors	Total	Inventors	Total	Inventors		
1990-1991	30	20	-	-	23	0		
1991-1992	72	2	-	-	21	1		
1992-1993	80	1	-	-	40	0		
1993-1994	61	0	27	0	29	2	5th year, evening c., Electromechanics	
1994-1995	82	29	16	1	30	2		
1995-1996	67	40	22	6	17	6	8	3
1996-1997	45	41	21	15	27	12	-	-
1997-1998	57	40	11	0	-	-	-	-
1998-1999	66	25	11	7	-	-	-	-
Total	560	198	108	29	187	23	8	3

Starting with the academic year 1995-1996, a "student creativity activating module" (SCAM) took shape. At its foundation there was the principle according to which "creativity" is a general function specific to the human psychic system. This module is based on the idea that every person is capable of creation and this creativity can be educated. In the limits of the inborn individual potential, partially activated, creativity can be modelled by priority training of one of its attitude or aptitude structural components according to the concrete objectives.

Mainly, a SCAM module has three courses, as follows:

- During the 1st year of studies - 2 course hours + 1 practical activity hour of Intellectual Work Technique
- During the 4th year of studies - 2 course hours + 1 practical activity hour of Psychology of Technical Creation
- During the 5th year of studies - 2 course hours + 1 practical activity hour of Basics of Technical Creation

The students who continue their studies at a graduate level attend courses of Creativity Applied to Technical Field or Creativity, Value Engineering and Marketing, according to their major.

3.0 QUANTITATIVE ANALYSIS OF INVENTIONS

The presented table synthesises the number of student inventors according to majors and academic years. There were also situations when patent demands were made by two students and very few cases in which they were made by three of them. The first two columns refer to the students major in general electrotechnique who studied Basics of technical Creation during the 5th year of studies. The 3rd and the 4th column refer to the students major in electromechanics who studied the Invention course during the 4th year. The last four columns refer to the evening course students for both majors. Global results for all academic years taken into account are the following:

- Total number of students: 863
- Authors of patent requests: 253, representing 29.3% of the total number of students

There were 586 male students with 191 inventors, that is 32.5% of the total number of students and 277 female students, with 61 inventors, that is 22% of the total number of students.

If we are to refer to the last five academic years with SCAM module, the results are obviously better as one can see below:

- 5th year, electrotechnics major, 317 students, 175 inventors, 55.2%.
- 4th year, electromechanics major, 81 students, 29 inventors, and 35.8%.
- 6th year, evening courses, 74 students, 20 inventors, and 27%.
- 5th year, evening courses, electromechanics major, 8 students, 3 inventors, and 37.5%.

In the process of technical creativity stimulation, the authors found out a surprising phenomenon. That is, about 10% of the students, who chose the traditional type of examination and did not obtain a satisfactory mark, preferred to become inventors.

According to the authors' opinion, being a unstandard course, the less significant results during 1994-1995 can be explained as follows:

- the subjects of the SCAM module are unconventional in comparison with the other technical subjects and the students thought they could approach them easily;
- the necessity of a certain period of time to evaluate the results of the initial training program and finding a solution in order to harmonise the two perspectives of the didactic activity, the technical one and the psychological one;
- the lack of trust in the individual technical creativity, caused by the over-evaluation of the patent notion;The necessity of an incubation period in student mentality triggered by the requirements of the subject as well as by their real possibilities that did not look obvious for the proposed goals but took shape by the passing from a generation to another.

4.0 THE ANALYSIS OF INVENTION FIELDS

In order to stimulate technical creation activity, students were not required to choose a certain field of research. During the first stage, we considered the most important thing for them was to invent a new technical object, to know how to describe it and how to emphasise the novelty of the solution, respecting the rules required by a patent. Most of the patent demands represented problems encountered by students during their technical, domestic, daily or leisure activities. Patent demands can be classified in percents according to the field they can be used in:
- Industrial field: 41.10%
- Domestic field: 25.15%
- Transports: 14.11%
- Sports: 6.74%
- Sanitary/Ecological field: 6.74%
- Agriculture: 4.29%
- Others: 1.84%

From the point of view of the field to which the constitutive elements of the pattern belong, the situation can be presented as follows:
- Mechanical field: 48.46%
- Electromechanical field: 22.08%
- Electric field: 20.85%
- Electronics: 4.29%
- Mechatronical field: 1.2%
- Others: 3.06%

In the authors' opinion, the fact that most of the inventions made by the students of the Electrical Engineering Faculty can be considered to belong to the field of mechanics, has at least the following essential explanations:
- The training program of the first two years of studies contains 5 subjects in the mechanical field - the electrotechnics training starting since the 3rd year of studies- and the field somehow imposes itself in the mentality of the future engineer.
- The concrete characteristic of the mechanics objects and the fact they are the most used in industry and daily life, leads to a habit of operating with them rather than with elements belonging to other fields (especially the electrotechnics one that has a highly abstract profile).
-Most of the necessities of daily life demand mechanics or electromechanical solutions.

5.0 CONCLUSIONS

The results of our project prove that students' creative potential is very important and an appropriate program enabling its stimulation and consciousness can lead to extremely encouraging results.SCAM refers to students who are not specially selected for technical creativity. It enables them to become conscious of their potential and to live the creation experience effectively triggering the motivation for subsequent creations.

We suppose the results can be even better through the employment of new methods of stimulating students, especially during practical activities. It is also necessary that the professors teaching these subjects should be creators, themselves, having psychological as well as technical training.

The entangling relations among the quantity, quality, and cost of higher education – a dilemma confronting private colleges and universities in Taiwan

C H CHEN
Tamkang University, Taipai, Taiwan

ABSTRACT

Private colleges and universities in Taiwan are facing a dilemma on the quantity, quality and cost of higher education. Due largely to the government's liberal policy since the late 80's, their number has more than doubled, and consequently the university student population has grown rapidly. However, there is a side effect: the quality of undergraduates, most notably at private universities, has steadily declined. Generally small campuses at private institutions became congested, degrading the quality of academic environments. Moreover, the rapidly rising cost of commodities and mounting payroll for faculty and staff, etc. have made the university operation more costly. One question stands out: if a university wishes to remain competitive, should the quality or the quantity of students be rated as higher in priority? Many universities then began to realize an urgent need for quality management measures of some sorts. The "Total Quality Management" program pioneered by Tamkang University in 1993 is delineated and its impact evaluated in this paper.

1. INTRODUCTION

Education in Taiwan is traditionally considered to be the best means by which one can achieve his/her career goal and advance his/her social status. Admission to colleges and universities through the government-authorized "joint entrance examination" had been highly competitive until the late 80's when the government began to loosen its grip on higher education. The number of colleges and universities has surged from 41 in '89 to 84 in '98, an increase of more than 100%, with the total undergraduate population expanding from 222,000 to 408,000, a remarkable increase of 84%! (1). The steady increase in the number of universities and colleges is also accompanied by a step-up rise of the admission acceptance rate, up from 34% in '89 to 60% in '98 (1) (Figure1). This is a clear indication that entering a university is much easier for high school graduates now than before.

2. THE DECLINING QUALITY OF COLLEGE UNDERGRADUATES

Many educators in Taiwan have been alarmed by a steady decline in the quality of college undergraduates in recent years, characterized by lack of yearning for learning and over-all poor scholastic performances, especially at private universities. The trend, according to an unofficial consensus, is firstly attributed to high admission acceptance rate resulting from admission of too many of those who were otherwise rejected a decade ago, and secondly to an easygoing and leisure-seeking life-style of youngsters in affluent Taiwan. Now the question is that what steps are needed to reverse the down-turned trend? Conferences/seminars on innovative teaching and educational reform were called for to address the problems and map out the strategies. But how much has the situation been improved since then?

3. STUDENT ENROLLMENT: QUANTITY VS. QUALITY

If an excessive growth of university's student population has indeed contributed to the decline in the quality of undergraduates, then will the school authorities willingly reduce the student enrollment in order to better the student's quality? The answer is most likely 'yes' for public institutions, but 'no' for their private counterparts. The reason is simple. While public universities are almost fully funded by the government, private institutions depend heavily on the tuition of students and little on governmental supports. Reduction of the student enrollment literally means a decrease in the revenue for private universities, and that would be certain to worsen their financial situations. Moreover, the rising cost of commodities and payroll for the swelling size of the faculty and staff has made the enrollment-reduction scheme more difficult to achieve.

4. UNIVERSITY'S OPERATION AND COST-EFFECTIVENESS

Many administrators of private universities then began to think of the means to cut down on some redundant expenses in order to make school's operations more economical. Some universities even quietly began implementing quality management measures of some sorts. Tamkang University (TKU), for example, a private and the largest comprehensive university in Taiwan with 25,000 students on a compact 20-hectre campus, pioneered in practicing the Total Quality Management (TQM) program in 1993. But, how effective is TQM in making the university's operations more efficient and economical? Has the program helped better the quality of education and particularly the over-all quality of students?

5. THE TQM PROGRAM AT TKU

The TQM program at TKU comprises the following measures:
- Office automation and computerization of library operation
- Installation of an across-campus computer network
- Enhanced instructional evaluation system
- Faculty reward system offering rewards for outstanding teaching and academic research
- Departmental evaluation based on its faculty's quality and research publications

6. THE TQM PROGRAM AND ITS IMPACT AT TKU

Statistics available are shown herein to help evaluate the impact of the TQM program at TKU (2), (3) (Figures 2, 3 and 4):
- TQM has seemingly expedited work process across the administrative ranks. The services provided for students and teachers are reportedly improved. Easier and quicker

registration by telephone, for example, is an evidence of improvement.
- Faculty reward system offers an incentive and due honors to those who excel themselves in teaching and/or research. Teachers are more conscientious about teaching methods as well as students' response and active on research work.
- While the student population at TKU grows steadily (Figure 4), its campus area has remained unchanged since it was founded in 1950. Congestion on the 20-hectre campus is evident as the campus area-student ratio downed from 20-sq. meter/person in '89 to 8-sq. meter/person in '98. The TKU administration's decision to acquire a 60-hectre land in eastern Taiwan is as the site for the new campus is a positive step in the right direction.
- Despite the university's policy of enhancing the quality and quantity of the faculty, the student-teacher ratio at TKU has hovered between 35 and 24 in the last decade (Figures 5 and 6), higher than the generally accepted ratio of 20. A bright spot, however, is that the number of undergraduates has tapered off in recent years while that of faculty members has been on a steady rise.
- The result of the author's 4-year experiment on the "effective and collaborative teaching" method to Civil Engineering sophomores and seniors at TKU is disappointing. The author's preliminary finding, which many colleagues concurred, indicates that the problem with the students at TKU is more complex than it appears. Firstly, the garden-like campus lacks an academic atmosphere, providing students with little stimulus for academic activities. Secondly, a large number of incoming freshmen at TKU are ill prepared for college life. Many engineering majors, for example, were either uninterested or of poor aptitude in mathematics and physical science, and it is only natural that they do poorly in most engineering courses. In addition, a widespread misconception that the college life promises a four-year smooth and easy sailing also contributes to their poor scholastic performances.

7. REMARKS AND CONCLUSION

In this paper the tangling relations among the quantity in terms of student enrollment, the quality of college undergraduates and the school's operational cost now facing Taiwan's private colleges and universities have been presented. Most of the financially disadvantageous private institutions of higher education in Taiwan are caught in a complex situations, struggling hard in order to survive. Many have come to realize that the quality management measures are urgently needed to keep the university's operational cost down on one hand and the quality of education at an adequate level on the other. TKU has implemented the TQM program for a number of years, and its outcome is mixed. The followings are some remarks and conclusion:
- Quality management measures are a positive step for today's money-pressed private colleges and universities in Taiwan. However, the TQM program alone is no remedy for improving students' scholastic performances.
- Faculty instructional evaluation system will never work without an active involvement of students. Students' learning habits, desires and motivation etc. should also be periodically assessed in addition to the faculty instructional evaluation to accurately reflect the true student-teacher relationship and the effects of students' learning.
- The atmosphere on school campus plays an important role in influencing the student's life-style. The administration and faculty should make a concerted effort in promoting on-campus academic activities such as international conferences on various academic studies, concerts or art festivals etc. Inducing students to engage in such activities will help fire up their yearning and motivation for active learning.

- Many quality college aspirants from middle- and low-income families in Taiwan shun a private university like TKU because of its high tuition. Providing them with more scholarships, exempted tuition or other forms of financial assistance will help attract more students of high caliber.
- Many students at TKU often find themselves in a field of specialization in which they are totally uninterested. Professors and the administration should actively seek and counsel them for a smooth transfer to an area in which their aptitudes and interests would more appropriately fit.
- The student-teacher ratio is always cited as an index of the university's academic standing. An adequate ratio should be based on the school's long-ranged goal (will it strive for academic excellence?), the orientation of the university (is it teaching- or research-oriented?), and consideration for cost-effectiveness. For a teaching-oriented private university like TKU, reducing the current ratio of 29 to 20 is recommended if it wishes to be rated as a quality university.
- The campus area-student ratio is another index reflecting the academic quality of the school. The congestion on the TKU campus should be eased when the construction of the new campus is completed.

As private colleges and universities in Taiwan are now confronting the tangling relations among the quantity, quality and cost of higher education, the choice is by no means a clear-cut one. However, what TKU learns from TQM provides an example that a private university in Taiwan, with an application of quality management principles and a quality academic program properly in place, can be cost effective and, more importantly, prosperous in the increasingly competitive academic world of the 21st century.

REFERENCES

1. Education statistics of the R.O.C., Ministry of Education (1998)
2. The University Records- Student enrollment, TKU dean of academic affairs' office (1998)
3. The University Records- Faculty and staff, TKU personnel office (1998)

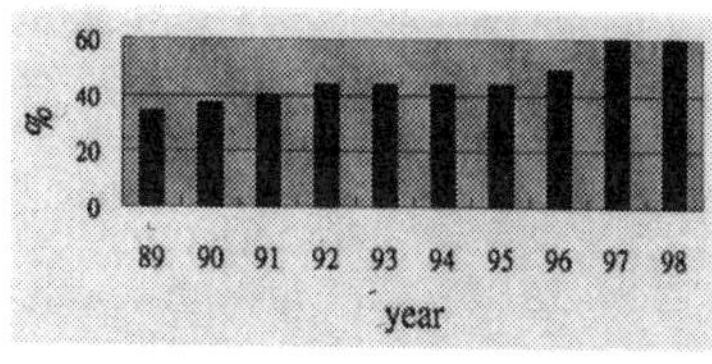

Figure 1: The admission acceptance rate at colleges and universities (technical colleges not included)

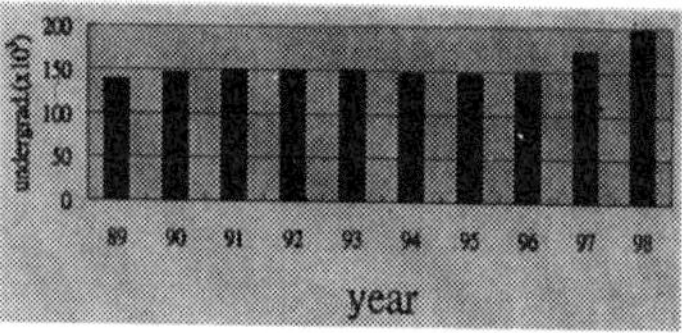

Figure 2: The undergraduate population at TKU (students at the evening school not included)

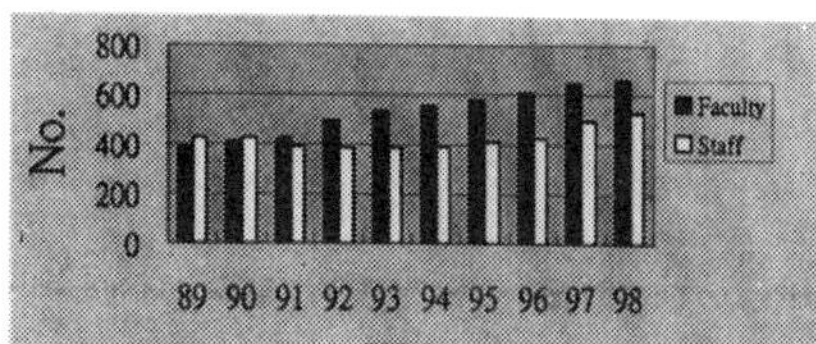

Figure 3: The number of faculty and staff members at TKU

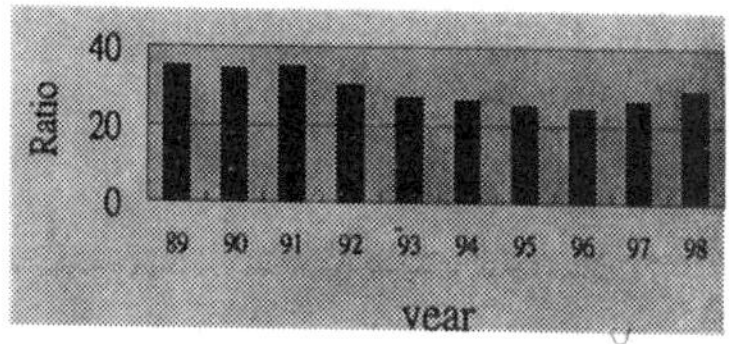

Figure 4: Student-teacher ratio at TKU

Sustainability – new paradigm of engineering education

Z CYWINSKI
Faculty of Civil Engineering, Technical University of Gdansk, Poland

SYNOPSIS

The mission of *sustainability* in civil engineering education is exposed as a major must today. Addressed to the *global village* needs of economy and the odds of engineering and information technology, it forms - together with the notions of *ethics, environment, aesthetics* and *heritage* - the philosophical and material representation of the society's *cultural landscape*. Thus, *sustainability* imposes a considerable effect on the present design and maintenance of constructions. Paper underlines the spiritual side of those aspects - having a crucial impact on the *sustainable development* and the life of people, as a whole.

1 INTRODUCTION

Sustainability originated in form of the *sustainable development* concept. The latter was derived in 1987 by the World Commission on Environment and Development (WCED) in its report "Our Common Future" (Brundtland Report) and later authorized by the Rio de Janeiro 1992 United Nations Conference on Environment and Development (UNCED) (1), (2). The relevant definition is the following: "Sustainable development meets the needs of the present without compromising the ability of future generations to meet their own needs".

Sustainable development relates widely to different spheres and practises various operations - locally and globally, thus becoming a very actual paradigm of engineering (3). Today, it is promoted by the world at large, to mention only the American Association of Engineering Sciences (AAES), or the World Engineering Partnership for Sustainable Development (WEPSD) with its partners: World Federation of Engineering Organizations (WFEO), International Federation of Consulting Engineers (FIDIC), and the International Union for Technical Associations (UATI). By developing a special network, WEPSD contributed essentially to the use of information technology in fostering *sustainability* (4). There is no doubt that such policy must earn its counterpart in the education of engineers - civil, in particular.

2 BACKGROUND

Until now, *sustainability* was governed mainly by the material needs of mankind. In author's opinion, it depends - largely - also upon the spiritual disposition of people. The western civilization is the effect of the well known phrase of the Bible: "Be fruitful, multiply, fill the earth and subdue it" (5); unfortunately, such recommendation could be understood as mastering nature at any price; eastern civilizations: Taoism, Confucianism, Shintoism, and Buddhism are here more moderate (5). Although the industrial revolution - the radical basis of the present technological development - appeared first in the west, today there emerged also here the belief that true development can not be limited to pure technical problems only; "human ecology" should be taken into account.

When addressing civil engineering, or even the more edged structural engineering, it is worthwhile to recall the relevant definitions:
- "Civil engineering is the profession in which a knowledge of the mathematical and physical sciences gained by study, experience, and practice is applied with judgement to develop ways to etilize, economically, the materials and forces of nature for the progressive well-being of humanity in creating, improving and protecting the environment, in providing facilities for community living, industry and transportation, and in providing structures for the use of mankind" (6).
- Structural engineering "is the science and art of planning, design, construction, operation, monitoring and inspection, maintenance, rehabilitation and preservation, demolishing and dismantling of structures, taking into consideration technical, economic, environmental, aesthetics and social aspects" (7).

According to those definitions, the civil/structural engineer must work within a milieu of many different aspects - all having substantial impact on the appearance of *sustainability*. A related educational approach is here necessary, as well.

3 SUBSTANCE

Sustainability is mutually and crosswise connected with *ethics* (8), *environment* (9), *aesthetics* (10), and *heritage* (11). All of them influence the appearance of the given *cultural landscape* which is the place that has been created, shaped and maintained by the links and interactions between people and their environment (12), (13).

Sustainability problems of civil infrastructure systems were discussed in (14). The "integration of organizational, financial, informational, managerial, and technical approaches" was investigated - concerning the interrelated life-cycle engineering, technology investment, performance evaluation, and project conduct and supervision. Environmental *sustainability* was specially studied in (15), with stress put on management and business practices, design technology, construction methods and equipment, materials, as well as - on public and government policy.

Different issues of *sustainability*, addressed to civil/structural engineering, are today vigorously introduced by the International Association for Bridge and Structural Engineering (IABSE) - by establishing a particular working commission (16) and by examining them, broadly, during the last 15th IABSE Congress Copenhagen 1996 and the recent IABSE Symposium Rio de Janeiro 1999. IABSE advocates the application of environmentally acceptable materials ("green products"), the use and re-use of products and structures ("stewardship"), the minimization of energy, waste, and pollution, as well as - life-cycle analysis and costing.

It is evident that all those particular activities of the civil/structural engineers must have an adequate foundation in their education. However, it is not easy to teach civil engineering *sustainability* by means of one particular subject. Therefore, that teaching must be conducted, more or less, within all the subjects of the civil engineering profession.

Another way bases upon a strong development of the students' intellectual power, of his creativity potential. This is possible to achieve by a major upgrading of their humanistic interests. The Faculty of Civil Engineering, Technical University of Gdansk, Poland, introduced recently the subject "Preservation of Heritage in Construction" which has been found adequate to deepen the students' intellect, considerably. Experiences in that matter, collected within the last two years, are presently very much encouraging. For a similar development within the community of professionals, in the nineties three International Conferences "Preservation of the Engineering Heritage - Gdansk Outlook" were organized in Gdansk, the last one in September 1999.

4 CONCLUSIONS

Sustainability (Fig. 1) became, surely, a very important aspect of the engineering practice - civil, in particular. Therefore, it must be intensively honoured in the present education of engineers. Hereby, upgrading of the students' intellect and of their interest for humanistic values, as well as - a broad promotion of *sustainability* within the world of professional engineers, are acknowledged to be substantial conditions for the respect and well-being of that idea in the present society, locally and globally.

It must be added that, to observe the needs of the *sustainable development* of mankind, the existing realities of the present world are extremely disadvantageous. However, without doing a visible step forward on this way today, a favourable development of future generation looks to be largely endangered.

REFERENCES

1. Golay A.: Sustainable development. *IABSE Structural Engineering International*, 6(1996), 3, 210.
2. Roberts D.V.: World engineering partnership for sustainable development • World engineering network. *UNESCO World Congress of Engineering Educators and Industry Leaders*, Paris 1996, Final Report, 139-143.
3. Quinn L.: Sustainability: Another new paradigm. *ASCE Civil Engineering*, 66(1996), 10, 6.
4. Foo Ah Fong: Sustainable urban development through information technology: The case of the city-state of Singapore. *CIB W89 International Conference "Construction Modernization and Education"*, Beijing 1996, Abstracts, 25-26; CD ROM, 6 pp.

5. Mac Isaac GAF., Morey N.C.: Engineer's role in sustainable development: Considering cultural dynamics. *ASCE Journal of Professional Issues in Engineering Education and Practice*, 124(1998), 4, 110-119.

6. *American Society of Civil Engineers Official Register*, ASCE, New York 1998.

7. *International Association for Bridge and Structural Engineering Membership List 1998*, IABSE.

8. Brennan R.A.: What are the ethical responsibilities towards sustainable development. *ASCE Journal of Professional Issues in Engineering Education and Practice*, 124(1998), 2, 32-35.

9. Cywinski Z., Kido E.: Environmental harmony of strait bridges. *3rd Symposium on Strait Crossings*, Alesund 1994, Proceedings, 91-98.

10. Cywinski Z., Kido E.: Bridge aesthetics - a call of the 21st century. *5th East Asia-Pacific Conference on Structural Engineering and Construction*, Gold Coast 1995, Proceedings Vol. 3, 1857-1862.

11. Cywinski Z.: Civil engineering heritage in construction education and research. *CIB W89 International Conference "Construction Modernization and Education"*, Beijing 1996, Abstracts, 19-20; CD ROM, 6 pp.

12. Cywinski Z., Kido E.: Rehabilitation of steel bridges for the 21st century - The problem of cultural landscape. *International Conference "Structural Steel Developing Africa"*, Johannesburg 1996, Proceedings, 63-70.

13. Cywinski Z.: Humanities & arts - Essential agents of contemporary engineering education. *SEFI Annual Conference "Humanities and Arts in a Balanced Engineering Education"*, Cracow 1997, Conference Proceedings, 22-35.

14. Aktan E., et al.: Sustainability of civil infrastructure systems. *International Workshop on Structural Health Monitoring*, Stanford 1997, Proceedings, 564-576.

15. Dreger G.T.: Sustainable development in construction management strategies for success. *CIB W89 International Conference "Construction Modernization and Education"*, Beijing 1996, Abstracts, 21; CD ROM, 5 pp.

16. Silman R.: Working Commission 7 - Sustainable Engineering. *IABSE Structural Engineering International*, 8(1998), 4, 327-328.

Fig. 1.   Author's recognition of structural sustainability related
to the cultural landscape of Gdansk

The methodology of research on the reliability of Socioecotechnical systems

J Z CZAJGUCKI
Department of Fundamentals of Technique, Gdynia Maritime Academy, Poland

SYNOPSIS

The aim of this paper is to present the methodology of research on the reliability of socioecotechnical systems in the light of formulated principles. Methods of carrying the considered research have been characterised. Remarks as to the systems of considered methods and research have been formulated. Conclusions are given at the end of the paper.

1 TERMINOLOGY AND BASIC PRINCIPLES

By system we define an entity consisting of real or abstract elements $\{e\}$ and of the relations $\{\rho\}$ between these elements. All the elements outside the system constitute its surroundings, but with a delimited area. The general designation of the system is $< e_1\, \rho_1\, e_2\, \rho_2 ... e_n\, \rho_n\, e_{n+1} >$.
In the definition of the reliability of a defined object we distinguish the reliability predicate, the reliability measure, the argument of this measure, and the conditions of existence of an object. Objects are various and may be treated as systems. A predicate is an expression defining the reliability of an object. Here are examples of predicates: the correctness of functioning of a defined technical object, the rightness of man's action, the ability of a defined object to fulfill tasks. The reliability measure most often is the probability of what is expressed by the predicate. The argument of this measure usually is time. The conditions of existence of an object are most generally the properties of an object and its environment.
The investigations of the reliability of objects are investigations of the relations between distinguished magnitudes, among which we rank the reliability of objects and their components as well as the properties of processes occurring in objects and their surroundings, in order to obtain models of these relations – most frequently mathematised. The investigations so understood will be called in a general way investigations of reliability relations. Usually we divide these investigations into model investigations (experimental and computer – so called simulation investigations) and investigations in real conditions – so called exploitation investigations. The investigations of reliability relations should be carried out according to defined methods. The quality of results of these investigations depends above all on scientifically founded methods of conducting these investigations. The created actions in methods and by the same research actions resulting from them are, in the author's opinion, subordinated to the following principles:

— the inference principle of operations, this principle expresses the logical sequence of actions,
— the principle of relativity of created models of reliability relations to the laws (and principles) of nature (expressed in physics, chemistry, biology and in other disciplines of science),
— the principle of concordance of formalisation of reliability relations with logic,
— the principle of mathematical transformation of reliability relations, a transformation which is relative to the properties of models of these relations resulting from the laws of nature,
— the principle of relativity of models of reliability relations to the needs of realising knowledge aims and utilitarian aims; the knowledge of causality relations in processes which cause damages, in a broad meaning, in socioecotechnical systems constitutes the main target of all the quoted kinds of investigations; if this cognition takes place the results of researches will be useful.
— the principle of entirety (completeness and consistence) in its component and relation in methods and investigations of reliability relations.

The author in the further part of his paper pronounces on the methods of investigating reliability relations in the light of the above principles.

2 ON INVESTIGATION METHODS OF RELIABILITY RELATIONS

We can distinguish two groups of investigation methods of reliability relations, namely:
1) methods creating researches on reliability relations,
2) methods of realising researches on reliability relations.

The first methods are creative and conceptual (conscious and subconscious). Deductive methods are an integral part of these methods. The second methods distinguished earlier are applied methods of the first ones. They include reductive methods (which comprise inductive methods). Besides, a third type of methods resulting from the two types of methods mentioned above; they are deductive and reductive methods. Well-known methods of research of reliability relations are parametric methods (**PM**) and nonparametric methods (**NPM**) of reliability exploitation research (**RMR**) of technical objects elements. These methods are described in many sources, and among others quoted in the work (**1**). Researches where these methods are utilised transform the probabilistic picture of events-damages of elements. In these researches the identification of functioning conditions of elements should be taken into consideration. The lack of information about these conditions makes the practical use of reliability functions of elements impossible. These conditions are being determined mainly by construction features (material, geometrical and strength) of elements, by load features (mechanical, thermal and others), strain and stress of elements and also by properties of technical object surroundings.

A less known method is the transformation method (**TM**) of the required reliability of a technical object into requested reliability values of its elements. This method permits the design of technical objects with the required reliability. It has been described, among others, in (**1**). Research carried according to this method is the design, simulation investigation (**SIR**) of reliability relations.

The methods of life estimation (**ML**) of elements submitted to the action of variables, stochastic ones also, of loads, mainly mechanical and thermal are well-known. These methods are applied in experimental researches (**ERL**) on the life of elements.

The literature where these methods are described is exceptionally rich and particularly literature concerning material fatigue, fracture mechanics and tribologic wear. Within the scope of these methods there is a great variety and particularly in the field of heuristic hypotheses of the course of processes causing damages and in the range of damage measures. In the light of this there is a need to create further scientific bases of the above mentioned experimental researches.

Let us try to create a system of methods (considered up to now) and it will appear that beside the distinguished methods, it is also necessary to take into account the methods of valuing technique (**VT**). By technique we understand all technical objects. The valuing of technique is a research (**VTR**) of a peculiar character. And the methods of this research are methods of assessing the risk of threats due to a broad interpretation of safety with which reliability is connected, and methods of optimising the reliability of technical objects. The system of methods permitting the research on reliability of technical objects is the following

$$< \text{VT } \rho_1 \text{ (PM} \vee \text{NPM)} \rho_2 \text{ TM } \rho_3 \text{ ML} >,$$

where the sign $\vee$ is the symbol of alternative and the other symbols are the designation of methods and relations; these relations have as a character to arrange methods with regard to themselves.

With reference to the already distinguished methods of reliability relations we should distinguish further indispensable research, so that a system of research on reliability relations might be created. If technical objects are considered in their relations with man (using or maintaining these objects) we have then to deal with anthropotechnical systems. The investigations of reciprocal relations in these systems as to their reliability in fact amount to investigate the reliability (**IRM**) of man according to defined methods (**MIRM**). The methods of such researches are known. They are, however, little developed. The base for creating such methods is psychology. The apprehension of processes taking place in technical objects, the perception and interpretation of these processes by man and his relation connected with it are dependant upon genetic characters, upon the level of education and the experience of man, as well as upon the conditions under which he acts. All that should be taken into consideration when investigating man's reliability.

In fact we have to deal with broadly conceived socioecotechnical systems composed of anthropotechnical systems and their surroundings i.e. people, nature and other technical objects (other than the considered ones in anthropotechnical systems). In the researches on reliability of socioecotechnical systems the above investigations will already take place and also the investigations of dangers (**IDS**) of the surroundings due to unreliability of anthropotechnical systems. These last investigations are carried according to methods (**MIDS**) coming from various sciences. An interdisciplinary integration of these methods is required in order to know, through research, the causality in various processes which, consequently, lead to endanger the surroundings in case of failure of anthropotechnical systems. Let us consider all the mentioned investigations as a defined whole which constitutes the following system of reliability research on the socioecotechnical systems

$$<< \text{VTR } \rho_4 \text{ RMR } \rho_5 \text{ STR } \rho_6 \text{ ERL} > \wedge < \text{IRM } \rho_7 \text{ IDS} >>,$$

in which we have designated the described investigations and the relations between them; these relations are connected with the results of investigations; the sign $\wedge$ is the symbol of conjunction.

If we take into consideration all the distinguished methods of research on the reliability relations, we have to deal with a consecutive system of these methods, namely

$$\ll VT\ \rho_1\ (PM \lor NPM)\ \rho_2\ TM\ \rho_3\ ML > \land < MIRM\ \rho_8\ MIDS \gg,$$

where ρ_8 is the symbol of relations of arrangement of the distinguished methods in relation to themselves; as to the other symbols they have been already described.

In order to concretise and guide the development of the above expressed systems of researches and methods we have formulated the following remarks which have the character of observations, propositions and postulates. They are:

1. Reliability of socioecotechnical systems which is, sensu stricto, the reliability of processes $\{X(\tau)\}$ occurring in these systems; $X(\tau)$ is a random function, τ – time.

2. Reliability predicates can be concretised, i.e. defined with the help of properties $\{c\}$ of processes $\{X(\tau)\}$.

3. Critical states (identified with unfitness or otherwise damages) of socioecotechnical systems can be defined with the help of critical (limit) values $\{c\}$.

4. The reliability of defined objects in systems can be expressed as to the states (fitness or unfitness) of these objects and we can express also this reliability as to the states which these defined objects cause in other objects of systems. For example the reliability of man can be expressed as to the states – effects which he causes in technical objects.

5. Modelling of reliability relations should be compatible with the principles expressed earlier. The main operations in this modelling are the identification of processes causing damages of the elements of systems, the creation of reliability relations with the use of suitable methods, the investigations of reliability relations, including analysis and synthesis of reliability relations, and the verification of created models. Deductive and reductive methods are considered as methods of the highest scientific and applied standard. These methods are worth while being created and perfected.

6. Modelling of reliability relations should be pertinent to the needs of designing, production and exploitation (utilisation, maintenance and liquidation) of technical objects treated as components of socioecotechnical systems.

The author intends, while presenting his paper, to illustrate the considered problem with many examples, particularly the models of reliability relations.

3 CONCLUSIONS

From the presented picture of the methods and researches on reliability relations in considered systems it results that:

- there is a need to create new methods in this domain,
- researches should be interdisciplinary and integral, they should particularly concern processes taking place in socioecotechnical systems,
- there is a need and even necessity, at this stage of evolution of civilisation, with an ample use of technique, to integrate scientific centres having a world range in order to meet research needs in the discussed field and also due to the costs of carrying researches.

„We need a new paradigm, a new vision of the reality, our former thinking, perception and values have to be changed" – F. Capra (**2**).

REFERENCES

1 **Czajgucki J.Z.** Methodical Approach to the Tranformation Process of the Required Reliability of Compound Technical Systems into Requested Reliabilities of Elements of These Systems. Proceeding 2nd International Conference on *Planned Maintenance, Reliability and Quality*, University of Oxford, 2-3 April 1998, Oxford, England, pp. 48-52.
2 **Capra F.** *The Turning Point. Science, Society and the Rising Culture*, 1987 (Polish edition by Panstwowy Instytut Wydawniczy, Warsaw).

Knowledge management or the management of knowledge – back to the future

D HAWLEY and **K A SMITH**
Sheffield Business School, Sheffield Hallam University, UK

SYNOPSIS

Knowledge has always been an important resource in organisations; it makes up the intellectual capital upon which the organisation's wealth is acquired, accumulated and measured. Knowledge transfer lines are the 'arterial highways' upon which organisational intellectual capital is transported. The operation of the 'arterial highways' is influenced by the organisational structure and social technology. An inappropriate structure, for environmental conditions, will inhibit the transportation and growth of intellectual capital and could impede organisational success. Knowledge is also the foundation of technological developments. The symbiosis of information technology, an appropriate social technology, and a better understanding of human capital are paramount for the development of intellectual capital.

1. A RESOURCE PERSPECTIVE OF KNOWLEDGE AND INTELLECTUAL CAPITAL

The resource base of an organisation is made up of tangible and intangible entities. It is the means by which organisations are established and operated.[1][2] It forms the basis on which the measurement of organisational success and worth is determined, usually expressed, in financial terms, and shown in financial reports as its 'net capital'.
Knowledge, in an 'information age', is a primary ingredient in the products and services organisations' buy, produce, add value and sell. It is – and always has been, albeight not always recognised as - an additional factor of production. Knowledge Management embodies those organisational processes that seek a synergistic combination of data and information processing capacity of information technologies, coupled with the creative and innovative capacity of human beings (human capital). Knowledge, as a component of intellectual capital therefore, is an important part of an organisation's capital structure, as is an understanding of the psyche of the 'human capital' who produce it. Consequently, the management of intellectual capital is an important economic task of individuals, organisations and nations.[3]
Illustrating intellectual capital is less problematic than providing a working definition. Illustratively, *"it is the sum of an organisation's intellectual property; employees' knowledge and skills; information and intelligence on customers, suppliers and competitors etc. together with that innate business nous instantly recognised but difficult to define, even by those who possess it"*.[4]

1.1 Intellectual capital defined

"Intellectual material that has been formalised, captured, and leveraged to produce a higher-valued asset".[5]

"Knowledge that exists in an organisation that can be used to create differential advantage" [6]

1.2 Knowledge in an around the organisation – a top and bottom perspective.

Implicit in the Klein and Prusak definition is that good ideas, initiatives and discretion is not only the preserve of those occupying high managerial positions, but that good ideas can be located in subordinate positions within the organisation. Furthermore, there is an implicit distinction between intellectual material, for example, patents and copyrights (intellectual output) and intellectual capital (intellectual resource). Intellectual capital being the raw material from which financial results are made.[7]

Information and knowledge are the thermonuclear competitive weapons of our time . Quality, reliability, innovation closeness to the customer has been suggested as 'critical success factors'. From industry to industry success is determined by those organisations who have the 'best' information, or who can wield the information most effectively. Knowledge has become a major corporate resource. Service industries and service roles predominate over manufacturing today, and knowledge is the raw material of their success. One recent survey showed that more than 80 per cent of the market value of ten leading Global 500 companies was 'intellectual capital'.

2. A DEVELOPING PERSPECTIVE OF KNOWLEDGE MANAGEMENT

Knowledge management, as developed by the 'classical' and 'scientific' organisation and management theorists (we later refer to as 'traditional theorists'', were professional engineers by discipline. They introduced and developed highly mechanistic concepts and principles, initially into manufacturing, which were employed in the processes of transferring knowledge from the 'skilled' to the 'un-skilled'.

2.1 Classical Perspective

emerged during the nineteen and early twentieth centuries that emphasised a rational, scientific approach in making organisations efficient operating machines. Problems arose in tooling the plants, organising managerial structure, quality and reliability of products and services, training employees and dealing with industrial disputes. Effectively, it was the way knowledge and skills were transported and managed in and around the organisation.

2.1.1 Scientific Management – emphasised scientifically determined changes in existing management practices that provided solutions to organisational productivity problems. [8]

2.1.2 Bureaucratic Organisations_– emphasised management on an impersonal, rational basis through elements such as clearly defined authority and responsibility. It became 'rule by the written word', supported by well defined policies and procedures.[9]

2.2 Emerging mechanistic organisational structures[10]

Organisational structures and design; management principles and concepts are not only manifestations of how the organisation plans and controls its activities, but is also a reflection of the organisation's approach to the way it manages its 'knowledge resource' or intellectual capital. The structures and knowledge transfer systems which have emerged from the 'traditional theorists' concepts indicate that the omnipotent, omniscient knowledge base resides at the 'top' of the organisation. Management governs operations and work behaviour by issuing strong, clear commands. Unilateral instructions flow downwards becoming more explicit at each successive level of hierarchy. Upward information, however, is often successfully 'filtered out' on its way to the 'top' of the organisation. What also emerges from this structure is a unilateral, autocratic system for transferring knowledge and information where opportunity for individual initiative, creativity and discretion are at best limited – at worst actively discouraged. The appropriateness of this type of structure and design is more suited to highly stable organisation environment relationships where the level of complexity, uncertainty and ambiguity is low. Such structures also restrict the knowledge content of the subordinate's work and the individual can become discouraged from being knowledge worker. It is the separation of conceptualisation (the role of management) from execution (role of the managed).[11]

2.3 Emerging organistic organisational structures

The dynamic, turbulent organisation environment relationship is characterised by high levels of complexity, uncertainty and ambiguity - probably more typical of contemporary society - where innovation, quality and creativity, knowledge about 'the customer' and quick strategic responses are key success factors. It requires an organisation structure enabling organisational processes to find a synergistic combination of data and information, processing capacity of information technologies, and the creative and innovative capacity of human beings that meet the challenges of a contemporary environment.

2.4 The Millennium Organisation

In an era characterised by knowledge as the critical resource for business activity, we are witnessing increasing hyperbole about the rewards emanating from the newest information technologies. Several providers, of the new technologies, for example, data-mining, archiving, intranets, video-conferencing web casting etc. are offering these as solutions and a panacea for the business challenges of the knowledge era. It has also been suggested that increasing investments in new information technologies somehow result in improved business performance.[12]

Electronic commerce is growing exponentially putting a premium on knowledge and on those who know how to manipulate it. E-commerce (roughly defined as a way of conducting, managing and executing business transactions with other businesses and consumers, using computer and telecommunication networks) grew from almost zero in 1995 to $26 billion worth of business in 1997. The Organisation for Economic Co -operation and Development suggest it could be worth $1,000 billion a year within the next four years.

2.5 Technology - the opiate of the (knowledge) poor

We posit that technological determinism is not a panacea for success. Organisations adopting this approach are seeking a form of utopianism that probably does not exist. Technology, in whatever form, is a 'tool' to be managed. We suggest that the organisation is equally reliant for success on its 'social technology'[13] as it is its hardware and software. If the organisation is inefficient and/or ineffective in the first instance, technology per se will not improve the situation. Both organisational structure and the technology employed must provide an effective information flow. It does not follow that an organisation full of 'brilliant' people is a 'beacon' of collective 'brilliance'!

2.6 The social technology of an organistic organisation structure .

An organisation's 'social technology' is largely determined by the environmental circumstances - measured on a continuum of stability to turbulence. It comprises the formal flow of authority in and around the organisation, the lines of communication, the division of tasks and activities and the arrangements for co-ordination and effort. It is the mechanism by which knowledge is processed as the primary ingredient in products and services organisations' buy, produce and sell.

In the highly turbulent environment of technological change and development, organisational information and knowledge transfer systems require a social technology which generates fluidity and flexibility. The findings of the Tavistock Research, is that the all encompassing knowledge is not situated at the top; that critical knowledge exists only at the 'tentacles' of the firm; that each functionary seeks out and interprets, for themselves, information necessary to perform their part of the task; non- restricted communication patterns resemble lateral consultations – sharing of knowledge – rather than vertical commands predominated.

2.7 Knowledge management: back to the future

We posit that knowledge management has existed in organisations for centuries. There could be an inherent danger in the concern to utilise and harness knowledge in the pursuit of competitive advantage, the desire to quantify, measure and control this valued resource will, paradoxically, result in the underachievement of the desired objective. In positing the concept 'knowledge management' one runs the risk of overemphasising the rationality of the human psyche. The history of the management enterprise to date and its failure to effectively manage its 'human capital' is an acknowledgement of the complexity of the individual as the 'manager' and 'the managed'.

REFERENCES

[1] Burns T., Stalker, G.M. *The Management of Innovation*, Tavistock Publications Ltd., (1961)

[2] Perrow C., *Organisational Analysis: A Sociological View*, Belmont: CA: Wadsworth, (1970)

[3] Edvinsson L., Malone M.S, *Intellectual Capital : the proven way to establish your company's real value by measuring its hidden brainpower*, Piatkus (1997).

[4] Stewart T.A., *Intellectual Capital: the new wealth of organisations*, Nicholas Brearly Publishing, (1997).

[5] Klein D.A., Prusak L., *Characteristics of Intellectual Capital*, Multi Client Programme Working Paper, Ernst & Young Centre for Business Innovation, 1994.

[6] Macdonald Hugh, Hour Futurologist, ICL, *Knowledge & Competitive Advantage* (cited in Stewart)

[7] Edvinsson L., Malone M.S, (OP.Cit).

[8] Taylor F.W., *Scientific Management*, New York, Harper & Bros, 1947.

[9] Weber M., *The Theory of Social & Economic Organisation*, New York, OUP, 1947.

[10] Burns & Stalker (Op.Cit).

[11] Braverman H., Labor & Monolpoly Capital: the degradation of work in the twentieth century

[12] Brynjolfsson B, Hitt E., *Three Measures of Information Technology's Contribution: Creating value & destroying profits*, MIT Sloan School, Cam. Mass. (1993)

[13] Burns T., Stalker, G.M (Op.Cit).

Life sciences, globalization, and engineering education

V HOPP
University of Rostock, Germany

What does life sciences and globalization mean?

Life sciences comprise an interdisciplinary science of nature, about nature, of mankind and its behavior, that means psychology and sociology. How mankind will behave in smaller and smaller living-spaces, is besides the basic needs of nutrition health and housing, the fourth problem we must solve in the near future facing (Fig. 1).

In favoured areas of settlement population density is increasing all the time. Areas of the resources for raw-materials, energy and nutrition are not identical with the locations for production and housing.

Globalization means to construct a network of storage, production-, transportation- and communication systems so that we can exchange and use supplies of water, energy, raw-materials, products and information without disruption and according to need. But we also need supplies of capital. We should remember that money is, in principle, a means of comparing the value of products, and value is only a way of showing the usefulness of products.

Life sciences and globalization depend on one another and force us to adapt production methods to shortage of natural resources and the regeneration of the environment. Research aims should be newly defined. The ability to regenerate means more than environmental protection it means health in the widest sense. Nature may be changed, it is not static. But nature must be capable of being regenerated.

Innovation is not only the ability of a nation or its society to adapt to worldwide changes in nature, technology and the economy, but also the ability to influence the process of change.

Population growth will give impetus to biotechnology, genetic engineering, environmental technology, nuclear energy a.s.o. As a consequence of this development our chances of survival will be greater.

These problems should be discussed openly and without ideological baggage or narrow-mindedness.

The creation of simple handicraft techniques should be an integral part of innovation in order that more people are enabled to participate in modern production processes.

Within the scope of globalization innovation in future must involve the setting up of new unsubsidised jobs. Innovation should not be restricted only to the fields of natural sciences, engineering sciences or of technologies.

Innovation potentials must be promoted and activated in the fields of economics, sociology and psycho-sociology.

Besides the requirements for sufficient nutrition and sustainable health it is necessary also to develop simple technologies and skills, which are cheap in production and which need no complicated maintenance. The new modern technology must be easy to use.

– *Responsibilities of the international companies*

It is a relatively simple matter for business to downsize and to cut tens of thousands of jobs in the pursuit of profitability, and for the state to be burdened with the consequent costs. However, to be able offer new job opportunities through innovative products, demands spirit, imagination and sense of responsibility.

The global spread of unemployment is the most fertile soil for revolutions, civil wars and wars. Politicians and top managers of international big companies should bear this in mind.

Globalization also means the interdependence of the various financial markets all over the world. That demands a high sense of responsibility on the part of the banks. They have to analyze the economics and not only look for quick profits.

The collapse of the financial markets in GUS (Russia) and in Asia were foreseen. The political background of the former USSR should teach us that the socialist systems could not operate with money. The governments of the following nations – e.g. Russia, Ukraine, Kasachstan and others – are still thinking in planned economic models. It will take one generation till they have understood and adopted a free market system.

One could foresee also the shadow of the disaster of the financial market in some Asian countries. This was only the greed for higher profit.

International engineering education

The world's population is increasing both rapidly and continuously. According to the latest statistics, there are currently around 6 billion people on our planet. That compares with only 1.8 billion in 1930. This dramatic increase illustrates the rapid developments mankind is facing in terms of co-existence and meeting basic needs.

However, the negative consequences of industrial production methods must also be taken into consideration.

The United Nations Conference on Environment and Development at Rio (UNCED) in 1992 and First Conference of the Parties (1st COP) to the Framework Convention on Climatic Change at Berlin in 1995, which was set up at the Earth Summit in Rio drew attention to this as did the World Congress on Population and Development in Cairo in 1994.

The tasks facing us can no longer be solved alone by a single nation or country. Nations must be able to communicate with one another. Professionals, including engineers, must work together and must be able to function as part of an international team. They must be able to speak the same language.

Engineering students must be prepared for and made sensitive to these demands during their university training. We need engineers who can think in terms of international, technical, social and financial relationships in addition to possessing expert knowledge. Apart from their native tongue, they must be proficient in at least two foreign languages (incl. English).

The world's engineers should have the same level of knowledge but different specializations

The spectrum of talents in the countries all over the world is wide and varied.

The scientific and technical innovation potential lies in the diversity of the talents, behavior and characteristics of the people in the different countries. This will be the power in future, if the people don't forget to work.

All nations have produced a lot of excellent and experienced engineers.

The Italians are wellknown as excellent constructors and builders of roads, bridges, highways and so on.

The French, Austrian and Egyptian engineers constructed the Suez-Canal (1859 - 1869) according to a concept of A. Negrellis.

In the 18th and 19th century the Spanish, Portuguese and British were famous nations of maritime trade. Therefore they need excellent engineers for shipbuilding.

The French engineer Alexander Gustaf Eiffel (1832 - 1923) is wellknown worldwide. He built the Eiffel-Tower which bears witness to his knowledge of static. The efficiency of textile engineering in the beginning of industrialization in England, France and Germany must also be mentioned. Chinese engineers can look back on a long tradition of construction of the Great Wall, making rivers navigable and the industrialization in the last few decades.

Indian engineers became famous for the construction of the Taj Mahal. The engineers of the United States are world wide well known for the construction of high-buildings, aeroplanes, computer-technology and so on.

If we describe the profiles of the engineering qualifications in the different countries of the world, then it is useful to describe the tasks and problems which they have to solve.

The comparison of the training time, the curricula, the number of courses and disciplines do not give the realistic situation about the training-quality and the standard of a degree.

The challenges, tasks and problems of the engineers in a country depend on the resources of raw materials, energy, climate and geological nature, geographical location, transport-infrastructure, materials and others.

Engineers in mountainous regions require different knowledge and experience than engineers near the sea.

In an area of earthquakes the houses must be built in an other manner than in a quiet area.

But all the engineers who work in different specialized fields should have the same level of fundamental knowledge in engineering.

The reputation of an engineer depends on his efficiency and experience during his occupation and not on his examination results.

If we compare the training qualification, then we must analyze the curricula, the teaching-contents of the disciplines, the didactic methods, the professors and so on, that's very complex.

In training we are not dealing with machines. Training is an individual, continuous long-term process that never ends.

The organization of engineering studies into three phases (Fig. 2)

A modern course of study in engineering that will fulfill all future requirements within the international countries could feasibly be organized into three phases.

- The <u>first phase</u> would consist of "general studies" dealing with the basics of engineering and the natural sciences and including a foreign language.

  After one year, a written or oral examination must be taken in every subject.

  General studies in engineering and the natural sciences include the basics of mathematics, informatics, geometry, physics, chemistry, biology, technical drawing and an introduction to the engineering disciplines which the student has decided to pursue, e.g. mechanical engineering, electrical engineering, construction, metallurgy, etc.

In this connection, the basic knowledge requirement in the biosciences is an important novelty. The biosciences are concerned to make future engineers sensitive to the laws of biological processes, selective optimization and evolutionary mechanisms and to show the relationships, similarities as well as the differences to technical processes.

General studies for degrees in engineering and the natural sciences should take up to 20 to 25 percent of the normal course duration.

- <u>The second phase</u> of engineering course work is then devoted to the subject-specific material. Since the basics of engineering and the natural sciences have already been mastered, phase two leaves enough time for the in-depth study of engineering. Yet the contextual components of this phase should always emphasize the fundamentals.

- <u>In the third phase</u> of this course of study, which only makes up 15 to 20 percent of the total, the student is to be given in-depth exposure to special fields, depending on his/her own interests. The dissertation is to be written during this phase. This gives the student the opportunity to demonstrate his/her own performance and thinking ability.

The qualification of good young engineers can be seen in their mastery of the fundamentals of engineering science and technology. It also exists in the ability to apply these basics to different problems in special fields.

Specialized knowledge is however best obtained at the workplace and not during university training. This knowledge is gained from topical subjects and its focal points often change rapidly. It is frequently short-lived.

In this context, the need to overhaul university curricula is applicable. This overhaul means doing without special fields while retaining the classical foundations of a field of speciality.

It is apparent that students conscientiously learn a great deal during their university education yet lack experience in problem-solving.

- *Proposal for foundation of an international University*

An international technical university should be set up. This university will be unencumbered by national traditions and can focus on the requirements of the global village of the future. The international Engineering Association (WFEO) should create a program to set up a new technical university.

The future is now. However, our education systems have not yet adapted to contemporary needs and developments.

This international university must develop suitable curricula. Financing could come from the World Bank until this university develop their own infrastructure and dynamics. The money invested would be sustainable development in the best sense.

offered as electives. Time will not then be spent on inert pieces of information and at the same time the student will not be deprived from the ability to pursue further information should the need arise. It is also recommended not to teach analogue and digital electronics in complete isolation from each other. Bissel [6] suggested a strategy for change in control engineering curriculum, two of the suggestions are closely related to the discussion here: one is to design our courses according to what we want our students to be able to *do* rather than we want them to *know*, the other is to "invert" the curriculum by bringing the aspects of digital techniques first.

The problem discussed here may find its parallel in other branches of engineering and probably could be addressed in a similar way, i.e. not only relying on comparisons of course offerings of other institutions but also by getting input from industry. This approach can contribute to maintaining the quality of the curriculum by assuring its relevance to the current needs of industry. It is intriguing to observe that the need to examine the relevance of an engineering curriculum was a concern of engineering educators since the turn of the century, it is not a new problem arisen because of computers. In 1917, Becker [7] wrote: "Engineering education is largely conservative, sometimes too much so, and the danger is that we cannot bring ourselves to discard a course that has been established for years. No matter how desirable or valuable a subject may be, the decision to keep it in the curriculum or to discard it, must depend upon the number of courses that can be given and upon the relative value of each ... To avoid personal bias, the engineer in practical work should be freely consulted ... New ideas are suggested frequently and we should be ever ready to make use of those that are worthy, either in whole or by adaptation, even though the introduction may cause the dropping of courses which we have hitherto considered necessary".

REFERENCES

1. A. M. Ibrahim, "Current issues in engineering education quality", 1st Conference on Life Long Learning for Engineers, Glasgow, Scotland, May 1999.

2. R. A. Pease, "What's all this technical reading stuff, anyhow?", Electronic Design, 33-38, August 4, 1997.

3. C. H. Small, "Mixed-signal methods shift gears for tomorrow's systems-on-a-chip", Computer Design, 31-42, October 1997.

4. C. H. Small, "Mixed-signal HDLs unite two worlds", Computer Design, 44-48, October 1997.

5. A. M. Ibrahim, "Prepared for the information age?", IEEE Circuits and Devices Magazine, 4-5, 1998.

6. C. C. Bissell, "Control Education: Time for radical change?", IEEE Control Systems, 44-49, October 1999.

7. A. J. Becker, "On the engineering curriculum", Bulletin of the Society for the Promotion of Engineering Education, 3, 112-118, 1917.

The continuing professional development of environmental engineers and managers

G R JORDAN
Department of Mechanical and Manufacturing Engineering, University of Portsmouth, UK

SUMMARY

The Integrated Graduate Development Scheme in Environmental Engineering meets this requirement of educating needed technical and managerial personnel . The paper describes the structure of the well established programme and discusses the future development of the scheme to reflect the increasing importance of environmental management. The scheme is adequately equipped for the purpose of Continuous Professional Development, C Eng requirements and fulfils the integrated programme for a Master's degree.

1. INTRODUCTION

The contemporary climate of increasing environmental legislation from UK and EEC countries and the concomitant dearth of appropriate technical personnel has resulted in the need for increased. Training and education... It is well established also that significant economic benefits can be achieved by managing the "environment". As a result, a major requirement now exists for managers and engineers who have both the technical and management knowledge of the issues involved.

The IGD Scheme in Environmental Engineering was started at the University of Portsmouth in January 1993 as the first step in meeting this need. Over the years this scheme has developed successfully with a track record of training and development graduate engineers from a wide range of industries. In this time, the scheme has progressively developed to meet the changing needs of companies and to reflect the impact of new legislation, changing technology and the public demand that the environment is important and will become increasingly so. The original fundamental basis of the scheme has remained, however, and this is the firm foundation upon which the present modular programme rests.

This paper describes the scheme as it presently exists, its development from the initial concept and the future role that it is projected to play in providing CPD in the field of Environmental Engineering. The linkage between technology and management is explored

2. THE IGD SCHEME IN ENVIRONMENTAL ENGINEERING

The scheme is based like other such IGD schemes on a series of modular courses each lasting around one week. There are now over 50 schemes all of which are composed of a series of modules, many of similar duration to the present Portsmouth scheme but increasingly variants have appeared aimed principally at relieving the time burden of employers. This problem is overcome by a judicious flexible block tuition to suit the employer and the attendees. Distance learning has emerged to reduce time away from work, and also equipping the student to read for a Masters degree. The Environmental Engineering programme has essentially maintained the 5 day format since its inception. 24 one week modular training courses are available consisting of 6 core modules providing the fundamental understanding of the subject as expressed in such subjects as Sustainable Development air and water pollution, resource husbandry and environmental legislation. The course is therefore broadly based reflecting the need for an understanding of the interpreted nature of a business's activities on the environment. This approach has proven to be attractive to a variety of industries who do not require a specialised scheme in single issue area. The award of a higher qualification requires study of the 6 core modules, 4 modules from the chosen theme areas and two options from the remaining 12 modules. At the same time, the Group activity, EIA and auditing must be completed. This is an essential element of the course as it provides direct experience and application in an industrial environment.
The EIA group activity which has increasingly become concerned with EM Systems and ISO 14001 and EMAS standards featured highly. Energy conservation and management, recycling of materials, waste management are other examples where management techniques linked to knowledge and understanding of the technologies involved have become an essential feature of the scheme. Project work focusing on the management of the technology as opposed to detailed technical solutions were included.
Discussions with both the industrial companies, the external examiners for the courses and the academic staff involved, resulted in a scheme outlined below.

3. ENVIRONMENTAL IMPACT ASSESSMENT AND AUDITING

The 4 week long module introduced by specialists on EIA and Auditing, forms an essential part of the scheme. It is a Group activity whereby delegates conduct an Environmental Review or Audit of a company's activities. The aim is to develop expertise in identifying the impact of a company's operations on the environment either as part of an EM System that the company may have implemented, as part of an internal system or as the initial stage in obtaining accreditation by an EMS accredited body.
While the topics will have been introduced in previous modules, it is at this stage that the subject is examined in depth and to ensure that before the delegates visit companies they are knowledgeable and fully trained. Company visits on a number of occasions to facilitate the Audit, which will culminate in the presentation of a Report of the findings and a stand-up presentation to company management is given by course students. A total of over 30 companies have been "audited" or "reviewed" by this process which is regarded as a vital part of the programme and where theory is turned directly into practice.

4. ACCREDITED PRIOR LEARNING

It is possible for those with previous experience or who have attended recognised training courses to obtain exemption from certain modules. A maximum of 4 modules may receive exemption provided that a strict process is followed.

Despite the advantage offered to delegates to obtain exemption from up to 4 modules it is surprising that out of the 60 or so attendees on the course to date, only around 4 have taken advantage of this facility, with its reduced commitment for study away from work. The reasons for this are not clear, despite bringing it constantly to the delegates attention. Perhaps the process of module accreditation and building of a portfolio of credits towards CPD is not well appreciated; alternatively, this may reflect a need for up-to-date information and experience.

5. ASSESSMENT

The performance of a delegate on each week-long module is carried out by completion of a mini–project or case study which integrates the knowledge gained into a coherent and structured response. There are no written examinations in the conventional sense In some cases, shorter assignments are required but generally this refers to the introductory core modules.

Each module with its attendant assignment receives a credit value and acts as an independent statement which then forms part of a CPD portfolio(?). It represents a building block as part of the Environmental Engineering MSc course, or indeed for any other course.

6. PROJECTS

A major project of 6 months duration carried out at the industrial company of the delegate or possibly at another agreed organisation is an essential requirement for these delegates working to obtain the MSc award in Environmental Engineering. This is a considerable undertaking and proceeds as a partnership between the delegate and his/her company and the University.

The final project report counting towards 40% of the full MSc programme. In all cases a viva voce examination is required where the delegate must defend his activities and these are conducted by the academic project supervisor, the external examiner and an industrial expert.

It is at this stage that the delegate must show the sum total of the taught element of the course in an activity of direct benefit to the company and showing in-depth understanding and analysis.

7. THE ROLE OF MANAGEMENT

Increasingly over the 7 years that the course has been presented there has been a trend towards increasing the "management" content of the scheme. Management in this context means environmental management involving EMS, waste, energy and materials management and not business management as defined by finance and human resources. The demand from companies has been such that after considerable discussion a new pathway for those wishing to emphasise the management aspects rather than entirely technical considerations has been proposed.

Two pathways are possible with a common core of modules as previous. Two degree routes may be studied with two awards; an MSc in Environmental Engineering and an MSc in Environmental Engineering and Management. The latter is possible by a carefully chosen programme which includes additional modules in Risk Management and Occupational Health and Safety. In addition content of the ELA and Auditing module will be changed slightly to emphasise the increasing need for understanding of EMS, Auditing & Review, ISO 40001 and EMAS. While these subjects were included in the original programme, the emphasis will change at the expense of EIA which is primarily concerned with the planning stage.

8. CONCLUSION

The IGD Scheme in Environmental Engineering is now well established and has provided a wide range of engineers and managers with the knowledge and understanding of the "environment" to enable them to make significant contributions to their organisation's activities and indeed their own careers. Emphasis however has been given to the qualification and obtaining the degree of MSc in Environmental Engineering. While this aim is important and will continue it is important that the short course and CPD element is developed more strongly. The structure is in place and with the introduction of the environmental management pathway, the opportunity exists to provide a spectrum of courses in the technical engineering aspects and those relating to environmental management. The opportunity therefore exists to provide a range of opportunities for engineers and others to obtain professional development in a variety of ways either to achieve CEng status or as one off training courses. The University for its part needs to respond further to the difficulty facing employees with regard to the time commitment required for a 5 day course. Various alternatives are possible, including a 3 day programme with significant pre and post course work, and increasing content of distance learning materials. Whatever method of delivery suits the market, the foundations have been established of a scheme to meet the future needs of environmental engineers, environmental managers and others.

REFERENCES

(1)   International Conference, Maintenance. Quality-Reliability. Cambridge 1995. April 6-7th. G.R. Jordan and M.R.I Purvis.

(2)   International Conference. Planned Maintenance Reliability and Quality. Oxford 2-3rd April 1998. G.R. Jordan.

Table 1 - Course Structure - Environmental Engineering

Core Modules
C1   Sustainable Development and Global Environmental Problems
C2   Modelling of Environmental Systems
C3   Materials Usage and the Environment
C4   Air and Water Pollution
C5   Strategies for Resource Conservation and Recovery
C6   Environmental Legislation

Materials/Manufacture
M1   Environmental Design and Materials Processing.
M2   Materials Life Cycle
M3   The Recycling of Materials
M4   New Materials, Resources and the Environment

Energy
E1   Energy Conservation and Management
E2   Thermal Plant Performance and Combustion Generated Emissions
E3   Air Pollution Control
E4   Environmental Control - Air Conditioning and Noise

Water
W1   Potable Water, Treatment and Supply
W2   Waste Water Treatment and Disposal
W3   Industrial Effluent, Minimisation, treatment and Disposal
W4   Water Supply and Sanitation in Developing Countries

E1A   Environmental Impact Assessment and Auditing - Group modular activity.

Option Modules
O1   The Management and Control of Municipal and Industrial Solid Waste
O2   Sustainable Development and Agenda 21
O3   Contaminated Land and Re-use

Industrial Project - 23 weeks duration.

Table 2 - Environmental Engineering
Environmental Engineering and Management - proposed.

Core Modules

C1 Sustainable Development and Global Environmental Problems
C2 Modelling of Environmental Systems
C3 Materials Usage and the Environment
C4 Air and Water Pollution.
C5 Strategies for Resource Conservation and Recovery
C6 Environmental Legislation

Environmental Management Systems and Auditing - Group Activity

Environmental Engineering Pathway	Environmental Engineering and Management Pathway

Theme Areas:
- Water
- Energy
- Materials/Manufacture

Choice - Theme Area of 4 Modules

EM1 - Solid Waste Management
EM2 - Sustainable Development and Agenda 21
EM3 - Environmental Risk Assessment and Management
EM4 - Occupational Health and Safety

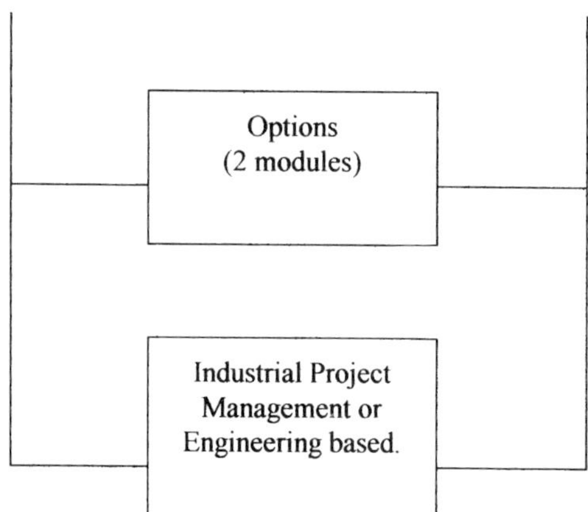

Optimization of teaching mathematics in a technical higher school

K KONOVALOVA
Department of Higher Mathematics, National Mining Academy, Dniepropetrovsk, Ukraine

ABSTRACT

The paper presents the substantiation of the approach to teaching mathematics to college-level students based on IDU which facilitates the development of the students' theoretical reasoning.

1. PECULIARITIES OF THE REAL TEACHING PROCESS

The process of teaching mathematics for students of a technical higher school has many aspects. We'll discuss one of them – psychophysiological. Necessity of its consideration is caused by the search of such approaches to teaching mathematics that would ensure a higher quality of perception, assimilation and reproduction of mathematical information by students. Students of a technical higher school are people of practical mentality. About 80 % of them prefer to use more often visual-active and concrete-figurative kinds of thinking in their lives. There exists a contradiction between a verbal form of teaching material presentation and students' reaction to a word. Therefore, definitions of mathematical notions, their characteristics, expressed by words, are perceived by students with great difficulty.

2. THEORETICAL CONCEPTS OF THE TEACHING PROCESS OPTIMIZATION

In our practice we proceed from the following characteristic of the process of thinking: the human brain processes any information by means of the staged system of functionally independent and subordinate to each other codes. First comes the code of sounds and symbols, then the code of words, the code of phrases and at last the code of meanings (1) and therefore we consider that educational information has to be presented in such a way that the lower stages of the coding system a priori provide its perception.

Besides, we use the theory of integrated didactic units (IDU) (2). The author treats the IDU as a unit of the educational process consisting of logically different elements having informational area (field) and proves that by using the given method in the teaching process, conditions are created for functioning of fundamental laws of thinking. In the psychological plan, the idea of IDU suggests that the previous and the subsequent chains of teaching material should have as many common exponents of information as possible beginning with the lower stages of the coding system.

3. METHODS OF FACILITATING THE PROCESS OF PERCEPTION AND ASSIMILATION OF THE MATHEMATICAL INFORMATION

3.1. METHOD 1. PRELIMINARY FAMILIARIZATION WITH TEACHING MATERIAL

Necessity of using the above-mentioned method is stipulated by the vital need of a person to know the whole picture of the development of events and to trace connections between specific event and the previous and subsequent events. In our opinion, it is expedient to familiarise students with the teaching material at the first lesson and to present it in the form of a multi-storied building – IDU-1 (Fig.1).

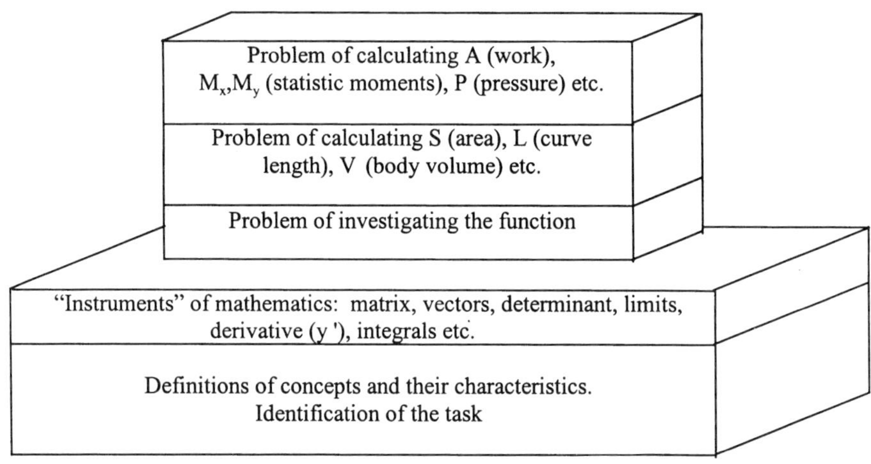

FIG.1. "BUILDING" (FRAGMENT)

Presentation of the teaching material should be based on the conceptual categories, which students know from their experience and to which they are accustomed. Let's quote one of the stories – commentaries to the drawing of the building: "Mathematics models and solves practical problems by its own means. Three types of objects can be distinguished among them: "instruments" for solving problems (placed in the upper layer of the foundation); problems themselves (placed on the floors of the "building"); certain definitions and characteristics of mathematical notions (placed in the lower layer of the foundation).Both instruments applied in the practical human life and "instruments" of mathematics are used in conformity with the problem a person wants to solve. The more complicated the problem is the more complicated instruments for its solution are. Ability to use the "instruments" implies assimilation of their definitions and characteristics, which in mathematics are reflected by hieroglyphic notes (formulae). There is a peculiarity in studying the "instruments" – each of the subsequent instruments rises from the previous one. It means that it is possible to assimilate the given "instrument" only after the "instruments" of lower stages had been learnt. The hierarchy of "instruments" and problems for the solution of which they are used are presented by IDU -2 (Fig.2)". This is the end of the story-commentary.

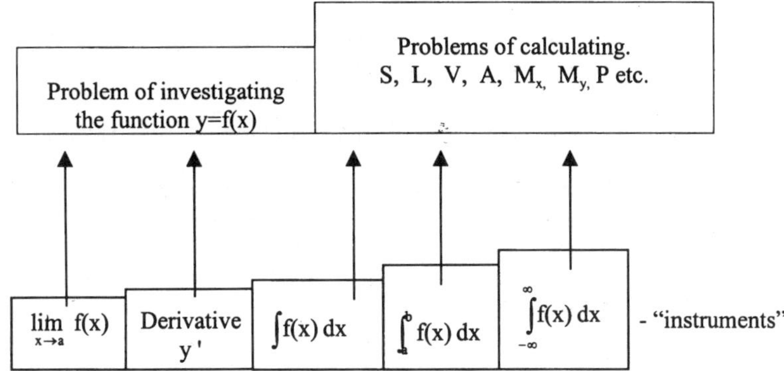

**FIG. 2. CONNECTIONS OF "INSTRUMENTS" AND PROBLEMS
(A FRAGMENT)**

What influence has the described method on the academic activities of students? First, introduced models of teaching material help students see it as an integrated system of mathematical notions, their characteristics and tasks. Second, students become aware that the process of learning should be continuous and it should proceed from simple to more complicated. Third, students experience psychological satisfaction from the possibility to see the ways by which they can learn mathematics from simple to difficult.

3.2. METHOD 2. MARKET CATEGORIES USED IN TEACHING MATHEMATICS
A civilised man understands well that any work he performs has a final product. In real life getting a job by a person is arranged in the form of an agreement on hiring in which a kind of job, a final product and payment are fixed. If the final product of the performed work doesn't correspond to the final product indicated in the agreement on hiring, the employee doesn't receive his wage or receives only a part of it.

Let's return to the mathematical activity of students of a technical higher school. Solution of any mathematical problem implies the final product, which corresponds to the condition of the problem. Problems and requirements to them form the system of setting tasks (SST) and the system of corresponding "final products" of solution (SFP). Because of poor reaction to a word students often incorrectly understand the essence of setting a problem and, therefore, substitute the given problem by another one. As a result of this, they receive the final product not corresponding to given requirements. A methodical and psychological problem exists: how to help students understand and learn the conformity of requirements to the problem and a "final product" of its solution?

We consider, it is necessary to cause the following associations with students: the call to get a job is a requirement to the problem. Soon the association begins to lead an independent life and before solving a problem each student asks himself the two questions:

- What job have I got?
- What final product must I receive?

These two questions induce students to apply the SST and SFP, trace the conformity of requirements to the problem and "final product" and, consequently, learn the material on a qualitatively higher level. Solution of this task implies application of corresponding mathematical "instrument", characteristics of which reflect formulae, that is equations. It is clear that left and right parts of formulae are interchangeable mathematical structures. A great number of those structures are called by us as a "bank of structures of the exchange fund (BSEF). Students' educational task is to remember structures of the exchange fund, increase their number in their memory, that is form the BSEF and at a moment's notice correctly exchange one structure for another.

Market categories –job, final product of job, bank of structures of the exchange fund – are perceived by students as natural notions of everyday life. Due to their use the teaching material is "humanised" in students' mind, therefore, its assimilation involves lower floors of the coding system.

Let's consider one more IDU-3 (Fig.3) which we use under the studying of the cognate area "Differential equations" the curriculum suggests . This IDU gives simultaneously the generalised comprehension of different kinds of equations of the given area and it also presents the material on the level of the code of symbols and words.

The first order equations	The second order equations
1. With the separable variables 2. Homogeneous 3. Linear 4. Bernoulli	1. $y'' = f(x)$ 2. $f(x, y', y'') = 0$ 3. $f(y, y', y'') = 0$ 4. $ay'' + by' + cy = 0$ 5. $ay'' + by' + cy = f(x)$

Fig.3. THE BLOCK-SCHEMES OF THE "DIFFERENTIAL EQUATIONS"

4. CONCLUSION

The use of IDU in the process of instruction creates favorable conditions for stimulating to resort to active operations: analysis, synthesis, comparisons etc. IDU are claimed to be a means of helping students to process the material of the study and thus to become aware of its systemic properties.

5. REFERENCES

1. Amosov N.M. (1965) Modelling of Thinking and Psyche. - Kiev, Publishing House «Naukova Dumka» 303p. (in Russian).
2. Erdniev P.M. (1978) Teaching Mathematics of School (From the experience of using the method of integration exercises). - Moskov, «Prosvechcheniye» 304p. (in Russian).

Methodological bases of engineering educational space

R B KVESKO, V G RUBANOV, and **T P GLUCHOVA**
Tomsk Polytechnical University, Russia
S B KVESKO
Tomsk Agricultural Institute at Novosibirsk Agrarian University, Russia

The methodological problems of engineering education are considered and the methods of engineering thinking formation are analysed in the paper. The correlation of the humanitarian and technical education, of personal and social factors, the role of humanisation of engineering education are analysed as well. The authors made an attempt to determine the criterion of engineering education, its forms and levels. The role of symbols in formation of an engineer as an expert and as a person, paradoxical character and efficiency of engineering education and engineering and creative thinking are considered.

Education and a human are inseparable. That's why it is impossible not to consider the problem of philosophical and methodological aspects of engineering education. Nowadays it is clear that we can't do without philosophical-methodological and sociological comprehending the engineering educational space phenomenon. Human environment and engineering are as important as human himself. Modelling, forming and developing a man in the process of education we should pay attention to the level of correlation between the human and environment and to their interrelations, to correspondence of needs, possibilities, aims, rights of a man with the quality and properties of society.

Using active methods in cognitive and educational sphere of engineering educational space it's possible to correlate the environmental requirements and the human inner world. In terms of engineering educational space one considers the degree and level of emotional, cognitive, educational, behavioural and social development and the activity of dynamic processes taking place within its subjects and objects.

Symbolic comprehension of the environment is necessary for cognition of reality. The symbolic knowledge form influences our world outlook, our intellect, our soul. Scientific ideas as symbols are the basic content of engineering education. A man is interested not in events but in these events symbols. The symbolic reality reflection is accompanied by awakening of our own understanding of processes and phenomena of the inner and surrounding world, it causes the human consciousness breakage, the interaction of the man and society. The increasing symbolic consciousness sounding is the engineering educational space magic, it directs a man to create, not only to change the world but to transform it with

"

the aim of saving nature, human, society. A symbol is a path of our urging to cognise the unknown, to create but not to destroy.

Modern level of understanding the problems of multi-level education has risen a number of methodological questions difficult for solution in terms of higher education. The education process formation under multi-level education conditions requires a theory of a student abilities development. The abilities developed in the education process represent a definite system within which some abilities actively interrelate with the environment, some are passive, some are neutral. One of the multi-level training goals is to transform potential abilities into actual ones. This can be done through formation of non-classical students' thinking. At present it's necessary not just to transform the education system but also to bring up a thinking man.

The education system transformation can't solve the problem of a new man formation without forming a new methodological approach to education system. The methodologization means to consider the education process as a cognitive activity system through a method. A method of science is just an educational potential and an important part in education is played by methods connected with non-classical thinking style - with understanding. Such methods should be the bases of all subjects teaching.

The method of understanding helps to create a scientific world picture and to master knowledge on the level of understanding. It's important not just to remember but to understand. We conceive scientific notions only in connection with the reality. The method of understanding is connected with the knowledge reflection and working out the non-standard principles of research and cognition. It includes such moments as: understanding the sense of actions and understanding the symbols sense and connection.

The material presentation by understanding method includes interpretation and illustration of the material and understanding different viewpoints on the problem. Interpretation is a main aim of understanding. It forms a man's conviction that he knows and understands the material. Only in this case understanding is possible and realisation of non-classical methods and forms of thinking of a student occurs.

Engineering education model can be a means helping to plan educational activity and control heuristic information. Having such model the educational process subjects are able to determine their activity, their aims, their life themselves.

The educational model of engineering education is a universal means of attracting the heuristic information for controlling and planning. It's possible to foresee these plans consequences before they begin to be realised in our life. This model creation enables to clear up the influence of different variants of the educational process subjects actions on education results, on the engineering education system efficiency, on the education level, on the education subjects activity and passivity.

The engineering educational space model consists of facts, suppositions, values connected with each other by definite relations which can be expressed mathematically. The information is introduced into the model and processed according the orders programmed. The model task is to describe the real educational process of a technical university and to predict future events influence on it. The educational models can give some advantages for better understanding the concrete educational goal as a result of the model designing and creation, of increasing the number of considered variants of future development of the engineering education by means of examining different situations, of time and finances saving.

The graphical model of the engineering education development made on the basis of the education history analysis shows its explainational, heuristic and methodological functions by interpreting the education past and present, by predicting its future. It's necessary, on our opinion, to take symmetry as the model creation basis for it's an arch-idea of the ordered educational space. It's this idea of order (the symmetry idea) which is the main one in an engineer's educational space, in his mind as the education subject. The symmetry in the engineering education as the educational process beauty is connected with the human harmony idea, with the unity of his needs, interests and real abilities in the world transformation. That's why, the symmetry principle is formational and substantial by creation of the engineering education process model.

As a result of the increased social determination of a subject actions and the aims of concrete individuals there appears the necessity of the humanitarian education system and structure, of creative persons developing. Education is the system of reproducing the essence of the society expressed in the intellectual potential. The society future depends on the level of the members humanity.

Maintaining the sequence of the humanitarization of higher technical education is very actual. The system of continuous multi-level education plays a very important part. The basic content of the humanitarian training of technical specialists in the engineering education process is the high professional level connected with self-reflection and practical realisation of the activity results and with the forming the responsibility for these results.

For extension of the humanitarian-technical educational space it is important to develop such human qualities which are beyond the pragmatic thinking of passive training and namely: imagination, reflection, value-personal orientations, activity, moral self-estimation. The person always tends to realise all that brings success and only success, that is most effective and successful.

Within the education space we deal with a person. The modern system of multi-level education makes it possible to form an appropriate multi-level structure of humanitarian-technical education. With the growth of social transformations the state of school becomes more complicated. The culture of the educational process subject becomes more and more isomorphic to the society culture, its individual existence is more and more connected with the socium fate, he is more socialised. A person as a representative of his individual culture determined by his inner life turns into a representative of the social culture. The cultural space of the society and a person becomes more and more isomorphic and synchronous. The person turns into an independent subject social-cultural action both his form and by his essence.

This global goal is achieved in the process of educational-methodical, scientific and educational activities of the teachers of humanitarian departments. Educational-methodical work is first of all connected with changes in approaches to the educational process itself. It should be economics and designs for the students to choose the methods of the student's inborn intellectual activities. Creation of mobile short courses sensitive to actual economic, social-economic and cultural problems favours this.

The formulated means of the information fixation are caused by striving to a strict formulation of the initial data, initial premises. Getting the objective result requires the application of the objective method. In this connection there increases the necessity in the formalisation for the humanities.

The use of the computer technique enables to create the visual image of the phenomenon on the base of the information characterising the object. This possibility enable to make sensible-visual base of the system of abstract characteristics. The application of the informatics makes it possible to carry out synthesis and the analysis, to make the data about the forms of connections between abstract definitions more concrete that exceeds the possibility of a human. A machine is able to give the object necessary for cognition in the form which is adequate to its other characteristics.

The transition from the abstract thinking to sensible-visual images is the basis for further development of the abstract knowledge by means of strengthening the connection between of the abstractions of a high degree of the community and abstracting and concrete-sensible perception. Due to this the possibilities of creating the conceptual systems connected with the images of sensible perception increase. The informational value of a sensible image in the process of cognitive activity of getting the economic and design knowledge is more and much more than the value of the complex conceptual system expressed in abstract notions. The dialectical connection between the sensible-visual and the abstract-conception images makes it possible to give the reliable support and enriches the characteristics of the system of scientific notions via essential ones for understanding their meanings.

A visual image is enriched by abstract characteristics and abstract notions are reflected in the form which is available for perception. This broadens the possibilities of giving and getting the knowledge. In this case we deal with the forms of united cognitive process in the economic and designs education.

A human being is not just a device for processing information. That's why, a very important factor for the development of the thinking culture of a cognition subject is the development of this emotional sphere by means of getting the humanities economic and designs knowledge. Reactions of a man are not simple and his joint relations with people and environment influences on his behaviour. A man has not strict determination his peculiarity is emotional experience. A man is a non-standard element of a very complex social system.

Pretoria Technikon's strategies to address the problems of inadequately prepared school leavers and the shortage of technicians and engineers in South Africa amidst the growing computer technology revolution worldwide

W J J KRUGER
Department of Mechanical Engineering, Pretoria Technikon, South Africa

ABSTRACT: The complexity of and sensitivity to change in educational systems world-wide are realities. All people live in the shadow of the information revolution which has been activated by the computer with its almost unlimited applications. In SA with its heterogeneous population and historical backlogs changes and adaptations are even more evident. As a developing country SA has high unemployment, and millions of people are either illiterate or products of an inadequate education system, aggravated by poverty. Pretoria Technikon has introduced bridging courses to redress the backlogs among school leavers and these have proved to be worthwhile.

1 THE NEW INFORMATION REVOLUTION IS NOW A NIGHTMARE.

Driven by the euphoria of the steam engine two hundred years ago the industrial revolution changed society beyond recognition. Today we are in the midst of the information revolution, driven by the mighty 'chip', with radical and far-reaching implications for the workplace and for mankind. Unemployment has reached a historic peak with almost one third of the world's economically active population of 3 billion without work. A major cause of this growing crisis is the technological revolution which has flooded the world. Fewer people are producing more and more.

The snowball effects of this revolution may well cause a world-wide depression. Another potential danger is that the wealthy, influential and computer-literate élite of the world may become the optimists who euphorically welcome the information revolution, while more conservative people (realists?) may become the pessimists who believe that man has brought the growing crisis upon himself by adopting a monster which may destroy him. There is a real danger that the information revolution will divide the world into two classes: the working class (the smaller component) and the unemployed class (the larger component). This is certainly a possibility to be considered.

2 TO BECOME PLAYERS RATHER THAN SPECTATORS EDUCATORS SHOULD PLAN THEIR STRATEGIES TOWARDS THE NEW REVOLUTION.

In 1998 a survey conducted by the Human Sciences Research Council in SA, to determine the level of knowledge of school leavers in Mathematics and Science, showed that SA had the lowest scores in a group of 22 countries (1). A change in our school system was therefore imperative. Our first attempt to address the backlog was to introduce 'mass education', but this resulted in alarming economic pressure on the state's frail economy. Learners are increasing at the rate of 4% per annum and there is already a serious imbalance between funds and numbers. Quality education and acceptable standards are therefore no more than a dream (2). The challenge facing us is to help young people to realize their highest potential so that they can end up in the 'working class'. There are, however, various problems which must be addressed:

2.1   Aging staff in academic institutions

Ten to twenty years ago well-trained educators were able to teach effectively up to the age of retirement (60). Now the scenario has changed completely because, with the revolution in information technology, human knowledge doubles every 18 months. As a result teachers who do not update their knowledge regularly, soon transfer only 'obsolete' technology and information to learners. Moreover, tertiary teachers who are near retirement are normally not keen to improve their qualifications or to update their knowledge through extra courses. We see this in increasing failure rates in subjects such as Mathematics, Science, Technology and Communication every year.

2.2   Available funds

Mass education of an acceptable standard is not possible without money, technological aids and logistical support. SA is already spending 22% of its budget on education. While the economic growth rate is about 0%, the imbalance between funds and numbers can only increase. The provision of schools and infra-structure is a government priority but success is limited by a lack of funds.

2.3   Abilities of school-leavers

Most of the school-leavers who apply for admission to tertiary institutions are very weak and not ready for further education. While admission requirements for first-year students remain high in order to maintain standards, there are fewer students who qualify for admission every year. There is no simple solution to the problem, but mass education and digital media, such as the internet, interactive television, computer-based education, video conferences and telematic education, are possibilities to be considered (3). These aids cannot replace teachers but can help to augment their knowledge and abilities. SA can use this world-wide revolution to address inequalities and backlogs.

3   DESPITE DISADVANTAGES INHERITED FROM PAST POLITICAL AND ECONOMIC STRUCTURES, SA EDUCATION CAN BE A SUCCESS STORY

Although SA is a developing country it is rich in terms of natural resources, except water. We urgently need to educate our youth to the highest level of their potential, but we have millions of uneducated, or poorly educated, people and an unemployment rate of more than 40%. The influx of immigrants (legal and illegal) from suffering neighbouring states, social and political instability and the increasing crime rate make the situation even worse. As an educator I feel that, if we refuse point blank to capitulate, we are committed to implementing different educational techniques for survival. Although 50% of our population are classified as being extremely poor we must do something to utilize our natural resources for the benefit of all our citizens. To achieve this we have to identify each Ramatueola Dansa, and give him our full support; then, who knows, he may become a person with the stature of Prof. Chris Barnard who performed the world's first heart transplant. Why not?

4   THE IMPLEMENTATION OF BRIDGING COURSES FOR STUDENTS WITH POOR RESULTS BUT HIGH POTENTIAL AT PRETORIA TECHNIKON.

The shortage of engineers, technologists, technicians and artisans is of national concern to government. Traditionally we have accommodated large intakes of students on level one with relatively high failure rates. Apart from the sad reflection on the standards maintained by our secondary school system, our own selection criteria had to be changed. The Bureau for Student Counselling, in collaboration with an established division in the Engineering Faculty, has developed a selection system which identifies students who have the potential to be successful, as well as those who are at risk. This has proved to be worthwhile. The test is done on application for enrolment and before registration. Students who perform well in the test are accepted and enrolled for the first level in engineering studies. Vast numbers of students do not make it, but the test shows that some of them have the potential for an engineering course. The rest of the paper will focus on the programme followed in 1997 to accommodate those students who were identified as having potential.

5.   THE COMPILATION OF OUR PRE-TECH PROGRAMME

To start this programme we asked ourselves what we had, where we could start and what we wanted to achieve. We kept in mind that these students often do not have access to a PC and that they often lack a culture of work and social ethics. Although some were indigent they were expected to make arrangements to pay for their own classes. We made use of some of our own lecturers but also of retired lecturers who have valuable experience. Through exposure to suitable role models and a sound institutional ethos during training we hoped to build an image designed to make them comfortable working in a multi-cultural society, and sensitive to diversity issues in this country.

The subjects we introduced to rectify their deficient background were: Mathematics, Mechanics, Technology, Engineering Drawing, Life Skills, Computer Skills and Communication. The syllabus for each subject was carefully planned and the standard was kept below first year level throughout. Ethics is an educational priority and we paid attention to it in the Life Skills programme. Popular topics in this programme were: motivation, setting of goals, time management, coping with stress, self-image development, assertiveness, communication, conflict management, work ethics, good behaviour, adaptability, code of conduct, preparation of a CV, job application strategies and job interviews. Though some students do not succeed, this course leaves them better prepared to fend for themselves in the open labour market.

Our test programme showed that most of these students have a weakness in communication skills, especially in English. The language course is therefore basically designed to enable students to think, speak and write clearly, to become conversant with the basic terminology in their chosen fields and to organize and present information logically. Consequently the main topics in the communication skills subject are: oral skills, reading skills, writing skills and corporate culture.

Teachers at historically black schools are often poorly trained and many are themselves unable to cope with the basic concepts in Mathematics. Consequently the syllabus for Mathematics consists of basic Mathematics in Algebra and Trigonometry only. Apart from the theoretical side the syllabus for Technology also makes provision for practical work like filing, arc welding, gas welding and the handling and mastering of different tools in Mechanical Technology. In Electrical Technology they are given the task of wiring a main distribution board with all the different devices in place, and some insight into the functions of each component. The Technikon is willing to use its resources in this way to mop up some of the spills of secondary schools.

What did we achieve? We started with a group of 33 students in semester one, 1997. During the semester we had an erosion of 6 students so that only 27 wrote the final examination. Thirteen students passed all their subjects, and being identified as reliable students, they were promoted to first year level in semester 2. We also had our normal intake during this semester and these students were used as a control group for comparison.

From the very beginning it was clear that the Pre-Tech group on average outperformed the control group, not only on the academic side but also in other areas such as, for example, their class attendance. Figure 1 gives a comparison between the academic performance of the Pre-Tech and control groups with respect to their predicate marks and final marks (class averages). Figure 2 gives a comparison between the percentage pass rate of the Pre-Tech group and the control group.

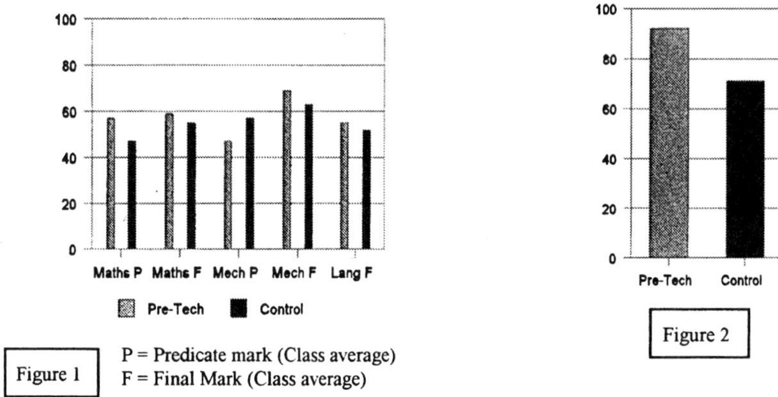

Figure 1

P = Predicate mark (Class average)
F = Final Mark (Class average)

Figure 2

Figure 2 shows that 12 of the 13 Pre-Tech students passed their first year to give them a pass rate of 92% while the pass rate for the control group was only 71%. The results of these students during their second and third years indicate that eventually 9 of the original 27 are going to get their diplomas within the prescribed time, a pass rate of 33,3%. Wonderful.

CONCLUSIONS: In view of these results we ran two Pre-Tech groups in 1998 and three in 1999, one each in Mechanical, Electrical and Civil Engineering respectively. Evaluation of Pre-Tech results seems to be a more valid criterion than final school results for entry into level 1. Implementing computer technology should be our next step. Identifying all the Ramatueolas in this country and making use of the positive aspects of computer technology will keep South Africa on the move.

REFERENCES:

1. Howie, S.J. & Hughes, S.A. *A HSRC document on Mathematics and Science Literacy of final-year school students in South Africa.* Pretoria, South Africa (1998).
2. Coetzee, C.E., Venter, J.H. & Voges, E.L. *A survey launched by the Department of Physical Science at Technikon Northern Gauteng, South Africa. The necessity of P.D. courses at Technikon Northern Gauteng, (1997).*
3. Green, M. *Transforming Higher Education.* American Council on Education, Oryx Press, Arizona, (1997).

John Arthur Roebuck PC, QC, MP (1802–1879): a social engineer of quality

R M LEDBETTER
Formerly Research Fellow, University of Sheffield, UK

ABSTRACT

He wanted an end to the House of Lords; he disliked political parties; he did not like Gladstone. Even more he wasted no opportunity in voicing his opinions. For a politician these attributes would go a long way to explain a lack of success or a short parliamentary career. For John Arthur Roebuck, however, they were no bar to an influential involvement in the affairs of this country for much of the half century between 1830 and 1880.

1   APPRENTICE POLITICIAN

1.1   India

John Arthur Roebuck was born in 1802, in Madras, where his father, Ebenezer, was a servant of the East India Company,which was building itself up as the dominant administrative, military and political force in India. In 1807 Ebenezer's wife, Zipporah, returned with their three Indian born children to London. On the day of their landing Ebenezer had died on a mission to improve the navigation of the river Godaveri. Barely five years old, Roebuck had lost his father and new influences began to shape the young mind.

1.2   London, and briefly Gumley

From 1807 until his teens Roebuck lived for most of the Napoleonic War years in London. He was the fifth born of six brothers and was the smallest and physically the weakest. Soon after arriving in London he sprained his right knee and in 1812 he aggravated the injury by a skating accident. The knee problem contributed to the weak health which dogged him through much of his life but it also meant that his education was mainly conducted at home by his mother and grandmother. English reading was the staple and he was given no Latin or Greek. His appetite for reading was unbounded and by the age of ten he knew Shakespeare by heart. He began to write under his mother's guidance. She had remarried and her new husband, Mr Simpson, was very kind but not very successful in his merchanting ventures. In 1813 Simpson tried his hand at farming in Gumley in Leicestershire.

1.3   Canada

The Battle of Waterloo in 1815 signalled a change for many Britons, not least the Roebuck-Simpsons, who were preparing to emigrate to Canada. They settled at Augusta on the Canadian north bank of the great river. John Roebuck grew stronger in Canada but could never match his brothers' physical exertions. Living for most of his time in a "dreamland" he continued to read and he taught himself French and Latin. By using his intellect he found that he could rule his family. Contemplating his distinguished scientific and literary ancestors he determined to return to England and to try his fortune at the bar.

1.4   Utilitarians

With fifty pounds in his pocket John Roebuck in 1824 headed for London. More important than the money was a letter of introduction to Thomas Love Peacock, Political Examiner at India House. Peacock in turn introduced Roebuck to his young colleague, John Mill. The young Roebuck was clearly impressed by the learning of the even younger Mill and he quickly became immersed in radical London concerns. He joined the Utilitarian Society founded by Bentham and Mill's father. If the bar had been Roebuck's ambition in 1824 by 1830 it had become a stepping stone to the greater bar of the House of Commons.

1.5   Radicals

Radical pressure for parliamentary reform had been mounting since the peace of 1815. The government's repressive policies only eased as economic difficulties began to lift in the early 1820s. In 1824 and 1825 the efforts of two notable radicals, Hume and Place, secured the Repeal of the Combination Laws which made it easier for workers to act together. Other radicals were pressing for prison reform, legal reform, removal of all religious and civil disabilities, state education, and many other changes but by 1830 it was parliamentary reform which was seen as "the first condition of reform in other spheres". A new monarch, a Whig government and serious riots in Bristol and Nottingham all prepared the ground for the eventual passage of the Great Reform Act in May 1832. For many in the middle and working classes it promised wholesale improvements. For Roebuck it brought a new career. Although he had been called to the bar political journalism was probably his main source of income. For some years he had been training himself for a life in politics, learning the art of public speaking and making for himself a "political scheme". In 1832 he was elected MP for Bath.

2   MEMBER OF PARLIAMENT

2.1   MP for Bath, 1832-1837. Education for the People

As MP for Bath Roebuck made it clear that he was "determined to advocate that which I believed for the interests of the people without regard to party considerations". He wasted no time in launching a barbed attack on the Irish Secretary's coercion policy. Also early in the session, in an "able and luminous speech", he requested a select committee to work out how a universal and national education system for the whole people could be set up. On this matter the government was prepared to listen and in 1833 they granted £20,000 annually for the purposes of elementary education. A much larger sum was spent on the royal stables in that year but today's education budget indicates the importance of this first acceptance of parliament's responsibility for the education of its people.

2.2   Pamphlets for the People

Outside parliament Roebuck was keen to educate people about political issues not least because he did not trust the newspapers to do it. Starting in May 1835 he published a series of pamphlets. They represent a valuable source of radical and utilitarian thought but their cessation in March 1836 was a blow to Roebuck. The pamphlets for the people had failed to galvanise radical opinion into supporting his views. They would have confirmed many people in their opinion of Roebuck as a firebrand who could burn friend or foe at the stake of his ardent principles. Some applauded him but too few of his intended audience bought the pamphlets to ensure their continuation.

2.3   The People's Charter

Although Whig achievements included reform of the suffrage, factories, education, town government, many working men had hoped for much more. William Lovett, founder of the London Working Men's Association sought the help of Place and Roebuck in drawing up a People's Charter containing six points:

1.	The vote for every man 21 years of age	1918
2.	Secret ballot	1872
3.	No property qualification for MPs	1858
4.	Payment for MPs	1911
5.	Equal constituencies	1885
6.	Annual parliaments	?

The dates, shown above, when the first five points of the Charter became law, are a good measure of how advanced the Chartists (and Roebuck) were in their political demands. The Chartists' first petition to parliament in 1839 was rejected but worsening economic conditions in the early 1840s swelled Chartist numbers and the movement's leadership passed from "moral force" to "physical force" Chartists like O'Connor. When the Chartists unsuccessfully presented their second petition to parliament in 1842 Roebuck denounced O'Connor as a "cowardly and malignant demagogue". Roebuck and the Chartists parted company.

3   BEYOND THESE SHORES

3.1   Canada

Lower Canada had been in a disturbed state since 1828. Roebuck , in 1834, had highlighted the problems there and in 1835 the House of Assembly for Lower Canada appointed him as their agent. The British government agreed to a committee of enquiry but this was not enough to avert a state of open rebellion by 1838. Although no longer an MP Roebuck was called before the bar of parliament to give his expert opinion of the Canada Bill. In 1840 Upper and Lower Canada were given legislative independence and by 1843 peace had returned.

3.2   India. MP for Bath, 1841-1847

In 1841 Roebuck returned as MP for Bath. For some his best speech in parliament concerned the First Afghan War (1839-1842), which was fought to protect India from Russian ambitions. He was backing the policy of the Tory government of Robert Peel. He had already praised the Prime Minister's income tax of 7d as a "really democratic measure". Not for the first time Roebuck's apparently Tory leanings caused dismay to his liberal supporters. In 1847 the final break with Bath came but other constituencies showed considerable interest in an advocate who spoke his mind both forcefully and eloquently. Glasgow might have secured him but fittingly it was Sheffield that claimed him as their MP in 1849.

3.3   The Crimea. MP for Sheffield

London was always Roebuck's real home but in March 1841 he was in York earning his living as a barrister. He wrote to his wife "I shall be in Sheffield late Saturday. York, however, is more pleasant than the smoke from 10,000 furnaces". When Roebuck became MP for Sheffield in 1849 the town was growing rapidly. In 1851 Sheffield manufacturers were proudly displaying their cutlery, tools and metal products to an admiring world in Joseph Paxton's Crystal Palace. The proud, self-confident town had a strong radical tradition and so had several things in common with its new MP. An early intervention in parliament for his new seat was Roebuck's vote of confidence for Palmerston's robust foreign policy. When Russia's equally vigorous policy against an ailing Turkey spilled into a great power conflict in the Crimea Roebuck defended Britain's involvement against Russia. The war, however, began badly for Britain and this led to Roebuck's most celebrated appearance in parliament in January 1855.

3.4   The Fall of Aberdeen's Ministry

When Roebuck asked parliament for a select committee to investigate the state of the army he barely needed all his oratorical powers to expose the failings of the government. Public opinion already needed much more than reassurance and when the government lost a confidence vote Aberdeen's Whig-Peelite ministry was at an end.   Palmerston, the only credible leader of a new government, had to let Roebuck chair the Sebastopol Committee which had wide terms of reference in its enquiry into the causes of Britain's poor state of readiness for war. Questions of army reform and administrative reform came hot on its trail.

4   SHEFFIELD OUTRAGES AND ROEBUCK OUSTED

By the 1860s mid-Victorian prosperity was improving the lives of many of the working classes and trade unionists were increasingly regarded as responsible citizens who deserved the vote. However, in October 1866, the house of a Sheffield saw grinder was badly damaged in a gunpowder "outrage". The local press and then the national press put the whole trade union movement under the spotlight and then in January 1867 a legal judgement threatened the safety of union funds. The government responded to these concerns by setting up a Royal Commission in February 1867. Roebuck was one of the eleven commissioners appointed to carry out the investigation. Three other commissioners were warranted to enquire into the Sheffield outrages. When the reports of the national and the Sheffield commissions were published they were generally favourable to trade unions and they led to the 1871 and 1874 legislation which considerably improved the status of unions. In the August of 1867 working men in towns were given the vote by Disraeli's Reform Act. When the newly enfranchised Sheffield workers voted in the strongly contested election of 1868 they helped to swing the result in favour of Mundella, a strong Liberal supporter of Gladstone.

5   THE FINAL YEARS

Although out of parliament during Gladstone's ministry, 1868-1874, Roebuck must have gained some pleasure in the passing of the Education Act of 1870 which took his scheme of national and universal education an important step closer to completion. Another cherished aim, the secret ballot, was fulfilled by this ministry in 1872. In his early seventies his sight began to fail and hearing became more difficult but his spirit for politics remained strong. In the 1874 election Sheffield crowned Roebuck's political career by choosing him again. His weaker health prevented him from making many appearances in parliament but his utterances were still forthright. To Queen Victoria he was a "true patriot" and she made him a Privy Councillor. He last visited Sheffield in 1879 to open a new asylum for the local Licensed Victuallers' Association - he had always strongly opposed temperance legislation. On the 21 November he apologised to the Lord Mayor of London for his absence at a meeting to promote the memory of his deceased friend, Sir Rowland Hill. Within ten days he too was dead.

Woundcare and infection control reliability – an engineering challenge

D C MITCHELL
Research and Development, Maersk Medical Limited, UK

ABSTRACT

It is well documented that micro-organisms are winning the infection war[1]. The era of guaranteed antibiotic effectiveness is over and those antibiotics that remain effective must be reserved for cases of greatest need. Totally new strains of antibiotics are not expected before 2005-2010. Wound care must therefor take advantage of alternative technologies to offer the best patient care programmes. Arglaes, a new wound care technology, offers effective bacterial control without resistance build-up, tissue reaction or other adverse side effects. Arglaes combines best engineering and medical practice to offer effective and reliable wound care with infection control of resistant strains of bacteria into the next millennium.

1.0 INTRODUCTION

It is no longer possible to guarantee that patients will remain infection free during their course of treatment in the hospital or community environment. Nosicomial infections (those acquired in a hospital) are now common occurrences not only in those patients who are severely metabolically or immunologically impaired but also increasingly in patients undergoing minor normally low risk procedures. The risks have further increased in recent years with the increasing development of strains of infecting organisms resistant to normal antibiotic therapies. The incidence of nosocomial infections resistant to vancomycin has risen from 0.3% in 1989 to 14.2% in 1996[2]. It is now accepted that continued widespread use of antibiotics must be reduced to preserve the efficacy of those drugs that remain active. The role of routine wound infection control must be taken on by alternative means.

There are many methods of attaining effective infection control but these have associated problems; high labour cost due to the need for repeated re-application, allergic reactions from the patient, limited spectrum of efficacy and of course the development of strain resistance.
One material, silver, has shown by its use and effectiveness over many years that it can successfully address many of the routine infection control needs but it still requires frequent re-application as it is rapidly neutralised by tissue fluids. Recent research has shown that silver in ionic dissociation is more effective than the silver itself, however, this again suffers from rapid elimination. If small amounts of ionic silver can be constantly made available at the wound interface then the efficacy of the antimicrobial action can be maintained. In addition the amount of ionic silver required is very small indeed. *Arglaes technology* offers just such an option.

2.0 ARGLAES TECHNOLOGY

Arglaes offers an alternative route to effective infection control without the need to resort to conventional topical treatments and the prophylactic use of antibiotic, *but what is Arglaes?*

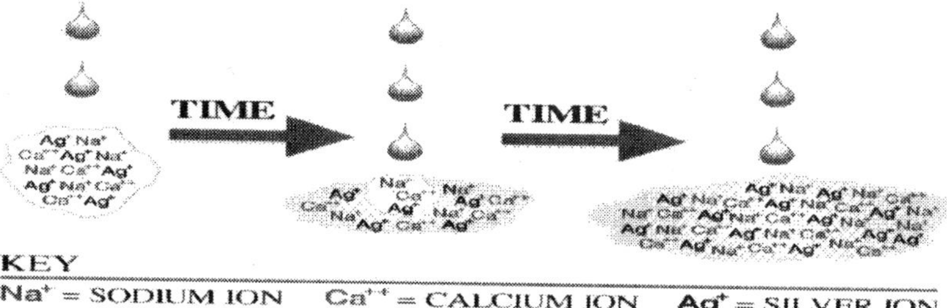

Figure 1 Arglaes dissolves in water or water vapour

Arglaes is an inorganic polymer of sodium and calcium phosphates fused in a furnace at high temperatures (in excess of 1000^0C) to produce a glass like polymer just as window glass is a fusion of silica materials. *Arglaes* has one very important property; it is soluble in water. This property combined with the ability of glasses to be seeded with other materials, in this case a silver salt allows the creation of a material capable of releasing silver as it dissolves. If the rates of dissolution and the threshold (moisture quantity) of initiation can be predetermined and set during manufacture then it is possible to create an infection control barrier dressing material of known performance and efficacy duration. *Arglaes* film dressings are the result of such a development, offering infection control without the risk of bacterial resistance build-up, but are they effective? Laboratory studies[3] have shown that Arglaes has a sustained antibacterial effect on a range of micro-organisms including methicillin resistant staphylococcus aureus. Staphylococcus aureus is the most commonly occurring organism infecting man. In all cases a sustained inhibition of bacterial growth was achieved. Figure 1 shows the result for staph. aureus inhibition over an extended period. The time studies were extended to cover a range of organisms. In all cases a sustained inhibition was observed over a seven-day period[4]. These studies and others indicate that a technology developed for other engineering and veterinary applications can have a beneficial effect directly upon the problems associated with bacterial control.

ORGANISM	24 Hours			48 Hours			7 days		
	Arglaes	Iodine (Povidone Iodine)	Penitility Control	Arglaes	Iodine (Povidone Iodine)	Penitility Control	Arglaes	Iodine (Povidone Iodine)	Penitility Control
Proteus vulgaris	11.3	12.1	+++	10.2	5.2	+++	9.3	2.9	+++
Acinetobacter bawnanii	14.8	6.8	+++	10.2	4.0	+++	10.5	3.8	+++
Enterococcus faecium	9.2	0.0	+++	8.8	0.0	+++	9.5	1.2	+++
Serratia marcescens	9.0	4.4	+++	7.5	0.0	+++	7.3	0.0	+++
Candida albicans	13.5	15.1	+++	13.1	8.2	+++	15.6	3.8	+++
Pseudomonas aeruginosa	9.7	1.4	+++	8.9	0.0	+++	8.0	0.0	+++
Proteus mirabilis	8.8	6.9	+++	8.1	5.7	+++	9.9	5.9	+++
Staphylococcus aureus	14.7	0.0	+++	13.5	0.0	+++	13.4	9.0	+++
Escherichia coli	7.5	0.7	+++	6.9	0.6	+++	6.7	0.0	+++
Enterobacter cloacae	5.4	8.0	+++	5.1	0.0	+++	4.8	0.0	+++
Staphylococcus aureus (MRSA, NCTC 12493)	20.0	10.6	+++	20.5	0.0	+++	20.3	0.0	+++
Klebsiella Edwardsi var Edwardsii	11.3	12.1	+++	10.2	5.2	+++	9.3	2.9	+++
Staphylococcus aureus (MRSA, NCTC 122.32)	13.9	2.6	+++	12.8	1.7	+++	12.5	1.5	+++
Staphylococcus epidermis	16.0	0.0	+++	15.7	0.0	+++	22.1	0.0	+++

Table 1. Sustained effect over 7 days

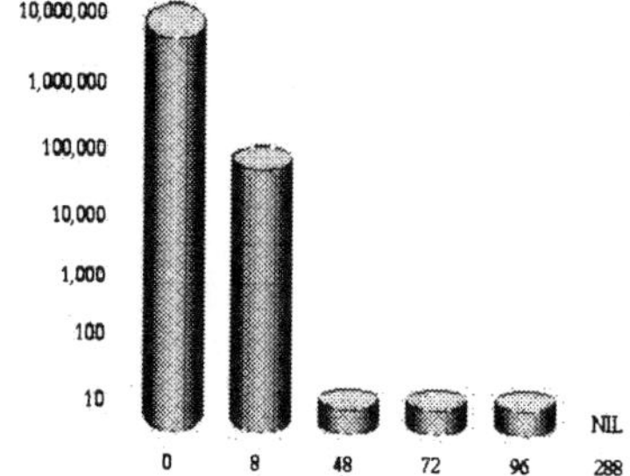

Figure 2. Colony count decreases with time

In addition to the laboratory studies for preservative and barrier properties, a full spectrum of toxicity and cytotoxicity studies[5] have shown that the Arglaes dressings do not exhibit any toxic effects when applied directly to wounds of many types. This technology when incorporated into woundcare dressings offers the control of bacteria in wounds on man without the spectre of 'super bug' development and spread.

In the UK and Europe extensive clinical evaluations have been undertaken studying a number of parameters. These have shown the dressings to be effective in reducing the incidence of acquired infection in a range of wound types. In one study, where *Arglaes* film dressings were compared to a placebo in a double blind randomised multi centre study, the results indicated that the percentage of patients with microbiologically confirmed wound sepsis in the *Arglaes* group was 2% compared with 8% in the placebo group[6]. A randomised multicenter crossover study against a commercially available film dressing showed the percentage of patients with microbiologically confirmed wound sepsis to be 3.6% in the *Arglaes* group compared with 11.3% in the competitor group[7] (Figure 4). In an evaluation of the *Arglaes* film dressing in a MRSA positive (heavy growth over an extended period) patient the growth of MRSA isolate was reduced to small to scanty growth in 8 days and to isolated scant growth in a further 24 days[8]. Further follow-up evaluations in 164 patients have shown similar results to those reported in the earlier clinical trials[9] with users reporting Arglaes to perform better than the dressings they normally use (Figure 3).

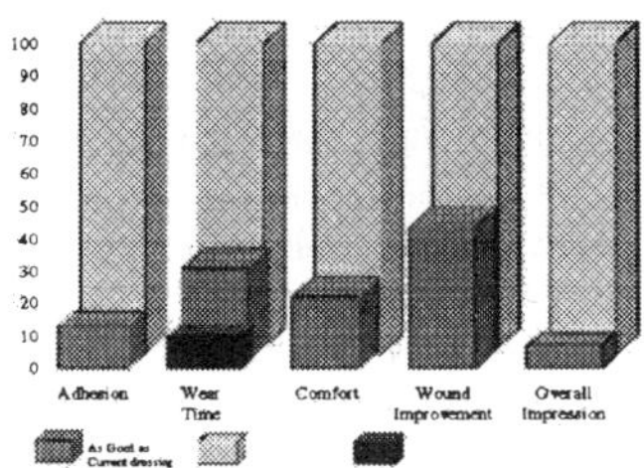

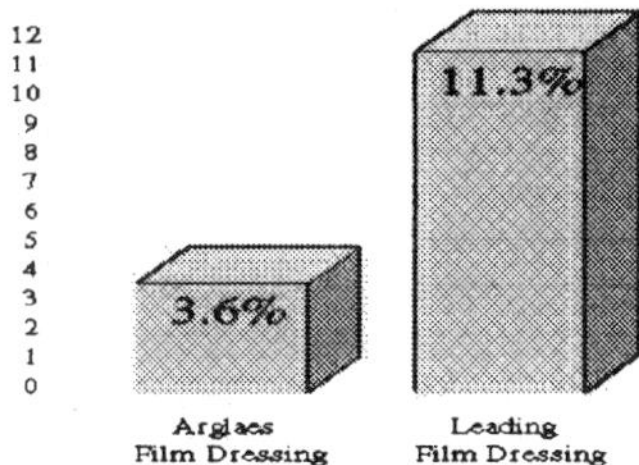

Figure 3. Better than, as good as or worse than existing dressing used.

Figure 4. Infection rate as a % of patients treated.

Two recent evaluations carried out at the Overton Brooks V.A. Medical Centre, Shrevport, USA[10,11], used Arglaes dressing on a number of cases where other dressings had not succeeded in promoting wound healing and where infection or the risk of infection was present. In all cases the beneficial effects of Arglaes as an anti-microbial barrier wound dressing were demonstrated and the wounds progressed rapidly to a healed state.

3.0 DISCUSSION

The need for the protection of wounds against the ingress of infecting agents is an essential feature of modern wound healing. This must now and for the foreseeable future be achieved without resorting to antibiotic therapy whenever possible if we are to preserve those that remain efficacious for use in essential cases. The most common infecting agent in man is staphylococcus aureus and this bacteria has learned by frequent exposure to antibiotics how to negate the efficacy of antibiotic therapy. In 1989-1991 the recorded incidence of infection from strains of staphylococcus aureaus resistant to the common antibiotics derived from methicilin was 1.5% of isolates examined by 1996 this had risen to 13.2% and continues to increase[12].

Bacteria have one further weapon in their armoury, they can transfer the knowledge of how to overcome antibiotics between different species of bacteria. In this manner a bacteria resistant to our most powerful antibiotic vancomycin (vancomycin resistant Eschericia coli) could transfer this ability to the most commonly occurring bacteria (staphylococcus aureus) to created a resistant super strain. Such cases have already been reported.

We must now look to combining our knowledge and expertise in chemistry, material science, engineering, manufacturing, bioengineering and medicine to ensure that the best patient care programmes can be provided at an acceptable cost offering safe and consistent quality infection free treatments well into the next millennium.

If in addition, the agent used can protect the environment and the carer when already infected wounds are dressed without the risk of further development of resistant strains of micro-organisms, then it must represent the preferred route forward. *Arglaes* technology heralds the next generation in woundcare offering the ideal wound healing environment with effective infection control beyond 2000.

4.0 REFERENCES

1. Souhrada l., Battle Against Bacteria. Materials Management 1997, 6(9); 31

2. National Nosocomial Surveillance System. Centres for Disease Control and Prevention. USA. 1996.

3. Gilchrist T. Silver In Infection Control. Tissue Viability Conf. Mar 1996.

4. Mitchell D. C. The Role of Bacterial Control in Woundcare Beyond 20000. Proceedings of the Symposium on Wound Healing : Science and Industry. St. Thomas Dec 3-5, 1998.

5. Studies Carried out as part of 510K Submission. Data on File, Maersk Medical Limited., Dec. 1996.

6. Double Blind Randomised Multi Centre Study of Arglaes Dressing v's Placebo. Data on File, Maersk Medical Limited., 1995/1996.

7. Randomised Crossover Multi Centre Study of Arglaes Film Dressing v's Standard Film Dressing Data on File, Maersk Medical Limited.

8. Holttum P., Case Study., Tissue Viability Conf., March 1996.

9. Multi Centre Evaluations of Arglaes Dressing v's Standard treatment. Data on File. Maersk Medical Limited.

10. Evans J.T. and Falconio-West M. Improved Healing of Homecare Patients `Undergoing Wound Therapy. Proceedings 13th Annual Clinical Symposium on Wound Care. Atlanta USA, October 1998.

11. Evans J.T. and Falconio-West M., Healing of Vascular Ulcers in Frail Elderly Patients. Proceedings 13th. Annual Clinical Symposium on Wound Care., Atlanta, USA. October 1998.

12. Speller D. C. E., Johnson A. P., James D., Marples R. R., Charlett H., George R. C., Resistance to Methicilin and Other Antibiotics in Isolates of Staphylococcusaureus from Blood and Cerebrospinal Fluid, England and Wales 1989-95. The Lancet 1997 vol. 350; 323-325.

CD-ROM training techniques of QRM

J M LEESON and **A F M OWENS**
Integrated Engineering Solutions, Sheffield, UK

1. THE NEED

The use of interactive CD-ROM based training packages can provide International Companies the opportunity to establish a common, and controllable programme that will ensure a minimum assessed capability of all it's employees. For companies with a diversified workforce, training to keep up to date with new equipment, systems, updates, service and maintenance can be daunting, as well as very expensive.

Many factors have to be taken into account – including:
- Who needs the training?
- When do they need training?
- How long is required to complete the training?
- Who runs the training? (Own Staff or External Provider)
- Where is the training best completed? (In-House or External)
- What is the skill level of the trainees?
- How much knowledge will be retained after the training?
- What temporary cover is required for staff attending a training course?
- What effect will this have on the business?
- What is the overall cost? (Staff, hotels, travelling, expenses etc., etc., etc.)

2. CONVENTIONAL TRAINING

In many instances where only a few people need to be trained in a specific subject then conventional educational establishments and training providers can usually give a satisfactory service. This can be cost effective and take place over a given period of time utilising the basic instructional techniques.

However due to this timed and structured approach the training may not take into account the difference between the individuals ability to absorb and retain all the knowledge being "thrust" at them.

It can therefore take longer for some people to understand certain aspects of technical detail than another person in the same group. The inevitable variable learning curve can create a difficulty for both the instructor and pupil.

Problems can also occur when the trainee returns to the company to carry out their normal duties when "a little knowledge can be a dangerous thing" – not only to themselves but also to their colleagues.

3. CD-ROM TRAINING CONCEPT

Most modern computers have integral CD – ROM drives allowing the user to view items such as brochures and catalogues, which are increasingly being produced in electronic format. This gives the user the ability to review the contents whenever they wish at their office desk.

With this capability in mind, ECONOMATICS, in conjunction with an International Company involved with Motion and Control Systems, developed, and produced a CD-ROM interactive training programme covering Pneumatic Technology. The resultant success of the programme, led to a further contract to produce a CD-ROM covering Hydraulic Technology.

The benefits of interactive CD-ROM technology to the client allows them to:
- Control the content of the training programme world-wide
- Ensure a common base of knowledge to all appropriate staff
- Identify which members of staff should complete the training
- Include "specifics" used by the company
- Control the time period to complete the course
- Carry out regular assessments
- Release updates whenever necessary
- Use selected sections for new staff induction (Company overview)

The benefits of CD-ROM technology allows the user to:
- Work at a pace they are comfortable with
- Work at a time and place that is acceptable to them
- Work methodically through the programme - which requires minimal computer knowledge
- Repeat sections or individual items that they are unsure of
- Move backwards and forwards through the programme to underpin their knowledge
- Take a refresher course at any time
- Use the information to develop their technical skills

3. CD-ROM FEATURES

There are near limitless features on today's CD-ROMS which provide an ever-developing role in training. The use of text along side an animated diagram or video of the subject matter allows the user to gain a far better and quicker understanding of the subject.

The features available are a mix of the following and include:
- Text (used to describe a subject in the least number of words)
- Photographs
- Static Graphics (2D and 3D)
- Animated Graphics (2D and 3D)
- Interactive Graphics
- Video Graphics
- Film and Video footage
- Student assessments
- Checks on student progress

- Test questions
- A final Master test (with results fd back to the parent company)
- Real world control interfaces

4. CD-ROM's in the area of Quality, Reliability, and Effective Maintenance Practice

The CD-ROM can be an invaluable source of information in all areas of Quality, Reliability and Maintenance.

The ability to be able to store and send a vast amount of information around the word on a CD-ROM, in any language, allows the user to induce a corporate standard for all their Maintenance staff. This ensures the quality of maintenance practices for their equipment is the same in Oxford as in Orlando or Okinawa.

Specific photographic and graphic support can emphasise a critical engineering point in the equipment that is the subject of the maintenance programme. Video can take the engineer through from the general operation of the equipment to the correct maintenance techniques and safety procedures required for a specific piece of equipment or system.

Video Graphics allow a more sophisticated way of showing how items work, how they are sectioned or how they can be taken apart and reassembled in simple stages. The CD-ROM approach can save many hours of instructor and engineer time if it has to be practically demonstrated in a workshop.

It would almost be inconceivable to consider repeating the exercise overseas, at the frequency required to ensure all engineers joining the company have the same knowledge base.

One CD-ROM programme could cover all these aspects at a fraction of the cost.

5. CD-ROM Project Management

To produce a professional programme the project must be managed from concept to final production by an individual who knows the complexities of the technical subject in question and can communicate and translate these requirements to a third party.

A CD-ROM training programme is best developed by an experienced Project Manager. Although he is not necessarily required to write any software programmes, as this is best handled through a specialist Multi-Media production company, the Project Manager must have an intimate knowledge of the various computer and video techniques that can be effectively used to support the programme.

The script for the programme is probably the most important aspect of the CD-ROM. This represents the foundation on which the whole programme is based. The text must be kept to a minimum, and used in close support of the graphics and video imagery which will cover the subject the shortest possible time.

Communication between the Client and the Project Manager, and the Project Manager and the Multi-Media Company must be crystal clear and precise, to ensure the end product meets the full requirements of the customer. This detail should be revisited throughout the development programme to ensure that the project achieves its original objective.

6. The Programme

The Programme will normally be divided into Modules, sub-divided into relevant sections covering individual subjects. Each section starts with an aim or objective and concludes with a summary. This ensures that the learner understands what is expected of them, and how this is to be achieved. The inclusion of a questionnaire at the end of each section and module underpins the knowledge gained from that particular section and module.

The Final Module is the master test, normally of 100 questions selected at random from all the questions in the sections and modules. Once into this Module the learner cannot exit, and is presented with a final score given as a percentage. The computer software enables the learner's manager to keep track of their progress, and the result of the final test.

7. Conclusion

Bodies such as the UK City and Guilds and The Business and Technical Engineering Council (BTEC), accept this method of training for accreditation, and in some instances it can be used in conjunction with obtaining a National Vocational Qualification (NVQ).

As training becomes more relevant to employees, especially in ensuring QRM, the CD-ROM method of instruction is becoming widely accepted and is very cost effective. It is particularly appropriate where a number of staff require training at different intervals, where the logistics and operating cost of a conventional training programme would generate large costs, and be complex to manage.

The versatility of CD-ROM Programmes allow them to be:
- Used at a Time, Place, and Pace to suit the individual.
- Viewed over and over again.
- Have obsolete or out of date information removed
- Have new material added
- Translated into any language.
- Be used by all company employees irrespective of their position - from the receptionist for company awareness, marketing for understanding of the product, sales to enhance their product knowledge etc. etc.
- Have the images produced into hard copy for publications, publicity, technical manuals

Once the Master CD-ROM has been produced literally 100's of CD's can be produced for under a pound (£1) each, possibly making it one of the most cost-effective methods of training World-wide.

J.M LEESON IEng. MIPlantE   A.F.M.OWENS MMS Euro.I.E.

CD-ROM laboratory versus internet laboratory as an aid in educational programme

L REFEROWSKI, R ROSKOSZ, and **D SWISULSKI**
Electrical and Control Engineering Faculty, Technical University of Gdansk, Poland

ABSTRACTS

Presented in this paper CD-ROM laboratory is a useful tool, which enables students to arrange better preparation for the tasks they have to perform in the university laboratory of electrical measurements. This CD-ROM enables them to perform, with virtual instruments, the same measurements and tasks that they will do in the university laboratory. With CD-ROM they have more time to think about certain problems they have met during the measuring session. Opposed to this solution is the Internet measuring laboratory. Indeed our students have powerful computers at their homes but only several of them have access to the Internet. Our CD-ROM laboratory is based on LabVIEW of National Instruments, but the successive measuring task can be perform with the computers equipped only in the Windows environment. In the paper are discussed the advantages and disadvantages of presented laboratory, practical solutions of certain stands as well as the conclusions, which can be drowned by the students after measuring session.

1. INTRODUCTION

In the last decade of the twentieth century it possible to notice the dynamic development of Internet Laboratories (IL) prepared by numerous teams from different technical universities and used by students for educational purposes (1, 2). These kinds of laboratories may form very convenient educational tool, in the case of so-called "weekend studies", now very popular in Poland. During their studies the students have contact with educational institution only on Friday, Saturday and Sunday and after five years of studies they obtain the BS (EE) diploma. But in order to take advantages of the IL it is necessary to have access to the Internet. Till now small group of our students fulfils this condition. Of course it is possible to obtain necessary access by modem but in contradistinction to USA in Poland as well in Europe local connections are payable and it is necessary to underline that the Polish telephone rates are rather expensive. In order to surmount these difficulties our team from TUG decided to elaborate the CD-ROM Virtual Laboratory. of Electrical Measurements.

2. GENERAL INFORMATION

The main assumptions of the CD-ROM Virtual Laboratory of Electrical Measurements (CD-ROM VLEM) are as follow:

- The CD-ROM VLEM must give the opportunity to exercise the same tasks and to solve problems that will be later realised by the students in the university laboratory (UL),
- The CD-ROM VLEM have to be used with PCs equipped only in Windows environment.
- The CD-ROM VLEM must not be treated as a substitutional university laboratory but as an aid tool for preparation to the tasks, which will be performed in the genuine university laboratory (UL). Our students, the future engineers, must have contact with real measuring technique.

The first and the second condition force to select the software, which will enable to create virtual instruments similar to the instruments, used in the university laboratory. Furthermore which enable to exercise them only in Windows environment. This condition is due to the fact that nearly all of our students have very good PC equipment at home but they have financial problems in buying special qualified software.

After consideration we decided to base our CD-ROM VLEM on the software LabVIEW of National Instruments, which we have used for last six years in our research laboratory. Its main advantages are as follow (3, 4):

- it has very comprehensive library with many examples and for this reason the productivity of designing adequate instruments is 5 to 10 times better in comparison with the software based on conventional programming language,
- its built in graphical user interface is intuitive in operation, simple to apply and as bonus, nice to look at.
- using Application Builder of National Instruments it is possible to create instruments which works properly with PC equipped only in Windows environment.

Practical solutions of our CD-ROM VLEM presented in this paper enable to understand better its idea and its conveniences.

3. PRACTICAL SOLUTIONS

The possibilities of our CD-ROM VLEM will be performed basing on the specific solution of measuring tasks. The first one concerns the measurements of electromotive force (EMF)

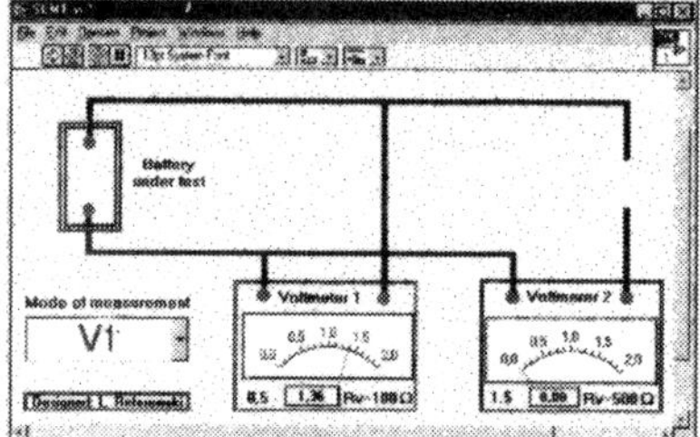
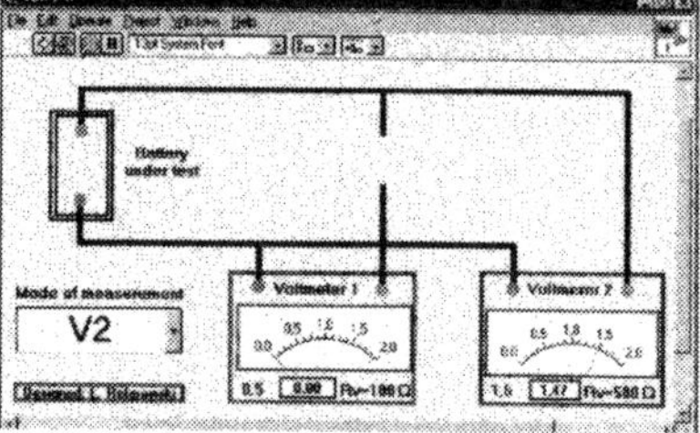

Fig. 1. EMF measurements by means of voltmeters with different inner resistance

of a battery using voltmeter and DC ptentiometer. The measurements realised by voltmeters are presented in Fig.1. In the first step of measurements the voltmeter V_1 is used. It has the accuracy of class 0,5 and resistance $R_1=100$ Ω, in the second step - voltmeter V_2 with accuracy of class 1,5 and resistance R_2=500 □. The difference between the indications of both voltmeters is □U=U_2-U_1=0.11 V (Fig. 1). This value exceeds the permissible error □U_p due to the classes of both voltmeters: □U_p=0.01+0.03=0.04V V. The best idea to verify the indications of these voltmeters is to make measurement using simultaneously both of them (*Mode of measurement V_1+V_2*). In this case the indications of voltmeters are equal U_1=U_2=1.34 V. Basing on this measurements it is possible to calculate the electromotive force E and the resistance r of the tested battery (the formulae 1 & 2).

$$E = \frac{U_1 U_2 (R_2 - R_1)}{U_1 R_2 - U_1 R_1} \qquad\qquad r = \frac{R_1 R_2 (U_2 - U_1)}{U_1 R_2 - U_2 R_1} \qquad (1\ \&\ 2)$$

They are as follow: E=1.50 V and r=10.3 □.

From the results of this experiment it is possible to state:

- making measurements it is necessary to analyse the errors of the method
- an instrument with better class of accuracy not always assures more precisely results of measurements

In order to verify the above measurements in the next experiment is used the potentiometer presented in Fig. 2. With this instrument the measurements are realise with the same order as in the case of the genuine potentiometer. In the first step it is necessary to adjust with the resistor the exact value of EMF of standard cell due to the ambient temperature. Then setting the mode switch in position E_n adjust supply current of the potentiometer using three knob resistors (*coarse and fine*). The galvanometer is protected by means of galvanometer's key *k* and switch *s*. The first one should be depressed cautiously and released immediately when the deflection of galvanometer appears. When an experimenter approaches to the state of balance and the deflection of galvanometer is small it is necessary to increase its sensitivity (switch position from *low sensitivity* to *high sensitivity*). It is also possible to shunt the galvanometer's

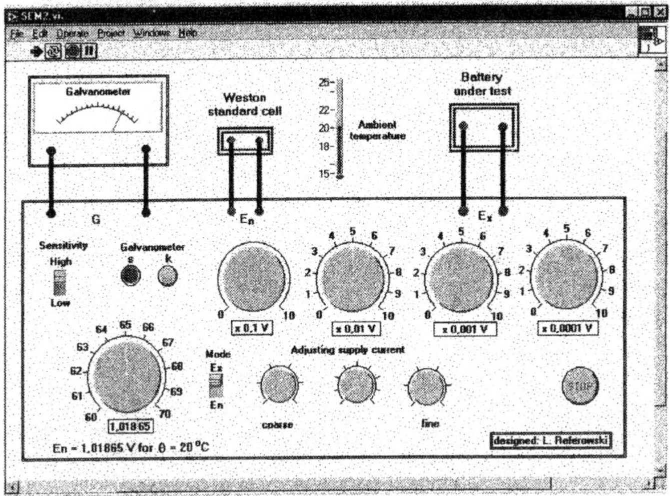

Fig. 2. CD-ROM direct current

key activating switch *s*, in this state the switch is blinking. When the supply current of potentiometer is adjusted it is possible to measure EMF of battery setting the mode switch into the position E_x. The manipulations with potentiometer are similar to these during the step of adjusting potentiometer's current but now the state of galvanometer's balace is obtained by adjusting four-decade potentiometer. The value of measured EMF is determined by reading the position of each decade taking into account the respective multiplying factor. In this case the measured value of EMF is E=1.051 V.

The measuring problem in Fig. 1 and Fig. 2 is one of the simplest of our CD-ROM VLEM but it explains well the idea of its using. During the conference there will be presented more complicated experiments e.g. autocorrelation measurements or vibration analysis.

4. CONCLUSIONS

Nowadays students' preparation for practical jobs in our Laboratory of Electrical Measurements is based on printed instructions or particular textbooks. In this case we have always problems concerning proper use of precise and expensive instruments especially when someone can damage them by incorrect manipulation. Preparation to the laboratory exercises based on CD-ROM VLEM enable:

- to understand better the laboratory task which will be realised during normal sessions in genuine laboratory,
- to have practice in manipulation which will have to be done as a job in laboratory, it is not possible to damage precise or sensitive instruments in CD-ROM VLEM as they are foolproof,
- to mark and think about certain measuring problems which could occur during measuring sessions.

Comparing our CD-ROM VLEM with Internet laboratory (1, 4) it is necessary to underline:

- The Internet laboratory enable to realise "on line" genuine measuring task in which the measuring signals are delivered by genuine transducers and are measured by "real" instruments. Students can also control different objects.
- The main disadvantage of the Internet Laboratory is that till now only small number of our students has possibility to profit from the Internet, as the costs of its use are rather high.
- CD-ROM VLEM enable students to prepare some preliminary laboratory tasks at home, afterwards they can realise them in the university laboratories in shorter time and in better way
- CD-ROM VLEM can be used by every student in his homework on condition that he has suitable computer at his home and it costs nearly nothing.

5. REFERENCES

1. K. Schilling, H. Roth – Virtual Laboratories to Perform Experiments with Remote Equipment, Proc. of 10th EAEEIE Conference, Capri, Italy, May,1999
2. L. Referowski, R. Roskosz, D Swisulski - Metrology Laboratory for Students using Internet Proc.of 8th EAEEIE Conference, Edinburgh, Scotland, June,1997
3. Gary W. Johnson – LabVIEWGraphical Programming, McGraw Hill, New York 1995
4. L. Carlson: Using LabVIEW to Reform Engineering Education, User Solution of NI 1996

Caf or decaf? – Impact of regular caffeine consumption on alertness, and mental and physical performance

P J ROGERS
Department of Experimental Psychology, University of Bristol, UK

Abstract: Recent research suggests that the net benefit gained for alertness and mental performance from caffeine consumption is much less than the 'felt' benefit. This is because significant fatigue is associated with overnight caffeine withdrawal, so that caffeine consumption through the day tends to merely restore (reinstate) and then maintain a 'normal' level of functioning (i.e., the level displayed by individuals completely free of the effects of caffeine and caffeine withdrawal). In contrast, high doses of caffeine have been shown to benefit certain aspects of athletic performance unconfounded by caffeine withdrawal, but the relevance of these effects for everyday physical performance is doubtful.

1   THE IMPORTANCE OF CAFFEINE

Caffeine is the most widely and frequently consumed drug in the world. Coffee and tea account for the vast majority of this intake, which has been estimated at an average of 70 mg of caffeine per person per day (1), although intakes in different countries and individual intakes vary widely (70 mg is approximately the amount of caffeine in a cup of instant coffee or a strong cup of tea). It follows, therefore, that even if caffeine has only a small effect on the well-being and functioning of individuals, its impact at a population level is likely to be very substantial.

2   MENTAL PERFORMANCE

Among its various effects on body and mind, mediated primarily by the blockade of adenosine receptors (2), caffeine has been shown to significantly affect both mental and physical performance. Consequently, caffeine consumption may be expected to have an important impact on performance in, for example, virtually all work situations. Generally it has been assumed that the impact of caffeine consumption is beneficial (except perhaps in cases of unusually high intakes, which can cause restlessness, nervousness, insomnia, etc.); however, recent re-evaluations of the research on the effects of caffeine on alertness and mental performance have questioned this conclusion [e.g., (2, 3)].

Results from placebo-controlled studies of the effects of acute caffeine administration have found caffeine to increase self-ratings of alertness, to improve mood, and to enhance psychomotor and mental performance (e.g., effects on tasks measuring tapping speed, simple reaction time, sustained attention, memory, and logical reasoning, and on simulated driving) relative to the placebo treatment [reviewed in (2, 3)]. The most obvious conclusion from this body of evidence is that caffeine consumption is beneficial. However, in the vast majority of these studies the study participants had a history of regular caffeine consumption, and they were tested on caffeine and placebo after a substantial period of caffeine abstinence (withdrawal). What this experimental protocol leaves open is the question of whether the results obtained are due to beneficial effects of caffeine or to deleterious, including fatiguing, effects of caffeine withdrawal. Furthermore, because significant fatigue is associated even with overnight caffeine withdrawal (4, 5, 10), it is possible that the psychostimulant effects felt by regular caffeine consumers represent only reinstatement of functioning. That is, caffeine intake merely restores mood, alertness and performance to baseline levels (i.e., the levels displayed by individuals completely free of the effects of caffeine and caffeine withdrawal).

One way to investigate this question is to compare the effects of caffeine in caffeine consumers and non-consumers (4, 10); however these groups are self-selected and therefore pre-existing differences might account for any differences found in responses to caffeine administration. Another approach is to test 'non-caffeine-withdrawn' individuals (6, 7), but again this is inconclusive, because in practice it is difficult to rule out residual caffeine withdrawal as an explanation for poorer performance in the placebo condition.

The best approach is to measure the psychostimulant effects of an acute caffeine versus placebo challenge in long-term (i.e., 'fully') caffeine-withdrawn consumers (L) compared with overnight-withdrawn consumers (O). A net beneficial effect of caffeine consumption would be demonstrated if acute caffeine administration led to significant improvements in functioning in O as well as L, and especially if caffeine brought both groups up to the same level of improved functioning. In contrast, the reinstatement hypothesis predicts that in the placebo condition L will perform better than O, and that acute caffeine administration will affect functioning only of O, bringing their level of functioning up to but not exceeding that of L.

More or less exactly this latter result was obtained in two recent independent studies, one using a letter recognition memory task (8) and the other using a long-duration, simple reaction time task (9). The data for this simple reaction time task are shown in Figure 1. What is striking is that the performance of the L group given placebo is almost exactly as good as that for the O group given caffeine. Moreover, there was, unexpectedly, a detrimental effect of caffeine on reaction time performance of caffeine in the L group, which might have arisen indirectly due to an adverse effect of caffeine such as 'jitteriness'. This is a frequently reported symptom occurring in normally caffeine-abstinent individuals given caffeine (10).

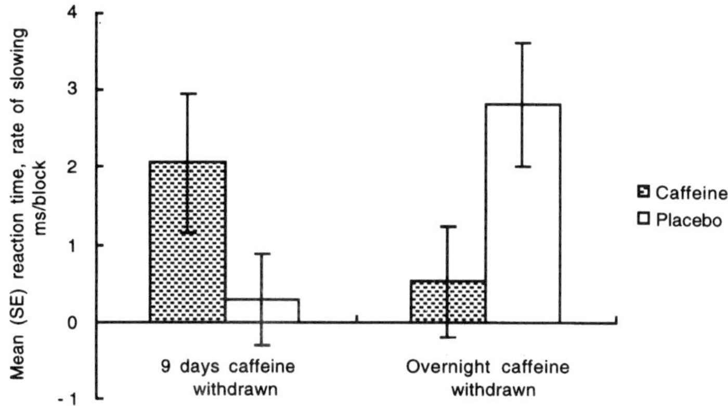

Figure 1. Effects of caffeine versus placebo on performance of moderate caffeine consumers (n = 40: 10 per group) on a 200-trial simple reaction time task after chronic and acute (overnight) caffeine withdrawal. The task was performed during test sessions repeated 3 times between 10 AM and 2 PM. Caffeine (85 mg) or placebo was administered 45 minutes ahead of each test session. The results shown are the rate of slowing of performance across 10 blocks of 20 trials collapsed across the 3 test sessions, for which there was a significant Withdrawal X Caffeine interaction effect, F(1,36) = 4.36, P < 0.05. The level of performance on block 1 did not differ significantly among the 4 groups, P > 0.1. (See ref. (9)].

An important feature of both of these studies on the effects of caffeine in relation to acute and chronic caffeine abstinence is that the acute caffeine challenge involved administration of several moderate doses of caffeine across a testing period lasting until early afternoon. This models the typical pattern of caffeine intake of regular caffeine consumers, and at the same time rules out the possibility that the failure of caffeine to benefit performance in the context of long-term caffeine abstinence, or to raise performance above 'baseline' in overnight abstinent participants, was due to the administration of inadequate amounts of caffeine on the challenge day. Nor can these results be explained in terms of a 'ceiling effect', since performance on the tasks was always below the maximum achievable level.

Recent evidence therefore strongly suggests that although regular caffeine consumers feel a benefit from caffeine intake, especially first thing in the morning after overnight caffeine abstinence, the actual or real benefits gained for alertness and performance are very small, or even negative overall (because of the repeated adverse effects of overnight caffeine abstinence). A small number of studies have claimed to show that caffeine administration can improve mental performance in non-caffeine-deprived individuals [e.g., (6, 7)], but as suggested above it is difficult to be certain that the participants were completely unaffected by caffeine withdrawal. In any case, it seems likely that the effects on performance of caffeine versus placebo measured in these studies were much smaller than those reported in studies of overnight caffeine withdrawn caffeine consumers. The only realistic way of testing unequivocally for a net benefit of caffeine consumption is to carry out further studies of the effects of caffeine in long-term abstinent caffeine consumers. It may be that under certain conditions, for example when fatigued by sleep deprivation or after eating a large meal (11), real benefits of caffeine consumption for alertness and mental performance will be demonstrated. Another situation in which the impact of caffeine consumption may be very important is during extended periods of driving. Driver sleepiness accounts for a substantial number of accidents occurring under monotonous driving conditions, and the rate of these accidents peaks in the early morning and early afternoon (12). Caffeine has been proposed as an effective countermeasure for this problem [e.g., (13)]. Again, however, because of the detrimental effects of short-term caffeine withdrawal on alertness and vigilance performance, the extent of this benefit is unclear. Indeed, rather than benefiting driving performance, the overall effect of caffeine consumption may be to impair performance on long, unbroken journeys due to the fatiguing effects of caffeine withdrawal.

3   PHYSICAL PERFORMANCE

Largely separate from the above research, there has been considerable interest in the possibility that caffeine consumption can enhance physical performance. Likely mechanisms for such an effect include direct effects on skeletal muscle, increased mobilisation and oxidation of free fatty acids (leading to glycogen sparing), and effects on the perception of fatigue and effort due to the psychostimulant action of caffeine (14). Confirming the view that caffeine can be a useful 'ergogenic' aid, many studies have found performance to be significantly improved by caffeine and, crucially, several of these studies have found performance enhancement to occur in non-caffeine consumers or in caffeine consumers irrespective of their length of prior caffeine abstinence [e.g., (14, 15, 16, 17)]. In other words, it appears that caffeine consumption can provide a real (i.e., net) benefit for physical performance.

The relevance of these effects of caffeine on physical performance for everyday activities is, however, doubtful. This is because the research has been carried out almost entirely on elite or trained athletes, and it has it used single doses of caffeine mostly in the range of 3-9 mg/kg, but sometimes as high and 15 mg/kg. To obtain the same amounts of caffeine from coffee would require the consumption of some 3 to 14 cups in one sitting. Thus, while caffeine in large doses (but still within the current limit allowed by the International Olympic Committee) appears to benefit physical performance to the extent that it can be expected to give a competitive advantage, particularly in endurance events, parallel effects at less extreme levels of performance have not been demonstrated. Indeed, at higher levels of caffeine consumption any advantage gained for physical performance is likely to be offset by adverse psychostimulant effects, such as 'jitteriness' and mental confusion (17).

4 CONCLUSIONS

The evidence reviewed above challenges the widely held view that caffeine consumption benefits alertness and mental performance. When the marked negative impact of caffeine withdrawal is taken into account, it appears that by drinking tea or coffee at breakfast the typical regular caffeine consumer is merely restoring his or her level of functioning to 'baseline'. Furthermore, it is doubtful whether subsequent caffeine consumption during the morning and the rest of the day confers much if any benefit above this effect. In contrast to studies the studies on mental performance, caffeine has been shown to affect certain aspects of physical performance unconfounded by caffeine withdrawal. However, benefits for physical performance have been found using high doses of caffeine given to trained athletes performing endurance and other elite sport relevant tasks, and consequently these results cannot confidently be extrapolated to everyday life.

REFERENCES

1 Gilbert, R. M. Caffeine consumption. In The Methylxanthine Beverages and Foods: Chemistry, Consumption, and Health Effects (Ed. G. A. Spiller), 1984 (Alan R. Liss, New York).

2 James, J. E. Understanding Caffeine: A Biobehavioural Analysis, 1997 (Sage, Thousand Oaks, CA).

3 Rogers, P. J. and Dernoncourt, C. Regular caffeine consumption: A balance of adverse and beneficial effects for mood and psychomotor performance. Pharmacol. Biochem. Behav.. 1998, 59, 1039-1045.

4 Richardson, N. J., Rogers, P. J., Elliman, N. A. and O'Dell, R. J. Mood and performance effects of caffeine in relation to acute and chronic caffeine deprivation. Pharmacol. Biochem. Behav., 1995, 52, 313-320.

5 Griffiths, R. R. and Mumford, G. K. Caffeine—A drug of abuse? In Psychopharmacology: The Fourth Generation of Progress (Eds. F. E. Bloom and D. J. Kupfer), 1995 (Raven Press, New York)

6 Smith, A. P., Maben, A. and Brockman, P. Effects of evening meals and caffeine on cognitive performance, mood and cardiovascular functioning. Appetite 1994, 22, 57-65.

7 Warburton, D. M. Effects of caffeine on cognition and mood without caffeine abstinence. Psychopharmacol., 1995, 119, 66-70.

8 James, J. E. Acute and chronic effects of caffeine on performance, mood, headache, and sleep. Neuropsychobiol., 1998, 38, 32-41.

9 Rogers, P. J.; Stephens, S.; Day J. E. L. Contrasting performance effects of caffeine after overnight and chronic caffeine withdrawal. J. Psychopharmacol., 1998, 12(Suppl.A) A13.

10 Goldstein, A., Kaizer, S. and Whitby, O. Psychotropic effects of caffeine in man IV. Quantitative and qualitative differences associated with habituation to coffee. Clin. Pharmacol. Ther., 1969, 10, 489-497.

11 Richardson, N. J., Rogers, P. J. and Elliman, N. A. Conditioned flavour preferences reinforced by caffeine consumed after lunch. Physiol. Behav., 1996, 60, 257-263.

12 Horne, J. A. and Reyner, L. A. Driver sleepiness. J Sleep Res., 1995, 4(Suppl.2), 23-29.

13 Horne, J. A. and Reyner, L. A. Counteracting driver sleepiness: Effects of napping, caffeine, and placebo. Psychophysiol., 1996, 33, 306-309.

14 Spriet, L. L. Caffeine and performance. Int. J. Sport Nutr., 1995, 5, S84-S99.

15 French, C., McNaughton, L., Davies, P. and Tristram, S. Caffeine ingestion during exercise to exhaustion in elite distance runners. J. Sports Med. Physical Fitness, 1991, 31, 425-432.

16 Graham, T. E. and Spriet, L. L. Metabolic, catecholamine, and exercise performance responses to various doses of caffeine. J. Appl. Physiol., 1995, 78, 867-874.

17 Van Soeren, M. H. and Graham, T. E. Effect of caffeine on metabolism, exercise endurance, and catecholamine responses after withdrawal. J. Appl. Physiol., 85, 1493-1501.

Main concepts of the elite oriented engineering educational system

G G ROGOZIN
Donetsk State Technical University, Ukraine

ABSTRACT

The paper presents the framework of the educational system and the main concepts of the effective methods in training engineers and technologists based on the new approach for making more active the students' thinking and intensifying the memory productivity during the lecture. Psychological and social aspects of the elite oriented educational system and their impact on the students' personality are described. Data are given about the fundamental tenets and the subjects of the humanitarian values in order to make the best use of the latter ones at lecturing the engineering courses. Recommendations are given about moulding the professional ethics and introducing the new technical terminology in the course of delivering the lectures. Conclusions are drawn concerning the main outcomes of reflecting the humanities and arts in the context of the elite oriented engineering educational system.

1 INTRODUCTION

The necessity of reflecting the humanities and arts at lecturing the engineering courses is called forth by the following objective reasons:

- World tendency in declining the human culture and ethical principles in the postindustrial society.
- Rise of psychological stress in society and faith loss in oneselves among the students in the states of the former Soviet Union.
- Isolation between the humanitarian and engineering courses.
- Feeling the needs of training the national technical elite in developing countries.
- Difficulties in achieving at the lecture some of the course objectives by conventional methods.

The latter item of the above enumeration connected with activating the students' brainwork (ASB) was dealt with in (1). The essentials of the new approach for solving the problem along the mentioned line are shown in Figure 1. As regards to co-ordinations of the associative information, the time exposition of the humanities and arts may not exceed 20 s (2). Integration of the humanities and arts into engineering education directly at lecturing the engineering courses with the proviso that the all methical recommendations are implemented and certain difficulties are overcome allows to obtain a number of positive results.

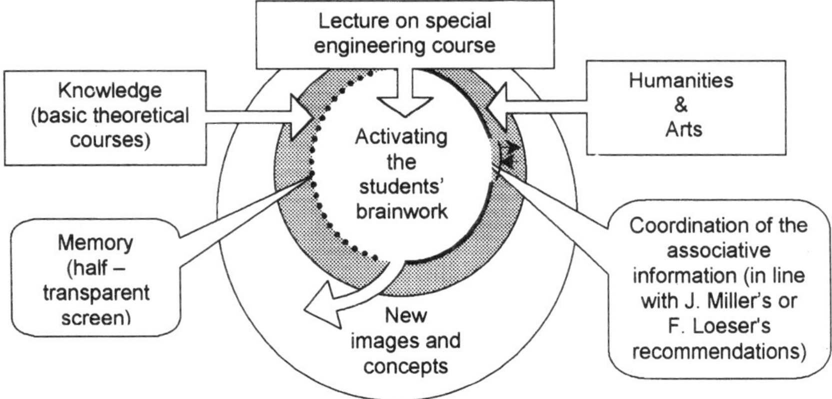

Figure 1. Information flow diagram of the method intended for ASB

The ASB method will also be available as a key to planning the educational system for training the technical elite.

2 INFLUENCE OF THE NEW APPROACH UPON THE MEMORY OF THE LECTURE LISTENERS

The next in importance is the matter regarding the influence of the ASB method upon the memory of the lecture listeners. The effectiveness of reflecting the humanitarian values at lecturing is readily apparent from the well-known laws of psychology. Analysis of the results being achieved from the view-point of memorizing the material set forth at the lectures allows to draw the following inferences:

- Contrary to the effect of Ebbinghause the application of the new approach enables to get the positive effect in the middle of the lecture.
- Being connected with interruption in exposition of some pieces of the lecture material, making use of the humanities and arts creates the so-called effect of uncompleted action. The latter one favourably influences upon the process of memorizing the fragments being expounded at the lecture, especially when students have not been given reasons clearly enough for setting up a lecture problem.
- If a lecturer gives grounds to doubt upon the piece of information in the field of humanitarian values the boomerang effect can be obtained that will lead to the negative after-effects.

Turning back to the effect of uncompleted action it is worthwhile to consider the analogy associated with applying the amusing and witty drawings in the columns of the popular scientific books. It has long been known that such a technique is destined for making more active the thinking of the reader. This mode clearly demonstrates the influence of the effect connected with interrupting the reading of the section being studied for improving the capabilities of the memory and perceiving the essence of the problem. In a like manner the positive result is attained by reflecting the humanities and arts at lecturing. It should be pointed out that the implementation of the new approach doesn't exclude the well-known traditional methods being used for memorizing the information being delivered at the lecture such as attraction of attention to the important material and imprinting it in the students' memory by using the relevant associations.

3 ELITE ORIENTED EDUCATIONAL SYSTEM

The educational system being described is proposed to solve the problems associated with overcoming the isolation between the humanitarian and engineering courses, moulding the professional ethics and the life-asserting aim position of the generation to come. Schematic representation of the mutual relations between the elements of the above system designed for training the technical elite is shown in Figure 2.

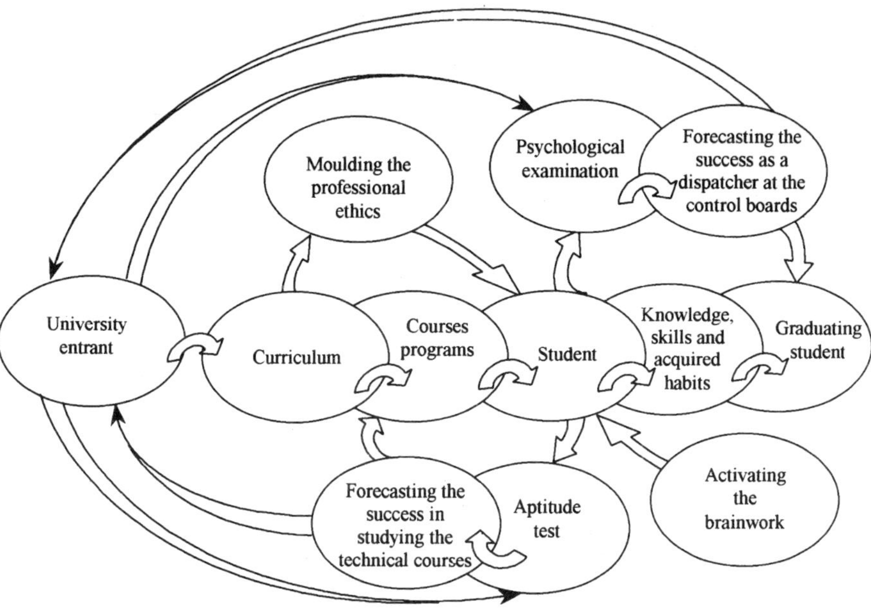

Figure 2. Moulding the elite orientation of the student's personality.

The great importance of the special testing program such as The Electrical Aptitude Test (3) must be taken as proved. The test developed in the Dept. of Electrical Engineering at The State Technical University of Donetsk (Ukraine) is meant for forecasting the successfulness of the graduates' professional activity in the capacity of an operator at the control panels in electric installations of the electrical systems.

The essence of the professional ethics problem is in clash of moral principles of the young specialists (the yesterday's graduating students) when carrying out their official duties in the context of the morals of their immediate superiors. The responsible decisions being taken in the professional situations ought to be in good agreement with the individual moral responsibility but not to be taken under cover of the common one or under the chiefs' responsibility. This principle follows from the A. Schweitzer's idea that everyone ought to have a reverent attitude towards the life (4). The mentioned problem being touched upon at the lectures, it becomes evident the deduction that all the important engineering solutions should be done not with account the strictly practical interest but, in a broad sense, the human ones and the man's individuality.

Introducing the new terms when delivering some special technical courses gives rise to difficulties for perceiving the essence of a lecture. The new technical terms call for additional explanations and definitions since, as a rule, these terms do not express the meaning of the latter ones precisely enough. Meanwhile the listeners' adaptation period to the new terminology can be significantly cut. Indeed, the international vocabularies allow to find more authentic view of subject. Inspection of the corresponding generally accepted terms adduced in various European standards done by their mutual comparison, is very useful for understanding and memorizing the new term.

In order to make the best use of the new approach for activating the students' brainwork it is advisable to take, as the starting points, some keynote concepts and subjects from the area of the humanities and arts. A wide range of examples involving in the detailed elucidation of the associative subjects being used for connecting the cultural and technical aspects containes information on ergonomics and human engineering, etymology of the terminology, ethics, religion, mythology, literary production, painting, architecture, music, etc.

CONCLUSION

On the way of implementation of the method being considered there are quite a few difficulties since it can't be formalized. The favorable results when employing the humanitarian values at lecturing the engineering courses can be achieved only under conditions when a lecturer continually takes a keen interest in the field of cultural wealth.

The main outcomes of reflecting the humanities and arts are as follows:
- Solving the problems associated with overcoming the isolation between humanitarian and engineering courses.
- Activating the students' brainwork.
- Making the favourable influence upon the students' memory at the lecture.
- Carrying out the favourable psychological and social effects.
- Rendering the positive educational effects.

REFERENCES

1. Rogozin, G., G., A new approach to making students' thinking more active when lecturing on special engineering courses. Proc. of Global Congress on Engineering Education, Cracow, Poland, 445 – 447 (1998).
2. Loeser, F., Gedächtnistraining, Urania – Verlag, Leipzig – Jena – Berlin (1976).
3. Pudlowski, Z. J. et al, An Aptitude Test and Associated Research on Basic Electrical Circuits. Sydney: E E E R G (1993).
4. Schweitzer, A., Das Problem des Ethischen in der Entwicklung des menschlichen Denkens, Frankfurt a. M. (1955).

The rehabilitation project of the Suleymaniye district

Z SAGDIC
Faculty of Architecture, Yildiz Technical University, Instanbul, Turkey

ABSTRACT

Suleymaniye is the most important historical part of beautiful Istanbul. Different cultures, such as christianism by Byzantium Empire, islamism by Ottoman Empire and however the modern movements of the Turkish Rebuplic has been entegrated period by period on Suleymaniye site.

Within 2600 years, Suleymaniye has been one of the only historical sites of the world, on where cultural, social and historical qualities have been living continiously.

The rehability project of this area was made by Proje Architectural Group from 1996 to the beginning of 1999. I was one of the architectural historian adviser and planner of this project.

The project consisted of analysis and sythessis phases about Suleymaniye site, 1/ 200 detailed rehability and design project phase and all under-structural details.
The aim was to preserve and transverse the cultural, historical, architectural and social qualities to the next generations.

Therefore, the paper will have the historical changes on Suleymaniye district and all of the project phases.

1.INTRODUCTION:

The phenemenon of urbanisation becomes visible when social identity
combines with culture through social relations. What makes cities
different from each other and what assigns them their identity is the accumulation of
culture. Thus, the past goes on existing without losing its identity and has the chance
of extending itself into the future.

2.DESCRIPTION OF THE SITE:

Istanbul is one of the most interesting and significant cities in the world with its
natural, cultural and historical values. The Historical Suleymaniye Site is one
of the most valuable and charming regions of the city of Istanbul. From 1550 to 1557
Architect Sinan designed the Suleymaniye Complex in the name of Suleiman the
Magnificent. Just after the construction of the mosque had been completed, the
district became the residental place of the Ulemas (Muslim theologians and

scholars), and finally the site took the name of Suleymaniye. The Suleymaniye
Complex comprises the significant Suleymaniye Mosque, the mektaps (primary
schools), the madrasahs (teological schools atteched to the mosque), darü'l hadis,
darü°°ifa (hospital), darü'l kurra, imaret/ darüzziyafe (temproray kitchen), hamam
(Turkish baths), kervansaray (inn), tabhane, arasta (commercial center/ shops in a
row) in it. As can be seen from the rich contents of the Suleymaniye Complex, from
the late16th century, the Suleymaniye Site was the heart of the city until the 18th
century, when the Pera Site became the second and finally the new center of
modernisation.

In the 20th century, under the modernisation movement the whole of Istanbul was
undergoing, it is clear that the site lost its centerilized structure and finally its
attraction.Istanbul, which is a beautiful city, now finds itself in a state where
historical sites and human patterns are degenarating day by day, which makes the
inhabitants of this city sad and worried.

The Suleymaniye District, which embraces buildings of all periods, has improved
and enriched itself throughout history and has aquired a prominent place in the
development of this mysterious city.

In all of the old cities, historical sites play an important role in providing
the cultural interaction of the new generations with the past cultures. For this reason,
it is crucial to protect and conserve them by turning them into livebale environments.
The Suleymaniye District, which still retains the typical settlement patterns having
street arrangements aligned with characteristic Ottoman-style wooden houses still
standing, was designated as a historical site in 1977.

Throughout the historical evolution of Istanbul, the Historical Suleymaniye Site
became one of the most important points of business and administrative activities.
However, the region has acquired a form of crowded structure with the decreace in
the density of business and administrative activities, under the limitations of
physical conditions imposed in the last decade. Today, not only with the decreasing
of commercial places on the region, but also with the present situation of historical
Turkish houses, the Historical Suleymaniye Site has a dramatical view. Taking into
consideration in the development of the dynamism involved metropolitian areas, it is
believed that the time for applying a rehabilitation project for the Historical
Suleymaniye Site came, that means to rehabilitating the region without giving any
harm to its basic structure.

This historical site is unique in that the magnificent Suleymaniye Complex built by
Architect Sinan constituses the most important triangulation point in Istanbul.
In addition to this complex, monumental buldings of the Byzantiene and Ottoman
periods can be found standing side by side. Among these are mektaps, dervish
lodges, saints' tombs, moseleums, coffee houses, and dwelling houses with bay
windows and jutting-out balconies etc., all of which represent the spatial organisation
in an urban unit in Istanbul.

Despite the fact that this historical urban pattern is losing its typical character day

by day due to fires, demolitions, functional transformations etc., it can still
convey of its original atmosphere.Looking back at its history, it can be seen that the
Suleymaniye District was a sophisticated settlement area with magnificent mansions
and kiosks having large and well-kept gardens. Most of these buildings were built of
wood which later became stone constructions. In time, with the increase in
population, the big mansions and kiosks were replaced by houses in order and
rowhouses.

Towards mid-20th century, there occured a sigficant charge in the settelment areas in
Istanbul. The small manufacturing businesses increased in number and began to
penetrate the dwelling sections. As a result, the old dwellers left their houses, which
were left empty or rented out to single men room by room.

Most of the buildings in this district which degenerated and rotted in time were
pulled down and replaced by badly-built concrete apartment blocks.Yet, in spite of
all this loss and damage, the Suleymaniye District still retains the
traces of the past cultures.

3.THE AIM OF THE PROJECT:

*"The Urban Design Project For The Protection Of The Suleymaniye District As A
Historical Site"* cosists of four stages as follows:

1. An analytical study of the Eminonu District;
2. Creation of the Design Project: Drawing up a conservation and organisation plan;
3. Designing the secondary units;
4. Presentation of the project.

1.   An analytical study of the Eminonu District:

The first stage consists of preliminary research, documentation, analytical studies
and synthesis of the results concerning the Eminonu district.The place and
importance of the Eminonu District on a macro-scale, that is in the metropolitan area
and also in the Historical Peninsula, have been assessed. The District has been
analysed and evoluated both in terms of its physical dimensions and of its social
dimensions throughout the historical process. Again the District has been studied and
evaluated in terms of infra-structure, transportation, environmental factors, and urban
development. The flora has also been established. And the whole site has been
photographed to provide data for the project. As a result of these studies, a synthesis
has been made taking into consideration the related economic and social factors.

2.   Creation of the Design Project:

Drawing up a conservation and organisation plan:
The second stage comprises the urban design project, the planning of the
relationships between the environment and transportation facilities, and creating

functions for buildings (like defining the commercial and the residental areas);
landscaping, and the preparing 1/200-scale urban design restitution project;
environmental organisation; different suggestions and drawings; taking appropriate
decisions on the guidelines.

The second phase is the creation phase. Thus, this phase has the suggestied
projects about the main decisions of the functions, transportation, landscaping, and
the decisions on restitution. Also, the urban-planning decisions are given according
to the main aims of the general design project and international rules.
Drawing on the results obtained from the above mentioned work and documents a
final general project is prepared for the whole of the Suleymaniye District.

3. Designning the secondary units:

The third stage involves work on the urban furnitures to give a new look to the
Suleymaniye Site. Also, in this phase, decisions about choosing all the materials to
be used in the site, such as the materials for the urban furnitures, street tiles, etc., are
given.

4. Presentation of the project:

The fourth stage includes the presentation of the project in general by means of
perspective drawings, and the detailed drawings were drawn and animations and
models were prepared.

4.CONCLUSION:

This project needs to be divided into several phases when it is got into application.
The Kirazli Mescid Street has been chosen as the pilot section. Starting from this
section, the project should be applied in stages. This vital project needs serious
intellectual and financial support in order to be realised.

4.REFERENCES:

1."Suleymaniye Sit Alaninin Koruma Amacli Kentsel Tasarim Projesi'nin Yorumu/
The Interpretation of The Renovation Project of The Suleymaniye District",
Barbaros SAGDIC, Mimarlik Dekorasyon, no:71, pp.52-53.
2."Suleymaniye Sit Alaninin Koruma Amacli Kentsel Tasarim Projesi/ The
Renovation Project of The Suleymaniye District", Barbaros SAGDIC, Mimarlik ,
Dekorasyon, no:71, pp.54-81.
3.Turk Evi/ Turkish Houses, Sedad Hakki Eldem, 1982, Istanbul.
4.Turk Hayatli Evi/ Turkish Living House, Onder Kucukerman, 1990, Istanbul.

The names of the bmp:
s1....The airphoto of the Suleymaniye District
s2....The map of the Europen side of Istanbul
s3....The graphic of the land-using criters in the Historical Peninsula

Scientific method and practical works in physics

L SAHRAOUI
Lab de Physique, Gouvernorat d'Alger, Algeria

ABSTRACT

Basic notions on scientific method and their application to practical works that consist in an introduction to experimentation are exposed. The incipient stages of physics instruction requires the presentation of physical experience pertaining to relevant phenomena. The primary objective of such an experience is to initiate students to contemporary techniques and devices. Furthermore the gradation of the apparatus is required. Measurement of physical quantities and compiling the relevant associated curves to determine laws and estimate uncertainties is a corollary of the main exercise.. This aspect must take into account what is called as the *"learning of learning of how to do"*.

I-RECALL OF ELEMENTARY SCIENTIFIC METHOD (1)

1.1 Experiment (2)

Scientific Method begins with observation of some part of nature or of a phenomenon. Einstein mainly built his Special Relativity after Michelson's 1882 observation of the constancy of light (3). Science appears as a result of observations and discoveries of that kind: discoveries of new physical phenomena, new investigation methods and detectors, etc. The object of science is the measurement by the scientist of different quantities that take place and their putting into order. The scientist must have a clear feeling of what to be experimented and already outline frames with which our world might have been built. To know if these frames describe reality, he has to experiment: we have to go on Mars or Pluto to know what is going on them. He chooses the subject to be studied, described and recorded, using words well defined for all physicists.

1.2 Bias (4)

Often bias are responsible for much incorrect work. Can one get completely rid of preconceived ideas or prejudices? Maybe not, because it is oft necessary in a certain meaning to have in one's head some hypotheses but that distort the least observation and experience: we have to overtake our own bias so as to try to grasp the physical event. That is why the scientist ought not to be afraid of critic of his colleagues and must accept checking and rehearsals, for the richness and force of science stay in that that a thing, once discovered, becomes independent from its discoverer.

1.3 An « essay » idea (5)

A real system at scales of the Universe ($\approx 10^{+140}\,m$), of man ($\approx 0.2.10^{+1}\,m$) or of particles ($\approx 10^{-12}\,m$ or less) is too complex to be understood in one time in exact and complete details by the human effort. Then analysis into different parts to be looked at, approximations and use of models are necessary. A model is a short cut to study a phenomenon. It does not describe reality but it is often sufficient.

After selecting the part to be studied and observed, a hypothesis, or *«an essay idea»*, is emitted on the nature and correlation of observations. Hypotheses may postulate relations between events or elaborate chains of cause and effect. The strength of hypotheses stays in the believing that it exists an order and simple laws in Nature. If two hypotheses fit physical facts and if one of them is simpler than the other, habitually, the simplest is accepted till a contradiction in the succession of the science statements makes it to be rejected. A hypothesis or an essay idea states a point of view of our mind, a temptation to understand the order of the things in Nature till its confirmation or its rejection by experimentation and tests. Great are hope and emotion of believing.

When a hypothesis has been devised, it becomes possible to apply formal logic rules and to deduce consequences. For the truth of a judgement, it is then important to distinguish between necessary and sufficient conditions called variables. Besides, there is no clear distinction between an experiment and a simple observation, but ordinarily in an experience, the observer directly implies himself in natural facts and makes conditions and events that are favourable to his purpose. In an ideal experience, one can give to variables all values he desires because verified predictions are often considered a serious confirmation of a hypothesis which not only must adjust facts that have contributed to its elaboration but also must fit itself to the whole body of the studied science: science truth is the absence of contradiction.

1.4 Cause and Effect (4), Search of causes

The basic method of science is generalisation or induction. The Induction pulls out « inferences » for an entire class of events on the basis of observation of few of their members. It makes us wait for the rehearsal of a phenomenon when all same circumstances repeat. Some of these circumstances occur to defect. Is it important? It is not always: the studied facts are sufficient and the probability notion enters into action. Oft induction refers to the recurrence or *«retour»* reasoning that is very fruitful and fertile because it contains an infinity of syllogisms in a unique formula from which it takes its force and power.

Strictly speaking, every event is unique and cannot therefore be observed a second time: e.g. sun or moon eclipses. Some events do show similarities. If one occurs, the second accompanies or follows it. The first is called as to be the cause of the second, which is called the effect. One aim of science is the prediction so as to control and utilise nature. It is closely allied to that of explanation. An explanation that does not increase man's power over nature is not to be considered in the same class as one, which does lead to an increased ability to predict or control future events or behaviours. Prediction and control are important aims of theory and science.

There are no fixed rules to search causes. The method of *«tacit agreement»* states that if the circumstances, leading up to a given event, have in all cases one factor in common, that factor may be the researched cause. The second is the *«method of difference»* which states that, if two sets of circumstances differ in only one factor and the one containing this factor leads to the event and the other does not, this factor can be considered the cause of the event. The third, the *«principle of concomitant variation»*, states that if variation of the intensity of a factor results in a parallel variation of the effect, then this factor is a cause. A kind of commonly used error of the *«circular reasoning»* is the use in a theorem demonstration, of points of the conclusion. Another difficulty can arise from confusing theorems and their converses.

2 PRACTICAL WORKS & EXPERIMENTS

2.1 How to attend and learn a lesson (theory and practice) (6)

The goal of the practical works is to custom the student to use, even unconsciously, the previous rules and to make him achieve standard reasoning in hope to help him to reach the required level of development of thinking. Before giving some ideas on *how* to attend practical works, here is a word on *how* to attend a lesson.

Within the lesson, there are in the mind of a student attending a lesson at least three simultaneous operations that may scatter his attention: listening, understanding, taking notes. It is the interference between these three operations that prevents him to follow correctly the lesson. From the point of view of the pupil and for a best concentration of his attention, it is important that he carries out a compromise and customs itself to take notice of the aim and the plan of the lesson, to good memorise in priority the main ideas, to remember the steps of the oft delicate calculations and demonstrations, to write notes to the full, overall on what is obscure, what is expected to be forgotten or what is worth a particular interest, to listen well to the teacher, to regenerate the lesson for absorbing it.

After the lesson, it is of great interest to teamwork, to learn the lesson the very day it has been done, to think of it and of the physical phenomena, the methodical approach, the used mathematical methods, the hypotheses and approximations, the conditions of formulae validity, and then to do again calculations (care to do literal calculations first then numerical applications), to learn by heart definitions, theorems, etc., to do a good use of the documentation, to well read the problem and the exercises, and do them again reminding the necessary knowledge and mathematics, without forgetting the physical meaning, the order of magnitude of quantities, their physical dimensions (7) in a generally four dimensioned frame .

2.2 How to attend practical works

Practical works and experiments are a necessity in Physics because without them no science is possible. So they should be given all the importance they need. Each teaching in Physics must start with an experiment to present the phenomena, to detect them, to proceed to calibration of the devices, to take the measurements of the quantities, to trace curves, to deduce laws, to perform uncertainty calculation. For obvious reasons, Physics must be summed up in presenting some well-chosen Practical Works, which seem good representative, by their power of initiation. Historical and fundamental experiments and phenomena should comfort this aspect. Our pupils should be initiated to the use of different devices and to the measurement techniques, and interpretation. They should be able to complete the lesson and prepare the rest of the program. To reach that goal, each pupil receives a booklet in which are reported the experiments to be done. Every week, he must a few days before prepare the work to be performed. The first thing to do is to read carefully the text of the experiment, so as to become impregnated with the *«physique de la chose»*. He should then notice the visible aim, e.g. *height and speed measuring*, the real aim of the visible aim, e.g. *energy conservation law checking*, the used principles, concepts and methods; if necessary, he must rehearsal the calculations, demonstrations, foresee the curves, reconsider the device utilisation and answer questions such as: did he reach the aim? what did he miss? what did he learn? conclusion? He completes the report he scrupulously wrote before, following the suggested master plan on the booklet, which generally indicates clearly aim, principle, list of operations, realisation, tables and measurements to be performed so as coming into the class-room the student has but to fill the empty tables, to trace asked curves directly if possible, and to give a conclusion. He must at last but not at least end by reading again his report before giving it to the teacher.

3 SOME FORMAL PEDAGOGICAL IDEAS (6)

3.1 Existence of latency time.
One must give time to the lesson to be etched in one's memory and be assimilated. That is what I call «*the minimum latency time* » which has been experimentally evaluated to approximately two years and is necessary to our pupils to perceiving the real physical meaning of the formulae he had learned: in the case of the Maxwell equations, it is the time necessary to understand really that every point interacts with its neighbour, in classical physics and space is a continuum till at least to atom scale. *"Latency time"* should be taken into account to optimise the teaching.

3.2 Distinction between the knowing and "learning of learning of how to do"
The by heart «*knowing*» is oft a usual sign of scholar evaluation, overall in primary or secondary school. This part of formation only develops *memory* for learning, storing and returning back on request. In that case, it suffices to consult a book or a dictionary to have the same result. Moreover, nowadays every computer can better store the list and give it back as and when you like. But one has to go further because studying physics demands more than memory and requires to strengthen other human brain abilities which are part of scientific method we wish to instil into the students' mind in physics. It is obvious that the development of new techniques and new pedagogical and scientific theories gives importance to *how* to make machines work but maybe overall to *how* to make machines that make machines work. To reach this objective, it clearly appears that not only should we teach the learning but also the «*Learning of learning of how to do* », i.e. the learning of how to grasp the hidden relationship between our universe and physical laws, for the intuition, the imponderable mental attitude, the *je-ne-sais-quoi* that allows a physicist do the link between a question and another far from it, make the presence of the teacher essential. The question is then what to do to structure the student mind and lead him think the solution of a given situation that makes him able to solve an entire class of analogous problems and moreover to solve unknown today and future problems: to-morrow techniques do not exist yet.
"One of the most cherished goals of a physics teacher is to teach students to think like a physicist…Such a lofty goal can be difficult to achieve" (8).

REFERENCES
1.A.Cuvillier, textes philosophiques, éd.A.Colin, Paris 1962.
2.F.Bacon, Novum Organum, 1620, Book I, Fowler Publishing, Oxford 1878
3. A.Einstein, Quatre conférences sur la théorie de la relativité, Gautiers –Villars, Paris.
4.E.B.Wilson JR, an Introduction to Scientific Research, Mc Graw Hill, NY 1952.
5. H.Poincaré, La Science & l'Hypothèse, Flammarion Paris 1912.
6. L.Sahraoui, Le Travail de l'étudiant, 2ème Congrès National de la Physique, Sétif 1996
7. J.Palacios, Analyse dimensionnelle, Gautiers –Villars, Paris 1960.
8. J.J.Prentis: Equation Poem ; AMJ. Phys, 64, 532 (1996)

Visual mathematics in art

C SINGER
Consultant, New York, USA

In the progress of this paper I have interspersed some reflections that we can collect into a general point of view on my art works. While the geometric style in the broadest sense can be classified in the art world as 'geometric abstraction', what actually takes place on the picture plane between form, space and color goes far beyond the limitations of such arbitrary labels. A complex underlying structure composed of straight lines, arcs, circles, elliptical sections, parabolas, hyperbolas, epicycloids, and spirals informs the viewer with a sense of controlled order and disciplined rational thinking reaching for a formalized mathematical perfection. In the final analysis, this geometrical structuring serves as no more than a foil for sensations for a more intangible and metaphysical (absolute) scope of vision. It is not the intention to present the viewer with a tasteful static equilibrium but to spark the imagination with suggestions of the powerful forces and processes operative on a universal scale. The work evokes ceaseless mobility, speed, time, space, growth, compression, tension, attraction, and the infinite nature of pure structural elements. The geometry provides the fundamental methodology for this dynamic expressive language. It is the personal vision that transforms this familiar vocabulary in mathematics into a unique transcendental syntax. If the circles were imagined on a larger scale they could represent the great circle that would bisect the celestial sphere or earth.

In understanding geometrical form there is classification of elements, functions, nonlinear approaches, and discussion of the accuracy of measurements, and of symmetry. There exists an internal configuration, and it is therefore possible to gain visual information from the order of the picture, its aesthetic qualities as well as the actual graphing and the pictures' corporeity. The latitude of geometrical method has many different examples of variations in the intensity of the forms and qualities: for example the dynamics of motion, gravity, flotation, and corpuscular unity. An intuitive design sense in conjunction with a practical mensuration is a major point. In this notion we see the continuance of a primordial condition that has carried from a first principle or element. Expressing a constant as the horizontal axis (longitude) in the graphing of the picture, and the variable (the varying intensity of quality throughout the picture) is the height on the vertical axis (latitude). The perimeter closes all of the formed figuration that represented the configuration or distribution of the varying qualities.

That a complex formalism is the foundation behind my paintings is a premise essential to the work in its constructible methods. Within the doctrines of the philosophy of mathematics, metamathematics includes questions of both semantics and syntax such as whether axiomatic systems are independent, consistent or complete. A transcendental extension field exists with applications to my art works. There is a duality of spatiotemporal reality between the physical and geometrical configuration that interpenetrate in superposition. Physical laws are simply applied to the geometrical configurations. The hyperbola can increase to infinity and therefore converge towards zero. The branch of the hyperbola bends more and more towards the x axis as the angle of the asymptotes become more and more obtuse. To look at far horizons we can imagine the far reaching parallel to our concept of nature and the physical world. In the totality of natural phenomena by enhanced approximations in the paintings we approach a system of reference x, y, z, t space and time by which these visual phenomena present themselves in agreement with geometrical rules and laws. H. Minkowski had introduced this fundamental axiom: *'The substance at any world point may always, with the appropriate determination of space and time, be looked upon as at rest.'* In painting it is my goal to capture a fragment of space and time although there is a duality where it is also rooted in singularity. In non-Euclidean space the geometry I have conceived consisting of the picture plane demonstrates an independent singular universe. Geometrical space and structure is adequately explained in that we consider points, curves and geometrical surfaces as minute particles, fine paths or thread, and thinly veiled surfaces.

Collinearity, (of a set of points lying in the same straight line) a recurrent theme is never coincidental but is an inevitable outcome, forced in the same way that it occurs in the classical Theorems of projective geometry, such as those of Pascal, Brianchon, Desargues, and others. These mathematicians recognized that spatial situations that produce collinearity were invariably the result of deep underlying geometric truths. The incidence of a point on a line is invariant under the projective rules. If three or more points are collinear along a line, then incident with a straight line, the images will thus be collinear. Therefore, the characteristics of incidence, collinearity, and concurrence are principle requisites of my work. So, the geometry focuses as to the overall collinear connective intersections and edificial detail of the inner organization.

Everything around us has a history or combination of historical influences is noted or emphasized. As an artist there are issues such as historic perspectives, ranges of influences, and importantly the sources of reference that are available to choose from which includes aesthetic and/or rational decisions. For my work in art, I have gone from art to mathematics or shall we say an integration of the two disciplines in tandem. I had never imagined that geometry would be a commodity or tangible product with economy. I have always considered geometrical paintings as windows or devices for contemplation and thought.

Clifford Singer, *Tractrix*, 1999, Acrylic on masonite, 36 x 24 inches

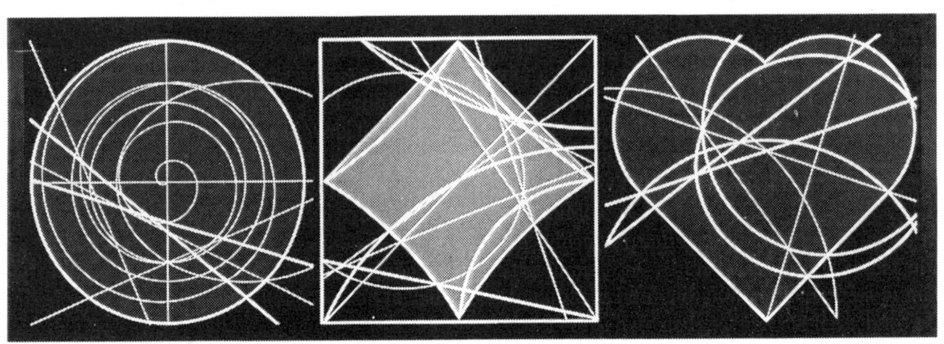

Clifford Singer, *The Geometry of the Circle, Square and Heart*, 1995,
Acrylic on Plexiglass, 30 x 92 5/8 inches

Clifford Singer, *Continuity*, 1999,
Acrylic on Plexiglass, 36 x 36 inches

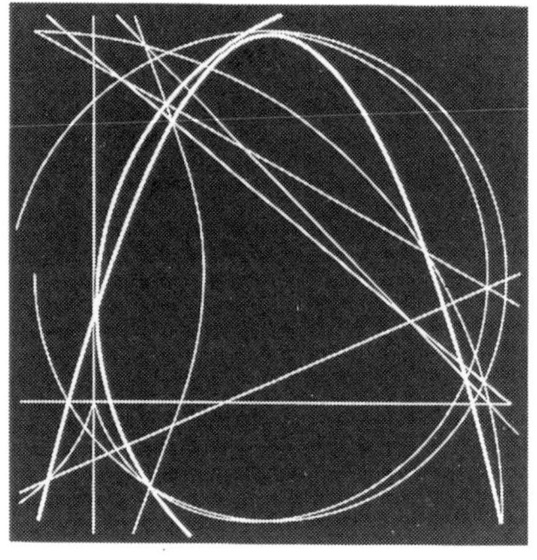

Clifford Singer, *Millennium*, 1997,
Acrylic on Plexiglass, 36 x 36 inches

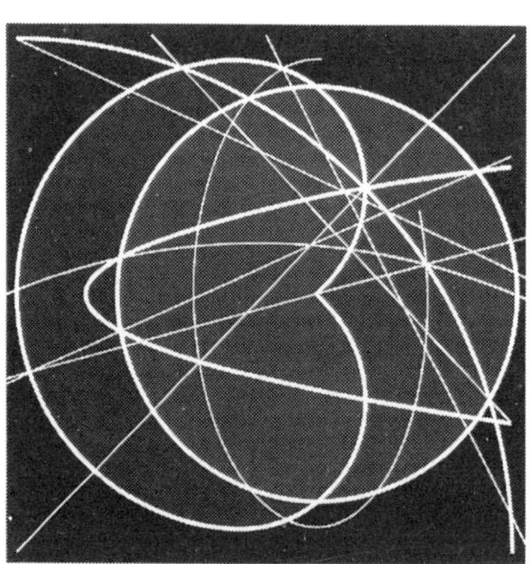

Clifford Singer, *Limacon*, 1998,
Acrylic on Plexiglass, 36 x 36 inches

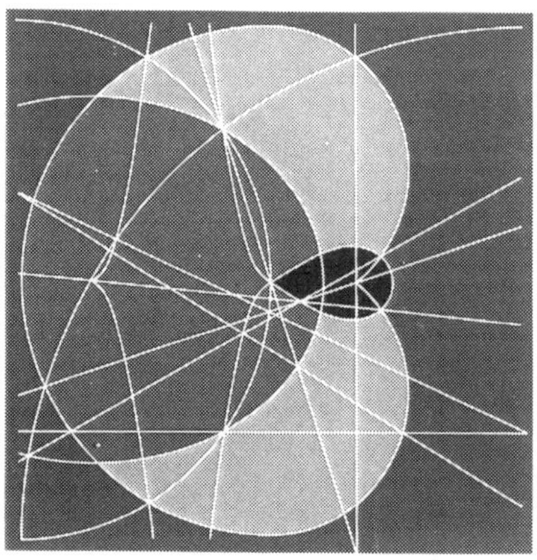

Clifford Singer, *The Geometry Lesson*, 1998,
Acrylic on Plexiglass, 36 x 36 inches

Effective teaching of circuit theory

K STEC
Institute of Theoretical and Industrial Electrotechnics, Silesian University of Technology, Gliwice, Poland

ABSTRACT

The paper deals with the problem of obtaining the best results in teaching of electric circuit theory. Three ways of teaching are combined: traditional lecture and solving problems by means of traditional methods, experimental work in laboratories and computer aided methods. Such combination, if applied properly, gives good results.

1. INTRODUCTION

Quality of teaching process is a very important factor having great impact on the future of any community. The world changes rapidly so people should be prepared adequately for the life in it. It concerns especially the world of engineering were the changes are more rapid then in other areas. We must be aware of the fact, that the professional life of an engineer is about 40 years. Students should be taught in such a way to prepare them for all their professional life. Is it possible with the rapid progress in science and technologies, with the acquired knowledge becoming quickly out of date? The answer is yes. If the teaching process is carried out correctly such result can be obtained. What we need are engineers that can learn new technologies, assess new ideas critically and are eager to find new better solutions for technical problems they have. Three way teaching used in Institute of Theoretical and Industrial Electrotechnics seems a good method of getting proper results in preparing future engineers for their work.

2. GOALS AND MEASURES

It seems that the only way to produce an electrical engineer that can solve problems properly moving with the times, is to support him/her with a good theoretical background concerning basics of circuit theory, electric field theory and use of computer [3]. The question arises if it is enough. The answer is - no, it is not. There is no way to become a good engineer without skills in carrying out experiments [4].
Students should
- be taught in such a way that would lead them to greater creativity,

- get accustomed to the critical examination of the experimental results ,
- get accustomed to finding correct solutions of the problems especially those that have not been solved yet,
- become eager to look for the new better methods and techniques,
- not to be satisfied with the knowledge they already have.

To obtain these goals the multiple-way teaching of circuit theory should be applied.

There is no doubt that good, up to date teaching should consist of three main components:

- theoretical background
- computer simulation of physical phenomena
- experiments (practice)

If these three compounds are adequately combined, we have a good tool to obtain mentioned above aims.

3. REALIZATION

3.1 Theory

The idea of combining three compounds of educational process has been shown in fig. 1.

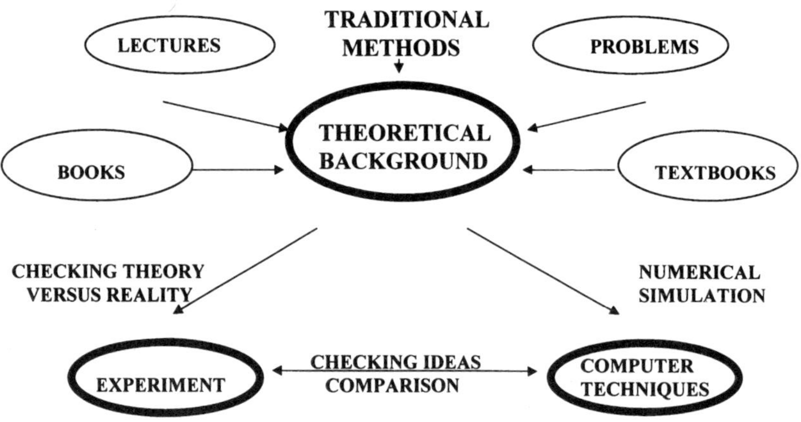

Fig.1. Mutual dependencies among the three compounds of educational process

The first of them i.e. theoretical background (TB) does not depend on the two others that means experiment (E) and use of computer techniques (TC). The knowledge acquired as TB usually is gained by means of standard traditional teaching and learning method such as lectures, reading books and textbooks and solving problems with the use student's mathematical skills. The very important part of this compound of the teaching process is student - teacher discussion during the lectures, lessons and consultation hours. The TB compound or its proper part must precede experiment as well as use of computer techniques.

Both TC and E can be used separately but the best results are obtained when both these compounds are used together complementing one another. CT used separately allows simulate by means of numerical methods equations describing circuit theory laws and rules and design models of electric networks and devices.

Use of computer simulation techniques for modeling circuit theory, field theory or other engineering problems results in better understanding of the theory. The possibility of getting a result without tedious calculations makes students much more eager to undertake computer experiments to find out the best solution. And now we must answer the question that is often asked: "If we can simulate nearly everything using computer techniques, is an experiment really needed? Is not it only waste of time and money?". Although many problems that have been solved earlier experimentally can be solved by means of computer simulation, experiment still remains very important. It concerns, especially such cases, when quite new problems are an object of research. The cases with no data for computer simulation or modeling. The cases when data can be acquired only by experiment. It should be noted that contemporary meters are not the simple tools that have been used, say 20 years ago, but are much more sophisticated and usually employ computer techniques. This was an answer relating to engineering practice. If teaching of future electrical engineers is concerned the proper combining of experiment and CT should be taken into account. Except of computer modeling students check all they have learned during the lectures and lessons experimentally. Of course they notice discrepancies between theory and experimental results being the consequence of the fact that theoretical equations are derived for ideal conditions i.e. ideal meters, ideal coils and other parts of circuitry. Reality is not ideal. Ammeter has its nonzero resistance, resistance of voltmeter is not infinitely large, ideal coils do not exist etc. One of the student's tasks is to decide what is the cause of observed discrepancies. He/she can make assumptions and then check their correctness using computer techniques. IF the results of computer simulation with the assumptions taken into account are similar enough to the results obtained experimentally, the assumptions are proved to be accurate. Students can use professional software, their own, designed themselves programs or educational programs designed by teachers (2). These computer educational programs are used for making the understanding of the discrepancies between theory and the measurement results caused by the use of non-ideal circuit elements easier. Very important part of teaching process are experiments undertaken to verify which of two or more hypotheses is correct (4). Such experiments teach student to examine all hypotheses and ideas critically. The stress should be put on correct argumentation and conclusions in the report from experiment.

3.2. Practice

3.2.1. Courses

What students are taught depends on the type of their study, whether they are going to obtain Master of Science in Electrical Engineering (M.Sc. EE), Bachelor of Science In Electrical Engineering (B.Sc. EE) or Bachelor of Science in Electronics and Telecommunication (B.Sc. ET). All students have three semester standard course of circuit theory but with different number of hours depending on the type of studies. B.Sc. students start their education in circuit theory on the first semester (lectures and lessons on solving problems). The semester later experimental work is added. M.Sc. students start their circuit theory course on the second semester and on the third semester they start carrying out experiments. M.Sc. students have additional one semester course, which content depends on specialization a student has chosen. Number of hours of circuit theory for different types of studies has been shown in the table 1.

Table 1. Circuit Theory

Course	Study	Semester	Hours/week
I	M Sc.	2	2L+2P
		1	2L+2P
II	M.Sc.	3	3L+2P+1E+1 CT
	B Sc.	2	3L + 2P+2E
III	M.Sc.	4	2L+1P+2E+1 TC
	B.Sc. EE	3	2L+2P+1E
	B.Sc. ET	3	2L+2P++2E
IV	M.Sc.	5	2L+2P
V	M.Sc.	6	1L+1P

L-lecture, P-solving problems,
E-experiment,
CT-computer techniques

Computer techniques mean simulation of electric circuits with the use of SPICE or by means of computer programs designed by students in PASCAL (1) or C. M.Sc. students have also lectures and computer practice on numerical methods so they can use it for computer simulation. Computer techniques are the part of circuit theory teaching program for M.Sc. students. B.Sc. student learn also SPICE, but as a engineering tool not as a part of circuit theory education. Of course they also use if for simulation of electric circuits.

4. CONCLUSIONS

Three compounds of circuit theory teaching program i.e. theoretical background, experiments and computer techniques together form irreplaceable educational tool for teaching future electrical engineers. If combined correctly, they allow to gain of such the abilities that an electrical engineer needs in his/her professional work. A skillful implementation of the teaching program based on these three compounds lets to form a professional that can be able to solve not only contemporary but also the future problems in spite of the quick changes in science and technology. The teaching program has been elaborated for circuit theory purposes but it should work also as an educational tool for teaching engineers of other specialties.

5. REFERENCES

1. Baron. B.: Garczarczyk, Z., Stec, K., Misiakiewicz, E., Wybrane zastosowania FFT w teorii liniowych obwodów elektrycznych, *XII Seminarium z Podstaw Elektrotechniki i Teorii Obwodów* , Gliwice-Wisła, Poland, vol.2, pp. 27-36 (1990) (in Polish)
2. Stec K.: Symulacja komputerowa jako narzędzie wspomagające w laboratorium elektrotechniki teoretycznej, *Zastosowania Komputerów w Elektrotechnice, Materiały Konferencji Naukowo Technicznej,* Poznań/Kiekrz (1996), pp.397-398 (in Polish)
3. Stec K.: How to Teach Computers for Electrical Engineering Purposes, Global Congress on Engineering Education, Krakow, September 1998, pp.280-281
4. Glinka T., Stec K.: Experiment as a Tool in Modern Teaching of Electrical Engineering, International Symposium on Theoretical Electrical Engineering , September 6-9,1999, Magdeburg, Germany

Sortal grammars – a framework for exploring grammar formalisms

R STOUFFS
Faculty of Architecture, Delft University of Technology, The Netherlands
R KRISHNAMURTI
Department of Architecture, Carnegie Mellon University, Pittsburgh, USA

Grammar formalisms come in a large variety and are commonly difficult to implement. To alleviate these obstacles to a more widespread adoption, a framework for developing grammar systems is needed that supports an exploration of alternate and varying grammar formalisms in a rapid prototyping way. We present such a framework based on a formalism for representational flexibility, named *sorts*. Sorts provides a component-based approach for building grammar systems, utilizing a uniform characterization of grammars.

1 INTRODUCTION

Grammar formalisms have been around for over 40 years and are found in a wide variety of disciplines and domains, to name a few, natural language, architectural design, mechanical design, and syntactic pattern recognition. Their implementations, however, have been mostly narrow-focused and sparse. In design, in particular, the expectation of grammar formalisms or similar rule-based systems to pervade design software has so far remained only an illusion. There are three main reasons for this. The first relates to the difficulty stemming from technical considerations of implementing grammars, which we addressed in an earlier paper R. Krishnamurti and R. Stouffs, Spatial grammars: motivation, comparison and new results, *CAAD Futures '93* (eds. U. Flemming and S. Van Wyk), pp. 57-74, North-Holland, Amsterdam, 1993.. The second difficulty pertains to ways of enabling designers to employ grammatical rules in a manner that does not impede their act of designing. In this paper, we consider a third difficulty that affects the rapid development, adaptation, and maintenance of grammar-based systems.

Grammar formalisms come in a large variety, requiring different representations of the objects being generated, and different interpretative mechanisms for this generation. Altering the representation may necessitate a rewrite of the interpretative mechanism, resulting in a redevelopment of the entire system. At the same time, all grammars share certain definitions and characteristics. Grammars are defined over an algebra of objects that is closed under the operations of addition and subtraction and a set of transformations. In addition, a match relation on the algebra governs when an object occurs in another object under some transformation.

Building on these commonalities, we propose a framework for exploring different grammar formalisms, based on a variety of algebras and match relations (or interpretative mechanisms). We base this framework on *sorts*, a concept for representational flexibility R. Stouffs and R. Krishnamurti, An algebraic approach to comparing representations, *Mathematics & Design 98* (ed. J. Barallo), pp. 105-114, The University of the Basque Country, San Sebastian, Spain, 1998.R. Stouffs and R. Krishnamurti, Sorts: A concept for representational flexibility, *CAAD Futures 1997* (ed. R. Junge), pp. 553-564, Kluwer Academic, Dordrecht, The Netherlands, 1997.. Sorts constitute a model for representations that defines formal operations on sorts and recognizes formal relationships between sorts. Each sort defines an algebra over its elements; formal compositions of sorts derive their algebraic properties from their component sorts. This algebraic framework makes sorts particularly suited for defining grammar formalisms. Furthermore, since sorts can easily be adapted and compared, these provide the basis for exploring alternate and varying grammar formalisms.

2 SORTS

Conceptually, a sort specifies a set of similar models; sorts combine to form new sorts under algebraic operations defined over sorts R. Stouffs and R. Krishnamurti, An algebraic approach to comparing representations, *Mathematics & Design 98* (ed. J. Barallo), pp. 105-114, The University of the Basque Country, San Sebastian, Spain, 1998.. In practice, elementary data types define *primitive* sorts, which combine to *composite* sorts under formal compositional operations defined over sorts R. Stouffs and R. Krishnamurti, Sorts: A concept for representational flexibility, *CAAD Futures 1997* (ed. R. Junge), pp. 553-564, Kluwer Academic, Dordrecht, The Netherlands, 1997.. For instance, an attribute operator provides for (recursively) subordinate compositions of sorts using an object-attribute relationship, in both a one-to-one and a one-to-many instantiation. For example, the sort of labeled points is specified as a sort of points, with one or more labels assigned as attribute to each point in the data form. The operation of sum allows for disjunctively coordinate compositions of multiple sorts, under many-to-one and many-to-many instantiations, where each sort may–but does not have to–be represented in the data form. As an example, a rule has both a *lhs* (left-hand-side) and *rhs* (right-hand-side) component, either of which can be omitted. Other compositional operations can also be considered, such as an array- or grid-like composition of sorts.

The definition of a sort also includes a specification of the operational behavior of collections, denoted as *forms*, of its members, denoted as *individuals*, for common arithmetic operations. This behavioral specification enables a uniform handling of forms of different sorts, on the sole condition that the universe of all forms of a sort is closed under the respective operations. Additionally, if a match relation exists that is a partial order relation on the sort's forms, a grammar can be defined over this sort. The simplest behavior that fulfills these requirements is a discrete behavior, corresponding to a mathematical set, where the part relation reduces to the subset relation and the operations of addition and subtraction correspond to set union and difference, respectively. A contrasting behavior offers the maximal element representation R. Krishnamurti, The maximal representation of a shape, *Environment and Planning B: Planning and Design* 19 (1992), 585-603.R. Stouffs, 1994. The algebra of shapes, Ph.D. dissertation, Department of Architecture, Carnegie Mellon University, 1994., where any element or individual contains infinitely many, not necessarily disjoint, individuals. This behavior applies readily to intervals over continuous domains, e.g., line or plane segments, or volumes. Primitive sorts have their behaviors assigned in order to achieve a desired effect, e.g., discrete behaviors for points and labels, an interval behavior for line segments, and an ordinal behavior for weights such as thicknesses or tones. On the other hand, a composite sort receives its behavior from its component sorts, based on its compositional relationship R. Stouffs and R. Krishnamurti, Sorts: A concept for representational flexibility, *CAAD Futures 1997* (ed. R. Junge), pp. 553-564, Kluwer Academic, Dordrecht, The Netherlands, 1997..

3 SORTAL GRAMMARS

Grammars are formal devices for specifying languages. A grammar defines a language as the set of all objects generated by the grammar, where each generation starts with an initial object and uses rules to achieve an object that contains only elements from a terminal vocabulary. A rewriting rule has the form $lhs \rightarrow rhs$, and applies to a particular object if the lhs of the rule 'matches' a part of the object under some allowable transformation. Rule application consists of replacing the matching part by the rhs of the rule under the same transformation.

The central problem in implementing grammars is the *matching problem*, that of determining the transformation under which the match relation holds for an lhs. Clearly, this problem depends on the representation of the elements of the algebra. Sorts offer a representational flexibility where each sort additionally specifies its own match relation as a part of its behavior. As composite sorts derive their behavior from their component sorts, the technical difficulties of implementing the matching problem only apply once for each primitive sort. New primitive sorts can be developed, distributed, and adopted by other users without any need for reconfiguring the system. At the same time, the appropriateness of a given grammar formalism for a given problem can easily be tested, the formalism correspondingly adapted, and existing grammar formalisms can be modified to cater for changing requirements or preferences.

4 EXAMPLES

References (1) and (6) give a variety of grammar formulisms. The specification is outlined in this section. Structure grammars (7) is set is represented by a set of pairs each consisting of a symbol. Each symbol in a set may have one or more transformations assigned as attribute. Tartan Worlds(8) that bestrides string and set grammars. Augmented shape grammars (4) (5) (9) is a finite arrangement of spatial elements form among points, lines, planes, or volume, of finite spatial dimension.

5 CONCLUSION

Technical considerations make grammar systems generally difficult to implement. Part of this difficulty also relates to the appropriateness of a given grammar formalism for a given design problem, and to the representational demands that grammar systems impose on users and developers. Together, these difficulties inhibit the rapid development, adaptation, and maintenance of grammar-based systems. Adopting an existing grammar system may present the user with a system that is not exactly suited for the purpose. On the other hand, it is unreasonable to expect every user to develop a grammar system from scratch or invest the time to analyze and adapt an existing system. Instead, a development environment for grammar systems based on sorts will provide the user with the ability to define a grammar formalism and explore its appropriateness for the problem at hand, then, to integrate it into a larger application.

REFERENCES

(1) R. Krishnamurti and R. Stouffs, Spatial grammars: motivation, comparison and new results, *CAAD Futures '93* (eds. U. Flemming and S. Van Wyk), pp. 57-74, North-Holland, Amsterdam, 1993.

(2) R. Stouffs and R. Krishnamurti, An algebraic approach to comparing representations, *Mathematics & Design 98* (ed. J. Barallo), pp. 105-114, The University of the Basque Country, San Sebastian, Spain, 1998.

(3) R. Stouffs and R. Krishnamurti, Sorts: A concept for representational flexibility, *CAAD Futures 1997* (ed. R. Junge), pp. 553-564, Kluwer Academic, Dordrecht, The Netherlands, 1997.

(4) R. Krishnamurti, The maximal representation of a shape, *Environment and Planning B: Planning and Design* 19 (1992), 585-603.

(5) R. Stouffs, 1994. The algebra of shapes, Ph.D. dissertation, Department of Architecture, Carnegie Mellon University, 1994.

(6) J. Gips and G. Stiny, Production systems and grammars: a uniform characterization, *Environment and Planning B: Planning and Design* 7 (1980), 399-408.

(7) C. Carlson, R. McKelvey and R. Woodbury, Introduction to structure and structure grammars, *Environment and Planning B: Planning and Design* 16 (1989), 215-242.

(8) R. Woodbury, A. Radford, P. Taplin and S. Coppins, Tartan worlds: a generative symbol grammar system, *ACADIA '92* (eds. D. Noble and K. Kensek), 1992.

(9) G. Stiny, Introduction to shape and shape grammars, *Environment and Planning B: Planning and Design* 7 (1980), 343-351.

Computer in practical work to optimize quality knowledge

I M SYMONDS
Department of Electronic and Electrical Engineering, University of Leeds, UK

ABSTRACT

This paper describes the experimental use of computer based material during practical laboratory work, to optimize the quality of student knowledge produced. Findings from educational research are reviewed to identify what aspects of knowledge and it's acquisition provide the quality. The learning situation used in the experiment is outlined and the role of the computer based material within it. Those aspects of the situation and computer use which are designed to encourage a deeper learning approach in students are highlighted, as this factor is seen as one the most important in optimizing the quality of the knowledge produced. Finally, the preliminary results of the first experiment are evaluated.

1. INTRODUCTION

What is meant by the term quality when applied to knowledge, and what aspects of knowledge acquisition, if any, affect this quality. These questions, albeit expressed in a different way, have been the subject of educational research for many years. The focus of this research is now student learning and their approach to it. Many researchers have attempted to discover a pattern in the learning approaches adopted by students and the outcomes that they produce. Most have agreed that there is a hierarchy of learning outcomes, ranging from surface to deep, and it is through these outcomes that classification can be made. For example, Säljö (1) proposed a framework of conceptions of learning with five categories. They range from learning as the increase of knowledge to learning as an interpretative process aimed at the understanding of reality. Perry (2) suggested that the two extreme positions of surface and deep level can be recognized by the learner characteristics associated with them.

"Deep level	Understanding intention, comprehending principles, ability to transfer; interpretation of meaning.
Surface level	Learning to repeat, memorizing, rote learning, precise reproduction of content."

These characteristics, and the insights provided by the conceptual framework, can be used as a guide to the preparation of suitable teaching material. The opportunities provided by such material, and the tutor expectations embodied in them, will determine whether a student is enabled to reach a deeper approach to learning. The term quality can now be seen as implying the achievement of a deeper approach to learning, and hence acquiring knowledge in a form that can be used rather than just reproduced. The following example, by Sandberg and Barnard (3), illustrates the relationship between the levels very clearly.

"What do we mean by deep knowledge or insight? Deep knowledge or insight refers to the principles and rules underlying the observable facts. For example, knowing a physics law just through its mathematical formula, can be considered as purely factual knowledge. Only when the factual knowledge is backed up by knowing why the formula is expressed as it is, knowing why it holds, and knowing its conditions for use, the knowledge is considered to be deep."

The aim of the approach described in this paper was to provide a style of teaching through practical work, with the aid of a computer, that optimizes the opportunities for deeper learning for the student, and hence the quality of the outcome.

2. THE LEARNING SITUATION

One of the principles behind the design of this course was to build in opportunities for students to take a deep approach to learning. These took a number of different forms. The first was to require the student to apply theory and practice from other modules to the current problem. Methods of indicating dimensions on engineering drawings were applied, by the students, for the drilling instructions of their enclosure by the workshop. The sizes and placing of the holes being determined from information in the on-line catalogue for the components chosen. The practical work covered in the first session on attenuators provided information that could be applied to the attenuator design problem. Calculation of the required resistor values could be done using techniques covered in the circuits module. The context in which students apply their knowledge is also important as Biggs (4) noted.

"Deep learning is associated with activity. Students are more likely to make connections between what is being learned and past learning if they are active rather than passive."

This principle is embodied in what is called active learning, where an activity which allows the learner to apply theoretical knowledge in a practical situation is used. Hannafin and Land (5) found that student oriented systems using educational technology regularly provide such active learning opportunities. The learning situation of the work described in this paper is both active and technology related, using computer based material as an integral part of the total design and build exercise, undertaken by first year mechatronic students. All teaching material, except for a brief introductory leaflet which included the assignment criteria, was on the departmental web site forming a framework for the complete course. A week by week breakdown of activities was included with links to the main sections of the work.

The first section was a practical introduction to attenuators where circuits were built on prototype board and measurements made on them. Hypertext links to extra pages on the use of the prototype board and test equipment, were used to provide to support students with less practical experience. The simultaneous use of the computer during the practical session allowed tight integration between background theory and practice. This integration puts the practical work in perspective and helps to avoid the situation described by Boud, Dunn and Hegarty-Hazel (6).

"The absence of any objectives make it difficult for the student, who may read the experiment for the first time prior to commencing the laboratory work, to realize what is going on. Yet because the procedure is so clearly detailed, the student is able to complete the exercise and achieve a result. It is not difficult to understand why, in the absence of a clear reason for the activity, students see such exercises as cookbook or recipe work."

The presence of the computer during the practical experiment provided opportunities for better communication between students and tutor. The instructions for the tests included embedded forms that enabled students to enter the results of their measurements as they made them. The forms were automatically e-mailed to the tutor, a process that had two important consequences. The first was the collection of assignment practical results, and the second was the chance to e-mail feedback to the students before the next session. Another use of this communication was in the form of design questionnaires, the first of which followed the practical work in the first week. The student task was to determine the values of resistors to provide the attenuation required in the design specification. The question "how could you determine the value of resistors required?" was phrased in this way to allow the student choice of method. The practical work provided the starting values and typical circuit configuration, and the use of simulation covered in the second session provided an alternative to calculation if required. Other questions asked how the two values of attenuation were to be selected, and what could be done if the design values of resistance are not manufactured. Here is where the trade off between conflicting factors, inherent in all design work begins. One solution is to use the nearest manufactured values provided the resulting error is acceptable. Another is to combine resistors to form the value required. The final question of this questionnaire asked if the proposed solution would actually fit on the 40mm by 40mm circuit board to be used. On completion, the form was again e-mailed to the tutor providing a "snapshot" of the thinking behind the student design process. The second week introduced the use of circuit simulation and printed circuit board design. Both required the use of computer packages and had pages devoted to tutorials on the use of each package. The parts required to complete the enclosure were chosen from an on-line catalogue as mentioned earlier, but the choice was further constrained by the requirement that parts cost must not exceed £6. A second design questionnaire was included at the end of this week to investigate the choice of components made and the reasons for them. The parts requested were ordered by consolidating all the orders e-mailed by the students as part of their computer use. The drawings produced were used to drill the enclosures, and the students continued with the circuit design and simulation between sessions using the information on the web site supporting this course. The circuit layouts from the third session were processed and circuit simulations completed. The final session involved the building and testing of the final design to check compliance with the design specification.

3. RESULTS AND CONCLUSIONS

The results of the short course described in this paper are still being examined in detail, but some broad conclusions have been reached. The value of the improved communications between tutor and student was illustrated by the results e-mailed from one student during the practical attenuator exercise. The results were incorrect in a way which was consistent with the disconnection of one of the resistors in the circuit. This was pointed in feedback e-mailed to the student, together with the reading that should have been obtained. This prevented the subsequent design stages being based on incorrect data. The use of open questions in both task requirements and the design questionnaire resulted in a range of designs and design methods from the students. Some found the values of resistors by manipulating the simulation rather than calculation. Some designs rejected the twin attenuator dealt with in the practical session, and produced a single stage modified by a switch which still produced measured input and output values within the design tolerance of 10%. The results from the second design questionnaire showed that ease of manufacture and assembly were being considered in component choice as well as the obvious factor of price. The logistics of such a short course were very difficult and many students did not complete the final assembly and testing within the very short time of the course. Extra sessions made available in subsequent weeks were well attended, showing an amount of student interest and commitment to this style of course despite the large workload it produced for them. Much of the evidence for the adoption of a deeper approach to learning is from the design questionnaire results as well as in the final product working within specification. The results indicate that computer use as an integral part of the learning situation, even in the case of practical work, is not a problem for the students on this course. It is also clear that this is not an approach for a total open learning package as the supervisors were kept busy during the time-tabled sessions. Finally, the use of a computer during the practical work has enabled one of the ideals proposed by Rowntree (7) to be realised.

"If he [the teacher] wishes to influence, or even understand, the student's learning style and how he is developing, for example, in his capacity for problem-solving or for collaborative work with others, he cannot wait for the product to give him insights but must seek understanding while it is in production – intervening, where necessary, with guidance."

REFERENCES

(1) Säljö R. *Learning and Understanding: A Study of Differences in Constructing Meaning from a Text.* Acta Universitatis Gothoburgensis, 1982. Page 192
(2) Perry W.G. *Different worlds in the same classroom.* Improving learning: New Perspectives. P. Ramsden (ed.) Kogan Page, London, 1988.   page 145
(3) Sandberg and Barnard *Deep learning is difficult.* Instructional Science, Vol 25, 19 Kluwer Academic Publishers, Netherlands, 1997. Page 15
(4) Biggs *Approaches to the enhancement of tertiary teaching.* Higher research and development, 1989. **8**
(5)Hannafin and Land *The foundations and assumptions of technology-enhanced student-centered learning environments.*   Instructional Science Vol 25, 1997. Kluwer Academic Publishers, Netherlands, 1997. Page 190.
(6) Boud, Dunn and Hegarty-Hazel *Teaching in Laboratories*, SRHE & NFER-Nelson, 1986. page 37
(7)Rowntree *Assessing Students: How shall we know them?* Kogan Page Ltd, 1987. Page 138

Quality-reliability-maintenance of knowledge

C TANDY
Department of English, Da Yeh University, Tatsun, Taiwan

SYNOPSIS

Part 1, The Philosophic Search For Knowledge, explains why philosophers today are sceptical of our ability to **know** absolute Reality. Nevertheless, it makes sense to talk about practical knowledge for good decision-making. Part 2 says that historically human knowledge has evolved alternatively in a 'normal' or 'revolutionary' way. Then a third approach, 'stereovision', is introduced and advocated. Part 3, Practical Knowledge And World Betterment, articulates a number of ideas related to the quality, reliability, maintenance, and increase of knowledge. The final part predicts the integration of biotechnology and infotechnology, and our creation of transmortal transhuman offspring.

1. THE PHILOSOPHIC SEARCH FOR KNOWLEDGE

Descartes, the acknowledged founder of modern philosophy, attempted to gain knowledge or find out Reality by trying to doubt everything. Both his sceptical methodology and his dualistic findings failed. Later, Kant countered Hume's extreme scepticism, but Kant's own epistemic project proved less conclusive than he imagined.

Philosophers today are sceptical of our ability to **know** the World of Reality. All we have are appearances. We can compare appearance with appearance, but we cannot compare appearance with Reality. Thus it makes some sense to concentrate on more practical matters: In the World of Appearances in which we live, some decisions and actions are presumably better than others. All alleged 'facts' are not equal; likewise, all alleged 'values' are not equal. Perhaps we can learn to better evaluate and extent our 'knowledge' of facts and values and logics and decision-making.

Presumably it is our interaction with the (unknown) World of Reality that presents us with a World of Appearances. However there is no guarantee that either Reality or appearances are rational/ consistent -- *a fortiori* if we mean human rationality/ consistency. Moreover, Reality may be incommensurable with any humanly possible account of what is the case. (Medieval philosophers asked whether we could talk about God **only** by use of analogy.)

Nevertheless, we can postulate alternative presuppositions, values, logics, facts, narratives, and models, then engage in good faith dialog with others, to attempt to choose in the World of Appearances alternative courses of action that may be reasonable/ meaningful/ fruitful. At any point in time in the course of such 'never-ending' good faith dialog over many centuries or millenniums, it is possible to ask what the consensus is with reference to our apparent knowledge of the world in which we live. For example, it seems reasonable enough to me to accept as a working hypothesis that there is an interior world ('self') and that there is an exterior world (other persons and objects). In our World of Appearances we attempt to detect reliable patterns and live meaningful or fruitful lives. But over time our state of 'knowledge' changes, sometimes falsifying old 'knowledge'. Considerations such as these suggest the following stance with reference to the quality-reliability-maintenance of knowledge (i.e., 'knowledge' of 'facts' and 'values' and 'logics' and decision-making): 'There is much I do not know. Moreover, at least some of my present knowledge may need revision. Perhaps the general framework or basic models I use to make good/bad and true/false judgements may need revision or replacement. "While, however, they continue to withstand criticism and seem superior to available alternatives, it is sensible to stay with them." ' (1)

2. NORMAL, REVOLUTIONARY, AND STEREOVISIONARY KNOWLEDGE

Sometimes our knowledge apparently advances in a rather simple add-new-knowledge or delete-bad-knowledge way. But sometimes the basic presuppositions, models, stories, emotions, values, or worldviews we use change radically. As Kenneth Boulding and Thomas Kuhn have said, sometimes our knowledge structure changes in a 'normal' way, other times the change is 'revolutionary'.

Godel's incompleteness theorems suggest to me the possibility of stereovision – using a variety of paradigms simultaneously. (Loosely described, it seems Godel rigorously demonstrated there are truths that cannot be justified even in principle without changing to a new paradigm – yet the new paradigm will necessarily have the same kind of insoluble difficulty.) Using a variety of paradigms simultaneously (stereovision) may help with our decision-making even though it is not a complete methodology. 'Normal advances' in knowledge are relatively easy to explain, as they fit within the same general framework or universe of discourse. But 'revolutionary advances' in knowledge -- as well as the stereovisionary approach I have just advocated – are **not** well understood; surely these issues are worthy of research funding. Such new knowledge about knowledge would be non-trivial.

3. PRACTICAL KNOWLEDGE AND WORLD BETTERMENT

What other matters seem worthy of attention as we try to improve our World of Appearances (facts, values, logics, and decision-making)? Many of the ideas to follow are based on the ideas of Kenneth Boulding or are plausible extensions of his ideas. (2)

3.1 Quality and reliability of knowledge
- In addition to the 'knowledge about knowledge' research previously suggested, research is also needed to gain interdisciplinary general knowledge and knowledge of general systems. Such general knowledge, if of high quality, would add vital unifying 'glue' to our knowledge structure.
- Evolution, including evolution of language and of knowledge, is a **disequilibrium** system. Thus language should be open-ended, open to modification, open to learning new realities

and new fantasies. In the course of the evolution of knowledge, truth has a long-term advantage over non-truth because truth is not falsifiable.

- We need quality knowledge and social inventions toward establishing a non-totalitarian world at stable peace (e.g., development of quality knowledge for conflict resolution, conflict management, societal evolution, and economic development). Power, wealth, justice, and quality knowledge are in general good in the long run for everyone. How do we make everyone in the world wise and wealthy?
- For advancement of quality knowledge in the domain of science, good faith dialog and ongoing feedback (so-called "reality-testing") are vital. Science seeks abstract laws (repetition) while the humanities prefer inward uniqueness. The **right kind** of philosophic, historical, and social-science knowledge can act as a bridge between the sciences and humanities.
- Knowledge of how to make the world better has more long-term evolutionary potential than knowledge of how to stop the world from getting worse. Excessive concern with alleviating evils can blind us to the long-term evolutionary (variation-and-selection) processes necessary to nourish the advancement of quality knowledge. The advancement of knowledge, the mentality of research, tells us it is **not** true that 'there is only so much to go around'. Rather, new knowledge creates evolutionary potential and new resources.
- Success reinforces our confidence and too easily leads to arrogance. Failure produces learning/mislearning and too easily leads to bitterness. Inefficiencies and redundancies, with room for mistakes, can help us meet intellectual crises and advance our knowledge in revolutionary ways. Highly efficient knowledge systems are systems operating at a minimal or "survival" level, with little room for the dynamic evolutionary processes that can lead to revolutionary intellectual advancement. The inefficiencies of biological evolution gave us sexual reproduction and the inefficiencies of the renaissance-reformation period in human history gave us science.
- In the long story of evolution, it is not the fittest that survive. The fittest fit too efficiently into the old environment -- they do not survive in the new environment. Thomas Edison said that invention is 99% perspiration and 1% inspiration; in a given project, he would continue to try and fail and try and fail – as long as it took until he tried and succeeded.

3.2 Maintenance and increase of knowledge

- Humans forget and humans die – thus we may say that 1% of all human knowledge is lost each year and must be replaced. As humanity's store of knowledge increases, the maintenance function of education becomes more burdensome; it takes us, say, not 20 but 40 years to catch up to the present state of knowledge in our area of specialisation. Perhaps Benjamin Franklin was the last 'Renaissance Man'; polymaths are no substitute.
- We need to invent new physical and social technologies to cope with the information overload. If we had more, and more creative, educational economists, perhaps they could help. With more specialisations, our store of knowledge increases – but we sometimes forget our need for those who specialise in interdisciplinary knowledge, even interdisciplinary general knowledge.
- If humans destroy themselves in the absence of transhuman offspring, then we will have failed in the knowledge quest. Thus we should attempt to gain knowledge of how to flourish rather than die. For example, quality knowledge of how to develop a high-technology society that is environmentally friendly would help us get from here to there.
- Likewise, technology allowing us to flourish in multiple locations other than in the single biosphere of planet Earth seems wise. I understand this technology is already in hand or at least not beyond our grasp. Such large free-floating Earth-like extraterrestrial

communities (self-sufficient and able to reproduce other extraterrestrial habitats at a geometric rate) would have multiple benefits. For example, if land is freely available, then there is less chance of conflict over a specific piece of land. Likewise, differing cultural traditions can be realised rather than globalised. Since the evolution of knowledge proceeds via variation-and-selection, this appears superior to globalisation.

- The advancement of knowledge allows us to build artefacts to test our fantasies/ speculations. Not only does knowledge growth allow us to improve our knowledge of the past, but also to realise/ actualise our fantasies about the future. Pre-humans simply expanded to fill their environmental niches, but humans are able to create the environments into which they expand.

- Especially in matters related to the domain of social science, it is too easy to hold on to untested knowledge rather than test it and find it false. Some things are easy to test, some difficult, and some impossible (at least for the foreseeable future). The decisions we make today help determine the terms under which the advancement of knowledge and 'the never-ending dialog' may and may not proceed.

- Practical knowledge expands not into absolute Reality but into an accessible universe of discourse. This means, as we talk to our self and others, that it is sometimes necessary to make up the concepts and discourse as we go along. (Poets, perhaps, are good at this?)

4. THE FUTURE OF KNOWLEDGE

Our image of the future should be an uncertain one. Nevertheless, we may have more or less reasonable expectations. We engage in practical reasoning with the future in mind.

Biotechnology and infotechnology have produced wonders some of us equate to science fiction. Integration of these two fields (including the developing technologies of molecular nanotechnology) is easily predictable. Although the gestation period may be uncertain and difficult to map, I fully expect that there are transmortal transhuman offspring in our future. Presumably transhumans will live for millenniums and will experience emotions and develop values unknown and unknowable to mere humans. Living in robust worlds of self-controlled virtual reality is also beyond the experience of humans past.

A human's conscious mind tends to focus on only one thing at a time and to interpret a single appearance as Reality. But the stereovisionary approach to knowledge, a difficult skill for humans to master, will be child's play to transhumans. Just as human knowledge superseded the knowledge of pre-human organisms, human knowledge will be superseded. Perhaps the scientific resurrection of all dead persons who have ever lived will someday become a practical political project. To what extent our transhuman offspring may be able to explore beyond appearances and into Reality, I do not know. In any case, I wish them and us well, as we continue our 'never-ending' quest for the good, the true, the beautiful.

REFERENCES

1. E.A. Burtt, **In Search Of Philosophic Understanding** (New York: New American Library, 1967), 176.
2. Charles Tandy, **Educational Implications Of The Philosophy Of Kenneth Boulding** (Ann Arbor, MI: UMI/ Bell & Howell, 1993), 7-280.

Teaching technical writing to Malaysian undergraduate engineering students through authentic learning

L E ZEITZ
University of Northern Iowa, Cedar Falls, USA

ABSTRACT

Introducing an authentic learning paradigm for teaching English to undergraduate engineering students in Malaysia, this study created a corporate simulation in which the students were divided into working teams. Students wrote interdepartmental reports as a process to create a product from needs assessment through production and to marketing. This format provided a learning environment characterized by high student involvement and initiative.

1 INTRODUCTION

Success in the engineering field is not solely based upon a professional's expertise in critical thinking and creative problem solving. Professional success also requires engineers to perfect oral, written and interpersonal communication (1). While universities and engineering firms do not generally train engineers in business technical writing (2), firms benefit when their engineers can effectively explain their ideas to their superiors, subordinates, clients and other professionals. This ability to explain clearly and precisely plays a major role in completing projects and promoting self-advancement in the field.

This is the philosophy held at the Institute of Technology Tun Hussein Onn in Johor, Malaysia. Throughout their programs in mechanical, electrical and civil engineering, the engineering students are required to take courses in English to improve their abilities in writing reports, managing meetings and overall communication skills.

The conventional style for teaching this writing class has been for students to read about report writing from textbooks (i.e., feasibility studies, investigative reports, and progress reports) and then write reports. The whole class is posed with the same problem, provided certain parameters, and students are asked to write reports which fulfill those parameters.

While this conventional system is time tested and has provided writing skills for today's engineers in the field, the major problem is its contrived nature. The students have no connection with the stated circumstances except to please the teacher. It also fails to use learning research in addressing the learning needs of engineering students.

Felderman and Silverman (4) found that a considerable mismatch exists between the common learning styles of engineering students and the traditional teaching styles of the engineering faculty. Most engineering students are visual, sensory, inductive, active and global learners. Most engineering education, however, is presented in an auditory, abstract (intuitive), deductive, passive (reflective) and sequential manner.

These research findings were used as a premise for the curriculum used in this study. An active curriculum filled with sensory input, inductive learning and allowing students to build their own understanding in an authentic learning environment.

2 METHODS

This study took place in an English 2 course for undergraduate engineering (mechanical, electrical and civil) students. The English 1 course involved the fundamentals of writing including analyzing the audience, identifying subjects and topics, and the mechanics of sentence and paragraph structure. English 2 was designed to provide the students with an arena for their language skills through report writing, conducting meetings, and overall business communication.

The present curriculum provided a workplace simulation arena. The students were organized into a faux-corporation, the PDQ (Pretty Darn Quick) Corporation, for building air conditioners (a premium industry in Malaysia). This scenario provided a realistic corporate environment for learning as well as a venue for the expertise of the air conditioning engineering students in the class.

The students were presented with the problem: Most homes in Malaysia are not air conditioned because the units are too expensive for the typical homeowner. Their task was "to create an air conditioning unit that was affordable for the individual homeowner". The whole corporation (class) brainstormed the possible solutions to the problem and the PDQ Corporation was then divided into eight departments: Project Management, Needs Analysis, Research and Design, Manufacturing Design, Purchasing, Accounting, Quality Assurance, Sales/Marketing, and Technical Writing." Obviously a physical air conditioner could not be manufactured in an English class, but a paper-based one could.

The project management team was charged to manage the process. At first, it was awkward because while they were trying to organize and instruct the teams in what needed to be accomplished they were constantly looking towards the lecturers for approval. Recognizing the problem, the lecturers "handed" the direction of the project over to the project management team. This placed the project management team in charge of the project in the eyes of students as well as themselves. Such a move was unprecedented in the Malaysian tradition. Typically, the teachers lectured and the students took notes. Providing the direction of the class to a group of students was novel for both the teachers and the students. The lecturers worked behind the scenes to ensure a quality experience and support the team by teaching them the intricacies of "being in charge."

Each team created a sequence of tasks their project would entail as well as a corresponding timeline. These were submitted to the project management team who correlated the teams' tasks chronologically so that the project could be completed in an 8-week period. The management team returned to the groups and negotiated the timelines. A master plan was created and the management team monitored progress through weekly progress reports submitted by the teams.

Progress reports were submitted on a weekly basis. At first, they were paper-based. After the third week, a web-based form was created and they were sent electronically directly to the management team member monitoring progress. Since report writing was a goal of the course, a different member would submit the report each week. Another goal of the course was for students to plan, administer and report on team meetings. Each student managed a meeting and submitted the minutes for evaluation.

3 RESULTS

Perhaps the most dramatic result of this curriculum was the authenticity that the writing process assumed. The departments assisted each other in the process of developing the product and generating all of the reports necessary for interdepartmental communication. The Needs Assessment department surveyed potential customers to identify the market. Their results were reported to Research and Design who used these opinions to design an air conditioner. The specifications of the new product were given to the Production/Manufacturing department who designed its production line and the production schedule. The production team submitted the purchase requisitions to the Purchasing department to ensure that the parts would be available for building the units. All along, the Quality Assurance team identified the quality requirements and the places in the production line where they needed to be checked. Marketing developed a campaign for selling the new air conditioners. Accounting created a project budget and taught the groups how to create expense forecasts to justify their fiscal needs. Technical Writing examined other air conditioning manuals to begin design on the unit's documentation. Project Management monitored, guided and scheduling all activities to ensure completion by the final deadline.

Every class meeting was an active learning experience. Instead of lecturing, the instructors monitored the students' progress towards the ultimate goal of bringing a product to market. While no actual product was produced, a three-inch thick report was created that could easily lead to the actual production of a product. While no quantitative measurement could be made, the lecturer who had been teaching the course for many years noted that the reports were of a much higher quality than seen in the past and students were more involved in learning.

The marked difference in the classroom atmosphere was that earning grades from the teacher was secondary to creating reports that were acceptable and beneficial for interdepartmental communication. Interviews with students found that students felt they had "learned more than in the normal classes". They stated that not only did this format provide a factory-like environment that was valuable for later employment, they felt more involved in the learning process than in other classes. Some of the interviewees complained that they didn't learn as much English as they had hoped, but admitted that it was because they spoke Malay in their groups. While speaking English was encouraged in the classroom, meetings outside of the classroom could not be monitored and future courses should address this. The majority of interviewees stated that they enjoyed this form of learning and that they would like to see it continued in this course and used in other courses.   *http://www.uni.edu/zeitz/techwriting/*

PDQ corporation work groups

Group	Activities	Resulting Report
Project Management	• Create the project tool • Goals/objectives of the project • Create management instrument • Maintain timeline	• Weekly progress report (compilation of groups' reports) • Coordinate the creation of the final report.
Needs Assessment	• Refinement of needs identified • Survey of potential customers	• Investigative report on findings from survey. • Weekly progress report
Research and Design	• Analysis of feasibility of potential solutions. • Generate additional solutions. • Select solution/design product	• Feasibility report • Product design • Parts list for purchasing • Weekly progress report
Production/ Manufacturing	• Identify processes necessary for manufacturing • Design manufacturing equipment	• Feasibility report • Production line design • Weekly progress report
Purchasing	• Identify sources for parts	• Product analysis • List of vendors • Weekly progress report
Accounting	• Project cost of product • Negotiate budget with teams	• Cost Analysis • Budget • Weekly progress report
Quality Assurance	• Identifies quality methods	• Feasibility report • Weekly progress report
Marketing	• Identify the market(s) • Identify size of market • Project potential sales	• Market analysis report • Weekly progress report
Technical Writing	• Define audience for documentation • Outline documentation • Write sample of documentation • Test documentation on a focus group and report results	• Product analysis report • Documentation sample • Weekly progress report

4 REFERENCES:

1 Hoyt, B. (1989) Bucknell's electronic classroom: Exploring tomorrow's education today. Available: www.eg.buckness.edu/bu_eng/Spring95.htm.

2 Schillaci, W. (1996) Training engineers to write: Old assumptions and new directions. *Journal of Technical Writing and Communication*, 26(3), 325-333.

3 Felder, R. Silverman, L. (1988) Learning and teaching styles in engineering education. *Engineering Education*. 78(7), 674-681.

4 Schmeck, R. (1988) An introduction to strategies and styles of writing. In R. Schmeck (Ed.), *Learning Strategies and Learning Styles* (pp. 1-19). New York: Plenum Press.

Twist drill vibration at environmentally conscious manufacturing

G VARGA, I DUDAS, and **T CSERMELY**
Department of Production Engineering, University of Miskolc, Hungary

Abstract: Among other factors reliability of a concrete product depends on the production as well. Nowadays developing of the new technologies which decrease the unfavourable environmental effects of the traditional manufacturing processes and using them in the industrial production get even more into the focus. The development of environmentally clean manufacturing technologies requires careful and circumstantial analysis. In this paper the analysis of the one model analysis of the vibration can be found. The mathematical model based on the known solvation method of partial differential equation which is a good base model for examination.

1. INTRODUCTION

Reliability of a product conveys the concept of dependability, successful operation or performance and the absence of failure. The reliability of the product is its ability to retain its quality and to continue to meet the specification over time. If a product falls its specification before its designed life time, it will cause economic loss. Nowadays, the survival of any manufacturing unit will depend on the quality and reliability of its product.

Reliability deals with the interdisciplinary use of different different topics such as design, production, science of failure mechanisms, modelling of stochastic processes. Many factors and disciplines can be mentioned on this field, however we mention only some on Fig. 1. which was based on the work of TN Goh [8].

Fig. 1 Some factors and disciplines which have influence on relibility.

Drilling is one of the most important machining operation. Till this time the aim of the manufacturers was the use of the maximum value of coolant or lubricant. It was done in order to

get as high drilling performance as possible. However nowadays the way of manufacturing a little bit changed because of the environment protection and the wish of machining more economical manner [1]. The coolants and lubricants endanger not only the environment but directly the health of the production workers as well. The increasing attention to environmental and health problems is reflected in the increasingly stricter legislation of national and international standards.
For these reason it is necessary to develop the environment-friendly, waste-free technologies with minimum quantity of coolant and lubricants or with the complete elimination of cutting fluids „dry" machining technologies [2],[3].

As the introduction of every new manufacturing technology requires the deep previous analysis of several parameters so the introduction of environment friendly drilling as well. Among other questions the following one arrises: „How the vibration of twist drills will change?" So it requires to work out a dynamic model for the transverse vibration of drill bit. In the posession of the model the effect of different circumstances can be examined. In this article our aim is to present a model which contains the effects of drill bit length, diameter and material properties.

2. BACKGROUND

Many researchers have investigated the drill vibration and questions of stability [5] when machining with collant and lubricant. Fuji and his colleagues experimentally studied the effects of geometric parameters on transverse vibration in drilling. They developed a vibrational model which includes radial forces along the main cutting edge and the frictional forces due to the collision of the flank with the machined surface [5]. Tekinalp and Ulsoy [6] created a multiparameter model which model predicts the fundamental frequences of the drill. They took into consideration of cutting speed. Analysis of Tekinalp and Ulsoy [6] when using FEM gives a good correlation between experimental and calculated results. These models were valid for drilling with coolants and lubricants. However in the case when the drill has inner coolant ducts (Fig. 2) the rigidity and the second order inertial moments will change, so the vibration will be different.

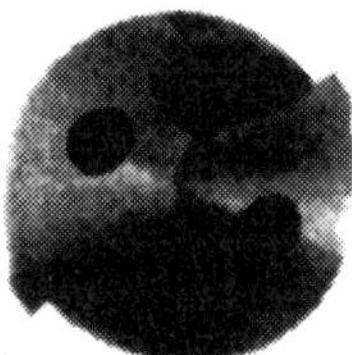

Fig. 2 View of a drill having inner coolant channels [9]

On the other hand in case of drilling having no coolant and lubricant the value of thrust and torque will change because of the increased friction factor, and it has some effects to the vibration as well. Our model is based on the Lagrangian equation.

3. MODELING

Suppose the drill as a stepped shaft. The shank is a cylindrical part of the drill. The working part diferring from the conventional drills, because it contains two twisted holes inside.
In this case the working part of the drill can be substitute with an another cylinder which diameter can be determined in the way which represents the same 2nd order inertial moment. The model considers the drill as a clamped-clamped support Fig. 3.
The examination of only l_1 and l_2 will be done, so the cross sectional areas A_1 and A_2 will be

different, and q is the generalized coordinate.

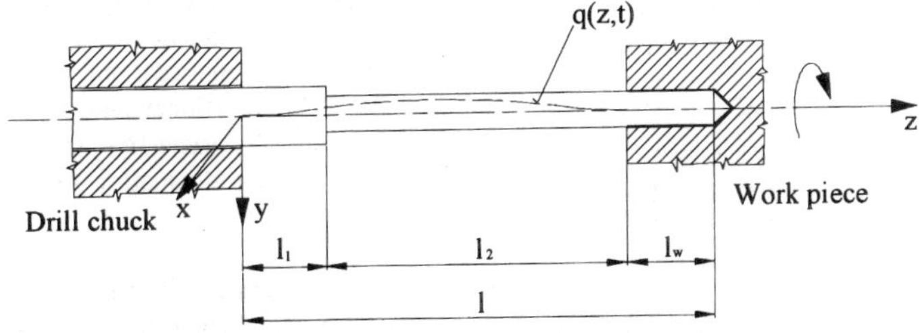

Fig. 3 Model of supporting of the drill

For the Δz infinitezimal length Fig. 4 shows the equilibrium of forces and momentums.

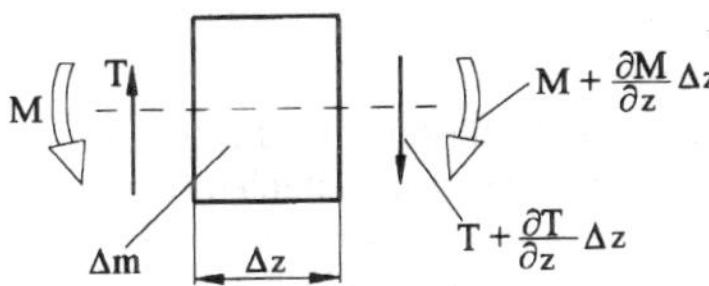

$\Delta m = \rho \cdot A \cdot \Delta z$

where: ρ - density of the drill material

Fig. 4

The equation of theorem of momentum:

$$\rho A \Delta z \frac{\partial^2 q}{\partial t^2} = -\frac{\partial T}{\partial z} \cdot \Delta z \,, \tag{1}$$

law of conversation of momentum for the x axis:

$$0 = \frac{\partial M}{\partial z} \Delta z + T \Delta z \,, \tag{2}$$

where: T - shear force,   M - moment

From (1) and (2):

$$\rho A \frac{\partial^2 q}{\partial t^2} + \frac{\partial^2 M}{\partial z^2} = 0 \tag{3}$$

can be obtained. Putting the $\left|\dfrac{M}{IE}\right| \approx \dfrac{d^2 v}{dz^2}$ approximation into equation (3) the following equation can be written, where v is the displacement of the centre of gravity of cross sectional area:

$$\rho A \frac{\partial^2 q}{\partial t^2} + \frac{\partial^2}{\partial z^2}\left(IE \frac{\partial^2 q}{\partial z^2}\right) = 0 \tag{4}$$

Respectively for the l_1 and l_2 length IE will not depend on the coordinate z.
The solution of (4) can be obtained by the Fourier-method:

$$q(z,t) = v(z)\tau(t) \tag{5}$$

where: q is displacement into direction y.
Substitution into equation (4):

$$\frac{d^4 v}{dz^4} - k^4 v = 0 \qquad \text{and} \qquad k^4 = \alpha^2 \frac{\rho A}{IE} \tag{6}$$

where: α^2 - const.

Solution:

$$v(z) = D_1 S_1(kz) + D_2 S_2(kz) + D_3 S_3(kz) + D_4 S_4(kz)$$

where:

$$S_1 = \frac{1}{2}(\text{ch } kz + \cos kz), \ S_2 = \frac{1}{2}(\text{sh } kz + \sin kz), \ S_3 = \frac{1}{2}(\text{ch } kz - \cos kz), \ S_4 = \frac{1}{2}(\text{sh } kz - \sin kz). \Bigg\} \quad (7)$$

are the Kruhlow functions.

The angular displacement:

$$\varphi(z) = k\big[D_1 S_4(kz) + D_2 S_1(kz) + D_3 S_2(kz) + D_4 S_3(kz)\big].$$

The moment:

$$M(z) = -IEk^2\big[D_1 S_3(kz) + D_2 S_4(kz) + D_3 S_1(kz) + D_4 S_2(kz)\big].$$

and the shear force

$$T(z) = IEk^3\big[D_1 S_2(kz) + D_2 S_3(kz) + D_3 S_4(kz) + D_4 S_1(kz)\big].$$

The D_1, D_2, D_3 and D_4 parameters can be determined from the boundary conditions. It can be done for the end of the first and second length, l_1, and l_2. In this way the displacement and moment can be calculated.

4. SUMMARY

The different forces caused different transverse vibration of the drill bits. Nowadays cutting theory based on shear plane is combined with other cutting theories. With these results the connection between torque, thrust and transverse vibration can be elaborated.

Vibrations of course play an important role in creation of hole quality. Increased vibration, when drill enters, can cause poor hole location accuracy, while in case of exit, excessive burr formation.

REFERENCES

1. I. Dudás, I. Tolvaj, G. Varga, T. Csermely: *Measurement applied at the experiments of environmentally clean manufacturing operations*, Proc. of ISMTII'98, Fourth International Symposium on Measurement Technology and Intelligent Instruments, Miskolc-Egyetemvaros, Hungary, pp. 487-492.
2. G. Byrne; E. Scholta: *Environmentally clean machining processes - A strategic approach*, Annals of the CIRP vol. 42/1/1993. pp. 471-474.
3. T. Cselle: *Is Eco tooling on the way in? The pros and cons of dry machining techniqus. Modern Cutting Tools Technology*, GÜHRING 20. Volume 14. pp. 6-8.
4. G. Varga: *Environment friend drilling-experiments by the use of factorial experiment design method*, Proceedings of the IXth International Conference on Tools, University of Miskolc,Miskolc, Hungary, 3-5th September 1996. pp. 567-575.
5. S. Ema, M. Fujii, S. Marui: *Whirling vibration in drilling, part 3.: Vibration analysis in drilling workpiece with a pilot hole*, Transaction of ASME Journal of Engineering for Ind., 1988, Vol. 110, pp. 315-321.
6. A. G. Ulsoy, O. Tekinalp: *Dinamic modeling of transverse drill bit vibrations*, CIRP Annals, 1984, Vol., 33, Part I. pp. 253-258.
7. I. Dudas, T. Csermely: *Vibration analysis of drilling tools, Proc. of ISMTII'98, Fourth International Symposium on Measurement Technology and Intelligent Instruments, Miskolc-Egyetemvaros, Hungary, pp. 499-503.*
8. TN Goh: *Special focus on Quality and Relibility in Manufacturing*, http://www.eng.nus.edu.-sg/EResnews/9905/p2.html
9. Precision Cutting Tools, GUHRING Cat., 5th ed., 1991, 078430/00005,801/9105-I-08,p. 108

Be part of the revolution !

Your Machines
Are deteriorating
but resources are tight
so how can CM be implemented

Our Products
Detect the faults
in time to take action
with a minimum of effort

More productivity ✔
Reduced costs ✔
Extended life ✔

There's a revolution taking place in Condition Monitoring (CM). More and more companies want to implement effective CM as part of their asset management strategy but fewer and fewer are prepared to pay the high costs associated with traditional vibration techniques.

Ask yourself what's most important to your business :

a)   Knowing which of your machines have problems and which are OK ?
b)   Knowing how rapidly defective machines are deteriorating ?
c)   Carrying out rocket science to know exactly what's going on inside defective machines (hopefully!) ?

If you are looking to answer (a) and (b) in the quickest time at the lowest cost and with extreme sensitivity and flexibility then ask for more details about our field proven instruments. You'll be in good company, they're in widespread use across the whole of UK industry including many blue chip companies and top-flight CM consultants.

HOLROYD
INSTRUMENTS

Via Gellia Mills, Bonsall, Matlock, Derbys. UK. DE4 2AJ

Tel : + 44 (0)1629 822060
Fax : + 44 (0)1629 823516
Web : www.holroyd-instruments.com
Email : sales@holroyd-instruments.com

HOLROYD INSTRUMENTS . . . .
. . . . rapid results
. . . . rapid payback

www.holroyd-instruments.com